Challenges for Health and Safety in Higher Education and Research Organisations

Challenges for Health and Safety in Higher Education and Research Organisations

Edited by

Olga Kuzmina
Imperial College London, UK
Email: o.kuzmina@imperial.ac.uk

and

Stefan Hoyle
Imperial College London, UK
Email: s.hoyle@imperial.ac.uk

ROYAL SOCIETY OF **CHEMISTRY**

Print ISBN: 978-1-83916-159-9
PDF ISBN: 978-1-83916-249-7
EPUB ISBN: 978-1-83916-281-7

A catalogue record for this book is available from the British Library

The Royal Society of Chemistry is a charity, registered in England and Wales, Number 207890, and a company incorporated in England by Royal Charter (Registered No. RC000524), registered office: Burlington House, Piccadilly, London W1J 0BA, UK, Telephone: +44 (0) 20 7437 8656.

Visit our website at www.rsc.org/books

Printed in the United Kingdom by CPI Group (UK) Ltd, Croydon, CR0 4YY, UK

Editor Biographies

Dr Olga Kuzmina (CMIOSH, CChem, CSci, CEnv and MISTR) is a Faculty Safety Manager at Imperial College London. Olga has more than 10 years' experience of working in manufacturing, research and academic environments. She routinely advises on the handling, storage and disposal of hazardous chemicals, outreach activities, and develops organisational policies and H&S training. She has 15 scientific publications and is a Committee Member of the IOSH Environment and Waste Management Group. In 2018 Olga was selected as one of the "40under40" outstanding H&S professionals.

Stefan Hoyle (MSc, CMIOSH) is the Head of Health and Safety in the Faculty of Natural Sciences at Imperial College London, with over 15 years of experience as a health and safety practitioner in a research environment. Stef started his career as a Medical Microbiology researcher then moved to a safety role to support frontline research. Stef is a Member of Expert Biosafety Group at Polo d'Innovazione di Genomica Genetica e Biologia (Polo GGB) and a committee member of the UK University Chemical Safety Forum. Stef's experience covers biological, chemical and laser safety, offsite working, computer health, outreach and events safety. Last year Stef's team received the Imperial College Provost Excellence award.

Challenges for Health and Safety in Higher Education and Research Organisations
Edited by Olga Kuzmina and Stefan Hoyle

Published by the Royal Society of Chemistry, www.rsc.org

Author Biographies

Philip Crosbie is a Solicitor specialising in both health and safety defence work and regulatory compliance advice. His experience includes acting for various organisations subject to investigations by the Health and Safety Executive and Local Authorities following serious injuries and fatalities involving employees and third parties. Philip will often work with clients from the outset of an incident, and guide them through the investigation process. Philip conducts his own advocacy where possible and has appeared in many Magistrates' Court and Coroner's Court Hearings.

Dick Hoeneveld is a researcher in the Safety and Security Science group at Delft University of Technology in The Netherlands. His research interests include nanosafety, safety management and intercultural safety. Dick has a background in Biochemistry, International Environmental Policy and Safety management. For over 25 years he advised Delft University on OSH matters. Dick is member of several national advisory boards, he is co-founder of the Lab Servant safety management system and winner of the EU-OSHA safety award. Dick's current research is focussed on sensor supported indoor evacuation, safety-by-design governance and homeland security innovation.

Monica Kanwar (BSc, MSc, PgD, PgC, MISTR, CMIOSH) manages occupational health and safety in higher education with over 10 years at a senior level. Monica has a degree in Biological Sciences, Master's in Environmental Engineering, a postgraduate diploma in Occupational Health and Safety and more recently a postgraduate certificate in leading through coaching. Her work undertaken on workplace wellbeing was shortlisted for two and won one of the National Safety and Health Practitioner Awards in 2013. She also chaired the reviewing panel for the University Safety and Health Association Leadership and Management of Health and Safety in HEIs (2015).

Challenges for Health and Safety in Higher Education and Research Organisations
Edited by Olga Kuzmina and Stefan Hoyle

Published by the Royal Society of Chemistry, www.rsc.org

Dr Derek Healy (Tech IOSH, MISTR, BSc (Hons), MSc, PhD) is a Health, Safety & Environment Officer at Brunel University London, he is also an HTA approved Designated Individual and Chair of the University Research Ethics Committee. He is an approved IOSH, and NEBOSH (GC) trainer and has vast experience in designing and delivering training. Derek began his career as a researcher in DEFRA's *Wildlife and Zoonotic Diseases* group at the Veterinary Laboratories Agency and acted as departmental safety adviser specialising on safety processes within and outside of the BSL-3 facility. Derek's publication list can be found on google scholar.

Dr Patrick O'Leary (BSc, PhD) is a Lecturer and Head of School of Chemistry in National University of Ireland Galway. Patrick has been in Galway since 2004 and his research group work in the area of organic chemistry and asymmetric catalysis. In 2008 he became safety coordinator in the School of Chemistry which involves advising the Head of School on safety matters, chairing the School safety committee and organising/delivering safety training. Patrick served on the University Safety Working Group for 10 years and acts as a contact point for the University on hazardous chemicals.

Dr Graeme Jones is a Senior Lecturer in Chemistry at Keele University and for the past eight years has been the Chemistry Safety Adviser. His career has spanned research, with over 50 publications in organic chemistry and chemical ecology; public science, with the BA Lord Kelvin Award and a NESTA Dreamtime Fellowship and a Guinness World Record for the largest model of DNA; and teaching, most recently the application of Team-Based Learning in Chemistry Education. In 2001 he won the Royal Society of Chemistry National Higher Education Teaching Award.

Diane Thomason (BSc Applied Biology, MSc Biomolecular Organisation, CertEd, DipNEBOSH) is a Faculty of Medicine Safety Adviser at Imperial College London. She has worked full-time in health and safety, mainly in universities and initially in biosafety, for 20 years following 10 years in research and five years as a laboratory manager. Her interests include safety management as well as technical aspects of laboratory safety. Diane is a former Examiner for the NEBOSH Certificate and Diploma.

Heather Combe has been a Faculty Safety Adviser in the Faculty of Medicine at Imperial College for 6 years and has worked at the College for the last 13 years. She has a degree in microbiology from the University of Edinburgh, and prior to her current role had extensive experience in laboratory management and technical support. Her particular areas of interest include biological and GM safety, laboratory ergonomics, outreach and work experience, safety training, and mental health.

Trisha Brown (BSc, PGCE) is currently the Head of the School of Medicine Secretariat at Imperial College London. She has over 10 years' experience of Medical Education and has been involved in major capital build projects for anatomy dissection, mortuary and specialist medical school facilities during that time. She holds an IOSH H&S qualification and advises the School on quality assurance of the student experience in a clinical environment.

Dr Peter Coulahan is currently Director of Safety, Insurance, Operational Risk & Compliance at University College Dublin. He has almost 20 years' experience advising on workplace health and safety in the University sector. Originally a field biologist, he has a particular interest in laboratory safety (he has acted as University Biosafety Officer during his time in UCD and has published on nanomaterial safety) and student safety during sporting activities. He currently lectures on a number of University occupational health and safety programmes, covering topics such as sports risk management, behavioural safety, risk assessment, machinery safety, electrical safety and occupational hygiene.

Dr Mehriban Ulusoy is an Associate Director of Environment, Health and Safety (EHS) at MSD in London. Mehriban has seven years of scientific experience in biotechnology and nanobiotechnology with nine scientific publications. After scientific research, she decided to pursue a career in EHS and has four years of EHS experience in academic and research institutes, and the pharmaceutical industry. Mehriban was selected as one of the "40under40" outstanding H&S professionals in 2018. Since 2019 she has been the EHS Lead for design and implementation of EHS management systems in the new MSD Discovery Centre in London.

Dr Sébastien Gillotin is a senior scientist and a team leader at MSD in London. Sébastien has more than 15 years' experience of research focusing on cell fate and gene transcription in British academic and corporate environments with work assignments in Canada, France, Germany and Israel. Sébastien has authored many scientific publications and peer-reviewed papers for several international journals. Sébastien is a representative member of the Environment, Health and Safety committee on GMO and laboratory safety and has extensive experience in creating and implementing risk assessments for new projects and techniques both in established and newly created laboratories.

Dr Vinita Jagannath (MBBS, MSc, PhD) is a scientist at MSD in London. She has more than five years' experience working in the field of cellular/molecular neuroscience. She routinely works in the laboratory planning and executing experiments to understand the pathophysiology of neurological disorders with a hope to find new therapeutics for the effective treatment of these disorders. She has nine scientific publications.

Dr Fiona Marshall is MSD VP Head of Neuroscience Discovery and Head of UK Discovery Centre. Fiona has 30 years' experience in drug discovery. She founded and was the Chief Scientific Officer at SoseiHeptares. Fiona has published over 100 scientific papers, won the 2012 WISE Women of Outstanding Achievement for Innovation and Entrepreneurship and the 2015 RSC Malcolm Campbell Award for chemistry. She is a Fellow of the Academy of Medical Sciences in 2016 and an Honorary Fellow of the British Pharmacological Society in 2018 and an Honorary Fellow of the RSC in 2019.

Tjasa Lepko is a research scientist at the MSD in London. She obtained her BSc degree in Biotechnology at the University of Ljubljana, Slovenia in 2010 and her MSc degree in Biology from the Technical University of Munich, Germany in 2012. She then started her PhD at the Graduate School of Systemic Neurosciences, Ludwig Maximilians University of Munich and Institute of Stem Cell Research at the Helmholtz Zentrum München, Germany. After completing her PhD in 2018, she moved to London to join the MSD to pursue research on the age-related diseases and contribute to the development of novel therapies.

Jeremy Presland is an *in vitro* pharmacologist with more than 20 years of experience within the biopharmaceutical industry with such companies as MSD, Schering-Plough, Organon and Pfizer. As well as extensive lab research expertise across a range of therapeutic areas and experimental techniques Jeremy has also held positions requiring management and oversight of EHS processes and procedures. Having worked in the UK and US and collaborated with scientists in Asia, the US and Europe, Jeremy is adept at working safely and effectively in global, multidisciplinary and multicultural teams.

Eimear Howley (MSc) is an Associate Principal Scientist at MSD in London where she leads a team of scientists. Eimear has more than 13 years' experience of working in pharmaceutical research laboratory environments. Throughout her career she has held numerous safety positions including Radiation Protection Supervisor in three pharmaceutical companies. In her current role she is an EHS committee member and also played a key role in the laboratory set up of the MSD London site.

Janet Brownlees is a Director of Neuroscience Biology at the new MSD site in London. She leads a group of biologists working on the mechanisms associated with Diseases of Ageing and neurodegeneration. Janet received a BSc (Hons) and PhD in Biochemistry from the Queen's University of Belfast and has held previous roles in King's College London, Glaxo Smith Kline and LifeArc. In these previous positions she has been very involved with safety leadership being a member of various Safety and Genetic Manipulation Committees, gaining strong background knowledge on the safety requirements to set up a new drug discovery research site.

Jane Hubble (MRSC) is a Laboratory Manager at Cranfield University. Jane has more than 25 years' experience in contract, industrial and research laboratories in the environmental field. She holds an MSc in swimming pool disinfection, alongside NEBOSH qualifications in Health & Safety and Environment and is a Biological Safety Advisor. As part of her role Jane develops and implements safe working policies, routinely advising on the handling, storage and disposal of hazardous chemicals and use of microbes. Jane leads on the Technician Commitment at Cranfield. She was Highly Commended in two categories of the Papin Prizes in 2019.

Heather Sheeley (BA (Hons.) MSc, MBIOL, CIOSH, FISTR) has more than 40 years of experience in laboratory research and health and safety across government, private and academic environments. Working internationally, she has an interest in sustainable and effective safety solutions to protect employees and communities. Heather has participated in numerous projects included laboratory design and multinational projects and reviews. She is a recognised trainer and presenter on topics especially related to biosafety and sustainable laboratory design. She has nine publications in journals and books and contributed to many international laboratory safety manuals and guidance documents.

Dr Alice Gallagher is currently a full-time Safety Adviser/Specialist Biological Safety Adviser at the University of Glasgow. An employee of the University for 40 years, she started as a trainee technician aged 16 progressing to senior technician. After completing an MSc and PhD she became a research fellow and has many peer-reviewed publications. Many years in research involved a safety component to the various roles culminating in her final career choice in safety and biosafety. She is a Grad IOSH member having completed a Diploma in OHS practice and a member of the Institute of Safety in Technology in Research.

Dr Jane Blunt spent 20 years as a safety professional, following a career that had consisted of research in chemistry, engineering materials and welding. She spent 15 years in the Department of Physics at the University of Cambridge where her remit covered all aspects of safety. She was responsible for developing policy, training, inspections and audits as well as advising on all matters of safety. These included outreach activities, electricity, chemicals, physical hazards, lasers, magnets, radiation and cryogens. She was a Chartered Fellow of IOSH and is a Chartered Engineer.

Dr Tanya Aspinall (CMIOSH dipNEBOSH MIIRSM MISTR) is the safety and risk manager at the Manchester Institute of Biotcchnology. With a degree in Biological Sciences and a PhD in molecular parasitology, she started her career as a medical researcher. Following an accident at work, she retrained as a safety and risk professional, and now provides strategic and operational leadership, support and professional advice, to the MIB Director and senior leadership team in order to support research, teaching and training requirements, and to ensure legal compliance and safe working within the Institute and wider University community.

Robert Derbyshire (MBA, DipNEBOSH, BSc (Hons), MISTR, CMIOSH) is a University Safety Coordinator for the central Safety Services Unit, part of the Directorate for Compliance and Risk at the University of Manchester. Robert has extensive knowledge in construction, maintenance, workshop machinery, statutory inspection compliance, accident investigation, noise and mercury assessment. Robert is passionate about "proportional" health and safety.

Louis DiBerardinis (CIH, CSP, MS) is the Director of the Environment, Health and Safety Office at MIT with BSc in Chemical Engineering from Northeastern University and MSc in Industrial Hygiene from Harvard University. He was an industrial hygiene engineer at Polaroid Corporation and an industrial hygienist at Harvard University Health Services. Mr DiBerardinis is a visiting lecturer at Harvard University School of Public Health. He co-authored "Guidelines for Laboratory Design: Health and Safety Considerations" and edited the "Handbook of Occupational Safety and Health". He served as chair of the ANSI Z9.5 subcommittee on Laboratory Ventilation from 1984 to 2006.

Tolga Durak is MIT's Managing Director of Environment, Health and Safety (EHS) and the responsible official for EHS programmes for all MIT academic and administrative units. He holds a BS in mechanical engineering, an MS in industrial and systems engineering, a PhD in building construction/

environmental design and planning, and a professional engineer licence. During his academic tenure, he held engineering and public health faculty appointments. In addition to occupational safety and health, he has research interests in building sciences with emphasis on performance, energy efficiency, retrofitting solutions, systems integration, integrated control strategies, and smart building materials.

Dr Surrinder Johal is currently the Director of Safety at Imperial College London and has over 30 years' experience in health and safety within the commercial analytical industry and higher education. Surrinder qualified as a microbiologist and started her safety career when she was appointed LGC's sole Biological and Microbiological Safety Officer in 1987. Her expertise expands across designing & establishing new laboratories and services, distribution & warehousing, small scale manufacturing, forensics and analytical laboratories across Europe, Asia and USA. She has represented the UK in international technical committees (including ISO) and in Europe (MICROVAL, EQUASE and BSI).

Ana Pedrero-Llamas is the Safety Deputy Director and Biorisk Manager at Imperial College London with over 20 years' experience in managing Health and Safety for multinational analytical and biotech companies. Following her training as a chemist, she became a forensic scientist and her expertise includes biotechnology manufacturing, diagnostics, applied research and due diligence risk profiling and risk management. She has hands-on experience in regulatory compliance and auditing in the UK, Germany, USA and South Africa. She was an active member of the AIRTO WG and is the current chair of the Francis Crick Health and Safety Audit Committee.

Charles Harrison is a Health and Safety Advisor working at the University of Exeter in Devon, England. Charles has had a varied career in higher education. After graduating from the University of Southampton with a BSc in Chemistry and an MA in Education he stayed on and worked in chemistry education, outreach, and technical roles before arriving at health and safety. After an eventful and educational start at Southampton Charles moved to Devon and specialises in chemistry and laser safety. Charles is a member of the Royal Society of Chemistry and has contributed to Chemistry World magazine.

Julia Cotton (MIIRSM, MRSPH) is the Lead Auditor for Imperial College London where she leads a team in auditing the research, teaching and support functions, as well as provides health and safety advice, writes Codes of Practice, and establishes safety management systems. Julia has worked in universities and research institutions for over thirty years at the bench, as a lab manager and as a safety advisor, and over the years has experienced enormous changes in the attitudes and practice of health and safety. Julia particularly enjoys mentoring potential safety personnel and has helped kick-start a number of safety professional's careers.

Foreword

Twenty-five years ago, in 1995, I started my career in occupational health & safety, that is to say, that I passed a written and practical examination that tested my knowledge and to some extent my understanding of what would be expected of me if I was lucky enough to be employed within a health & safety role. I mention this because what was missing in terms of my understanding was a guide, a 'how to', perhaps a Haynes type manual that takes you through a step by step of how to use the tools or, in this case, the knowledge you have gained in the real place of work, whatever the place of work. I quickly learnt that there is no such manual, like life, it is best learnt through experience. Over the next 25 years I worked within several industries, training, and education establishments, moving from practitioner to manager, director and business owner, gathering further and higher qualifications. The steps taken would always include my previous experience and my knowledge, however each sector and place of work offered differing challenges in terms of the specifics.

One of the authors of this book, Dr Olga Kuzmina, currently Faculty of Natural Sciences Safety Manager at Imperial College London, whom I have known both as one of my students of Occupational Health and Safety (OH&S), and now as a peer, mentioned to me that while there are many books on the subjects of OH&S covering most sectors, she could not find any that addressed the subject specifically in terms of higher education, universities and research. Olga had taken the initiative along with Stefan Hoyle, Head of Health and Safety in the Faculty of Natural Sciences at Imperial College London and 30 other contributors to provide a 'how to manual' for this sector. Between them, they have shared their expertise, knowledge and more importantly, their experience of applying health and safety management specifically within places of higher education and research.

Challenges for Health and Safety in Higher Education and Research Organisations
Edited by Olga Kuzmina and Stefan Hoyle

Published by the Royal Society of Chemistry, www.rsc.org

This is an important publication, it offers those who have health and safety responsibilities within higher educational and research environments and those who are seeking opportunities to gain employment within the sector the opportunity to learn from those who have applied their skills and knowledge to the unique and complex challenges in this innovative sector of the economy.

This publication is of great interest to me. As a health and safety professional at Chartered membership level with the Institution of Occupational Safety & Health (IOSH) since 2005 and holding Fellowships with Chartered Management Institute and the Institute of Leadership & Management, I continue to study. I find that I am most interested in the personal accounts of others who have met and overcome OH&S challenges and the different approaches depending on culture, location and personal approach. This manual provides me with these insights, and I would recommend it for all OH&S practitioners as a useful guide to consider in their sectors.

Mike Watson CMIOSH Dip RSA FCMI FInstLM

Preface

In 2018 we advised and actively participated in the relocation of the research activity in the Department of Chemistry, Imperial College London to a new building on a new campus. This required considerable effort to ensure processes were safe before, during and after the relocation. For example, planning for the disposal of five tons of legacy chemicals and the relocation of thousands more as well as the training of all staff and students moving into the new building. The opportunity was taken to review the health and safety management in the Department and identify opportunities to further improve the safety culture. The relocation went smoothly and as a recognition of our work we received The College Provost Award for Excellence in Health and Safety in 2019.

During the relocation process we wanted to learn from practical examples of best practices from other universities. We discovered that while there are resources for other industries, and research papers on individual accidents and approaches to safety management in the higher education sector, there is no single source of this valuable content for research organisations. This encouraged us to contact colleagues in other organisations across Europe and the US to coordinate compiling examples of best practices of application of health and safety in the higher education sector, to enable us all to learn from each other.

At the same time we contacted the Royal Society of Chemistry (RSC) publishers who were encouraged by our idea and agreed to publish our book. It was highly satisfying that the information from us and our collaborators will be shared with students, researchers and scientists, managers and safety professionals in academia and industry around the world.

Challenges for Health and Safety in Higher Education and Research Organisations
Edited by Olga Kuzmina and Stefan Hoyle

Published by the Royal Society of Chemistry, www.rsc.org

We created a questionnaire to gain anecdotal evidence about the current situation in research labs and academia with regards to the perception of health and safety. We asked our co-authors to include as many examples from their own practice as they could to illustrate the best practices and difficult scenarios that they have had to overcome. When editing the book we read through the presented information which sometimes was reassuring and sometimes eye-opening.

The initial deadline for drafts submission coincided with the COVID-19 pandemic. Many health and safety professionals prioritised their duties and got involved with advising on urgent queries regarding COVID-19 research and drafting new workplace guidance. This delayed but not prevented the publication of our book.

After reading the book we created a list of potential improvements to implement within our organisation and we hope readers will be able to do the same. We believe our book will serve as a helpful guide to higher education and research organisations to enable improvements in health and safety culture while ensuring researchers and students are working in a safer productive environment.

Contents

Challenges for Health and Safety in Higher Education and Research Organisations
Edited by Olga Kuzmina and Stefan Hoyle

Published by the Royal Society of Chemistry, www.rsc.org

CHAPTER 1

Introduction to Challenges for Health and Safety in Research

O. KUZMINA*[a] AND S. HOYLE*[a]

[a]Imperial College London, South Kensington Campus, SW7 2AZ, UK
*E-mail: okuzmina@imperial.ac.uk, s.hoyle@imperial.ac.uk

Post COVID19 Note

This is the first time I have been involved with writing and helping to produce a book. All was going well; chapter drafts were being sent to us by the various authors and we enjoyed reading about experiences and the challenges faced in our sector (we hope you do too!). Our day-to-day work was busy and my team were focused on providing great advice and support to our academic colleagues. We were preparing for various challenges and reviewing how we could improve and provide staff and students with skills to apply principles of risk assessment to their work. Then on 18th March our university was effectively closed, and all work was stopped due to COVID19.

The effect of this was acute – labs having to be safely shut down in a very short period of time, and stressful for all involved due to the logistics of being able to achieve this while also thinking about the potential impact on work, life, family and our future.

It is June 5th and in the UK the lockdown is being eased. Olga and I must admit that for the last few months this chapter has been left to gather dust while we worked remotely to support our Departments in their decision making, and provided advice on assessments for work on COVID19 related projects. For the last two weeks I have returned to our campuses in London to help with planning for the gradual reintroduction of staff and students into our buildings and for work to start again in this 'new normal'.

Challenges for Health and Safety in Higher Education and Research Organisations
Edited by Olga Kuzmina and Stefan Hoyle

Published by the Royal Society of Chemistry, www.rsc.org

Many of the chapters in this book have been written and completed before the COVID19 crisis, however I think the experiences are still relevant to practitioners in our sector and beyond, even with the introduction of social distancing and other COVID19 controls. The COVID19 work requirements will require us as a profession to adapt and be even more innovative in our approaches. We don't believe that higher education, and in fact most workplaces, will be able to return to pre COVID19 working for a while yet. Looking through the chapter titles our authors have provided, you can see that the challenges to operating successfully with social distancing and other strategies to reduce infection will require significant thought and planning. Therefore, we think we can confidently say that a revised edition of this book will contain reflections by our returning authors (and new ones) on how they and their workplaces adapted to COVID19 and how it influenced their approaches.

Crucially I would say that in my experience so far during this crisis, the work done by technical staff, admin support and estates to get buildings and procedures ready for a phased return of researchers has been magnificent. Academic leadership from our Departments and Faculty has been excellent, and we have all pulled together to provide support to each other and direction through these difficult times. We are on the cusp of having the first few researchers return to work on the 8th June, with more returning over the coming weeks. Currently the plan for our Departments is to achieve up to 25% occupancy of research buildings using cohort models or similar that allow lab-based researchers within a department to complete their lab work while remaining socially distant. Where this is not possible, other control measures must be used. Departments have 'COVID19 oversight' management teams to receive feedback from the researchers to ensure procedures reflect their needs. My team have produced 'return to work' induction materials delivered *via* online learning platforms to ensure researchers are reassured that buildings are safe to return to, and to provide information on the new procedures for them and their colleagues to work safely. Face-to-face inductions and training are moving online which creates additional challenges. The next large piece of work for us is to support our colleagues in teaching to plan for lab practicals for returning students in October.

All areas of health and safety described in this book will be impacted by the reaction to the COVID19 crisis, those of us in health and safety operations are in a unique position to provide advice and support to many operational areas. The changes to our working practices and behaviours will be profound, but we hope that it will lead to improved communication and trust between health and safety advisers, academics and students. Then, no matter what the conditions, we will achieve a safe working environment and a healthy balance between work and home.

1.1 Introduction

Research laboratories provide a unique environment of continually evolving work, varying levels of individual competence and increasingly shared workspaces between different projects and research groups. Health and safety management in research laboratories faces unique challenges to ensure that science and knowledge is advanced without being a barrier to scientific progress. Health and safety, rather than being a barrier, needs to be an enabler to these aims in these dynamic and challenging environments.

A recent review and critique of academic lab safety[1] highlights how little progress has been made over the past ten years to improve safety practices and culture in the academic environment. The aims of this book are to examine some of these challenges in detail, from the perspective of the health and safety practitioner, researcher and academic. This book is an analysis of health and safety management in the research laboratory environment across a broad spectrum of topics (predominantly within the university sector). As well as literature reviews and chapters on specific topics by individual experts we have revisited a survey, the results of which were first published in Nature.[2] It is interesting to note the differences seven years on and reflect on why these have occurred and, most importantly, what can be improved to support researchers and enable safe scientific progress.

What are the challenges that modern researchers face and how do these impact on safety culture and the implementation of a safety management system? Various barriers have been identified[2,3] or conjectured in previous studies. The shift in focus from theory, reflection and evolution of ideas to milestones, defined end points and real-world applications with monetary value has significantly changed the economic landscape at many universities. Academic and research careers are hard to achieve and maintain, with pressure from internal reviews, funding bodies, politicians, the public and business to ensure continual progress and value for money. Obtaining funding to do research is extremely competitive, even at the most prestigious institutions. Time is critical to ensuring work that is funded is then completed on time and within budget. Key stakeholders are the organisations hosting the research. They must ensure that their estates and infrastructure can keep pace with technical requirements, the need for flexible lab spaces and deliver a modern comfortable environment for study and work. The need for a physical space to be able to accommodate diverse research activities while achieving the lowest impact on the environment has never been greater, more challenging and expensive.

These factors have an impact on end user health and safety operations in various ways. From the need to 'take a shortcut' due to time pressure to meet a milestone, install an item of equipment in a less than optimal lab environment with insufficient cooling or extraction, to working in an overcrowded lab space. From the lab bench it can appear that the organisation or university is not managing its space with their scientific requirements or health and safety in mind, potentially leading to a poor relationship between researchers and 'central services' and poor safety culture. A recent review of what researchers think of the culture they work in by The Wellcome Trust[4] indicated that 'their working culture is best when it is collaborative, inclusive, supportive and creative, when researchers are given time to focus on their research priorities, when leadership is transparent and open, and when individuals have a sense of safety and security'. But too often research culture is not 'at its best'.

It is critical that health and safety in a research lab environment is practicable to those working in the labs and those that support the researchers (technicians, estates staff and contractors). Over the last 20 years there has been a definite shift of approach in health and safety from prescriptive top-down management to a consultative approach to facilitate this.

Team leaders are responsible for their group members and reliant on their PhD students, research assistants and postdoctoral researchers to ensure their projects are progressed. These individuals will have different technical competencies as well as different attitudes towards health and safety. Their previous training, either formal or informal will determine their personal attitudes and approach to safety in the lab environment. Those in the lab face the risks on a day-to-day basis and make critical decisions about risk based on their competence and the prevailing culture they are working in. They are constantly monitoring experiments and adjusting processes and procedures based on previous results. A key skill for researchers is to be able to identify when a risk has changed significantly and be able to either ask for help or devise a suitable control measure. How do researchers gain these skills? Is it trial and error? Or, is there a need to integrate risk, hazard and safety considerations into undergraduate degree courses formally and on a wider basis across the sector?

On a day-to-day basis, academic leaders are reliant on individual members of their research groups to ensure the work is done safely and the labs managed appropriately. The researchers are then reliant on the academic to continue to gain funding to continue research. This needs to be maintained to ensure work progresses, careers are successful and scientific knowledge improves. If the working environment is not optimal it is foreseeable that health and safety could be a secondary consideration and either ignored or bypassed.

The nature of cutting-edge research science in a multi-disciplinary setting (science, engineering and medicine) can lead to a focus by health and safety support services on bureaucratic solutions. This is driven by the regulatory landscape of the host country, and the need to comply with separate Regulations relating to individual hazards and risks. For example, in the UK a laboratory could be subject to multiple regulatory requirements depending on the hazards and risks they are dealing with. These hazards and risks are often fundamental to the research, restricting the ability to substitute hazards with those which are less hazardous. The lab worker normally experiences this as a large volume of paperwork, which from their perspective can be thought of as a barrier to completing research activity in a timely manner. The worker may perceive it as unhelpful and a way for the institution to 'cover their own backs'. Chapter 2 of this book examines how these risks should be managed from a legal perspective to ensure that compliance and a safe working environment can be achieved.

Technology has reduced barriers to international communication, resulting in increased opportunities for research and collaboration. Scientific research groups are international in composition. Individuals not only bring their personal experiences and attitudes to the workplace but also their cultural influences. The short-term nature of most research contracts requires scientists to continually seek further employment, either at home or abroad, so these cultural influences on the approach to health and safety are a significant factor when assessing health and safety culture within the workplace.

On the other hand, technology, and specifically combined software solutions, do not appear to have been able to be fully utilised. The challenge of creating a flexible system of interactive online assessment forms, health and safety training records, computerised lab notebooks and health and safety management software is difficult to achieve across a large teaching and research organisation with multiple disciplines. The researcher is then confronted with a fragmented system within their organisation that can be inefficient and confusing.

Are health and safety practices in research laboratories less stringent than in other sectors? In a manufacturing environment, process-related safety rules are followed. These rules may seem to be excessive and are often criticised by safety professionals.[5] The impression is that industry and manufacturing have a tighter control of health and safety due to adherence to requirements set by quality control or safety management systems, fewer changes in established processes, strict policies related to employment, and are more sensitive to the financial impact of delays to business activities. In 2019 the ACS President Elect of the American Chemical Society (ACS) ran an initiative on collaborations and safety, specifically "Bridging the (Safety) Gap between Academia & Industry".[6] As researchers move into industry and vice a versa there is an opportunity to identify what these differences are and whether they can be harnessed to improve safety culture in both sectors.

1.2 Survey Results Comparison

In 2013 Nature Journal published results from a survey among laboratory workers revealing that despite 86% of the scientists considering their lab to be "safe", almost half of them had experienced work related injuries.[2] We have conducted a similar survey and collected 427 replies from mainly UK based researchers (but also from some US, EU and Asian organisations). Approximately a third of the responses were from students, the rest comprised of postdoctoral research assistants, research fellows, technicians and other support staff. Scientists within industrial research development were also represented (Figure 1.1).

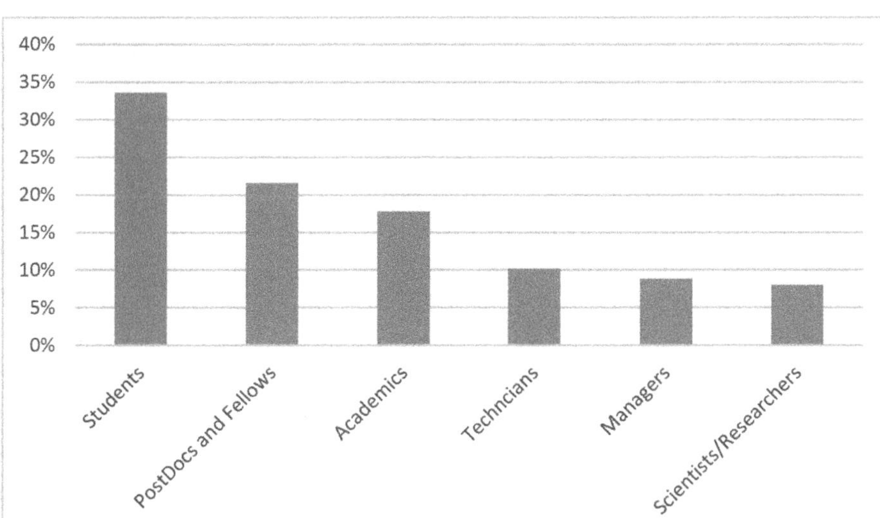

Figure 1.1 Representation of the responders of the survey by their position.

Like the 2013 survey, invitation to take part was by email and social media. Similarly, 86% of responders considered their lab as "safe" (Figure 1.2). The percentage of responders who indicated they had experienced some injury was 35% in our survey (46% in the 2013 survey).

The most commonly occurring injuries are lacerations/cuts/bites, thermal burns and needle pricks (Figure 1.3). In our survey, 14% of responders did not report their incidents to their supervisors. The willingness of an organisation to receive data on incidents is critical to establishing a good reporting culture. The organisation must review its resources for following up and closing out incidents. Clear guidance on what needs to be reported, as well as a simple, easy to use reporting mechanism is essential. Trained individuals completing the investigations, closing them out and identifying

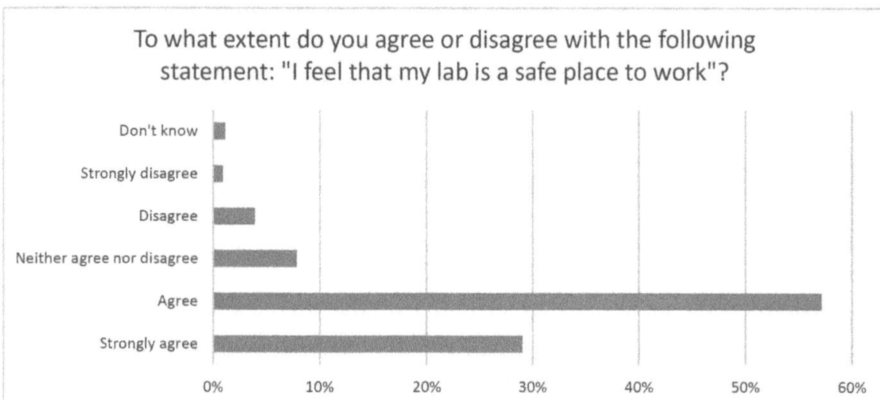

Figure 1.2 Perception of laboratory safety by the survey responders.

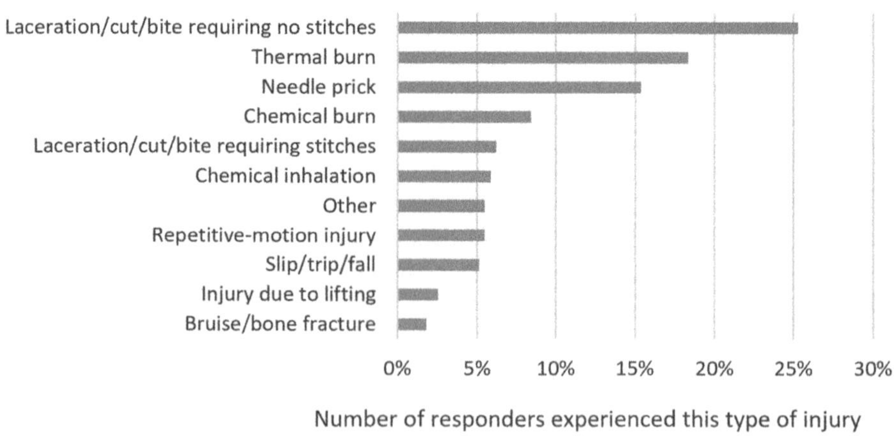

Figure 1.3 Types of injuries reported to be experienced by the survey responders.

recommendations that can be implemented to reduce risks of reoccurrence are also a key factor. The research scientist needs to be reassured that if they do report an incident that it will be dealt with quickly and efficiently.

In 2013 40% of researchers admitted that they had not received suitable safety training. Our survey indicated that less than 5% of responders had not received training on specific hazards in their work. The survey did not attempt to identify how useful the responders found their health and safety training.

It was previously indicated that two thirds of British scientists were using formal risk assessment templates; our survey showed the use of formal organisational risk assessment templates to be 70% in this sample. However, it is worrying that 2% of responders admitted not assessing the risks, and 27% conducted informal risk assessments. Risk assessment is the cornerstone of health and safety in the UK so strategies for ensuring the process is understood and applied appropriately are essential.

Figure 1.4 represents the opinions of the responders to selected statements regarding safety culture. While 89% of responders agreed that "Safety is paramount and takes precedence over all other lab priorities", almost one fifth of them also admitted that safety rules impact negatively on their productivity.

The majority of the responders admited that safety inspections improve safety compliance which is already known from other studies.[7] What is key is that the inspection process is perceived to be there for the health, safety and wellbeing of the researcher rather than institutional compliance. The majority of inspections are completed by health and safety officers, who can be seen as 'outsiders'[8] and therefore either not qualified or 'safety police', only listened to when in the lab. Some responders suggested increasing the number of inspections as a solution to improve safety in their lab. While this seems to be a logical suggestion, more inspections by safety officers, in our opinion, will not deliver the systemic improvements

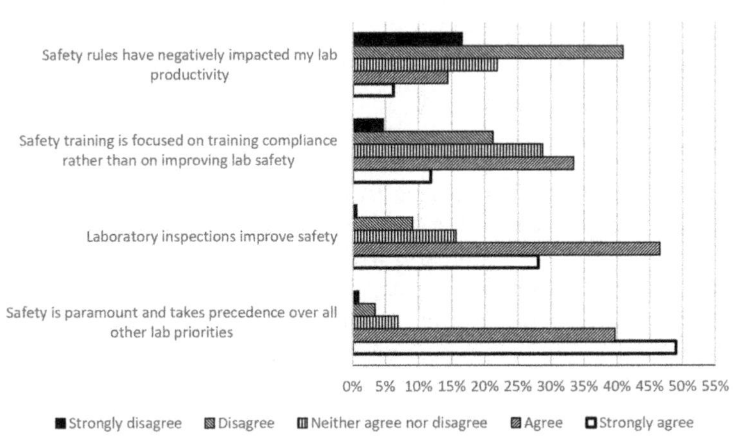

Figure 1.4 Responses to the selected statements about safety and its perception.

in lab safety that are needed. Current methods are taking a more collaborative approach, giving the responsibility for monitoring and inspecting to the researchers and their peers and ensuring that research leaders and senior management are taking part in the inspections with support from health and safety practitioners.

Of the respondents, 21% believed that they do not have any safety duties or were not sure what their safety duties are. At a time when health and safety in labs has a relatively high profile (due to serious incidents over the last 10 years) this is a worrying indication. Part of acting responsibly is to know what your responsibilities are. Then you can ensure your actions will not only keep you safe but also those around you. In the UK, the University Safety and Health Association (USHA) produced a Safety Leadership and Management Guidance document[9] for Higher Education Institutions (HEIs). This provided a template for all management and worker levels to understand their health and safety responsibilities. Potentially using this as a training tool, HEIs in the UK could improve individuals understanding of their health and safety responsibilities, depending on their position within the organisation. It is not currently possible to identify how many UK HEIs have provided training to not only senior management but all workers based on this document.

Almost one third of the responders for our survey were working in chemistry laboratories. Respondents from biological, physical and medical sciences also took part in the survey. Overall, we received a snapshot from experiences of workers from 17 disciplines, who worked in over 30 countries in many universities and research institutes.

We would echo the recent Nature Chemistry article[1] in which it is identified that to get more than 'indications' from these types of survey, health and safety practitioners and social scientists need to work together to devise suitable experiments and test hypotheses. Until then conclusions supported by data will not be possible.

The indicative responses we received have been collated into discreet categories, detailed below. Additionally, we collected the individual views on how safety could be improved at the end of this chapter in Appendix 1.

1.2.1 *Category 1: Physical Barriers*

Researchers identified that building infrastructure, ageing equipment, lack of appropriate equipment and space are serious barriers to working safely. Space constraints are often a reason why researchers neglect safety, particularly with the evolving nature of research work and research groups growing or moving in and out of universities on a regular basis. The work changes, but physical space does not. When new hazards are introduced this can result in engineering controls that are either too expensive (for example installation of a LEV system) or compromised to fit the space.

Lack of appropriate space also leads to overcrowded labs, inappropriate storage of reagents and waste, poor housekeeping and consequently more near-misses and accidents. The management of space and its contents is also an issue in research laboratories. Labs do not always have appointed managers or technical services to assist researchers in waste removal and housekeeping duties. These duties can often be neglected if the researchers are expected to complete these duties, partly due to a lack of time and partly due to attitude *i.e.* 'I'm here to do research, not manage the lab on a day-to-day basis'. This is a persistent concern which does not have enough attention in the literature. In 2006 Francisco Javier Penas and others[10] described the implementation of industrial standards when designing and starting up a laboratory for chemical engineering teaching. Their experience proved to be a successful example of setting high safety expectations when assessing the risks from every point of view. The detailed design of the research labs and considerations when starting a new research site will be discussed in Chapters 12, 13 and 14.

The costs of clearance of legacy materials that have accumulated in a lab space can be considerable and will normally have to be paid for by the host department or School. In 2017–2018 when the Chemistry Department at Imperial College London relocated to a new building over 5 tonnes of chemicals were disposed of. Legacy material introduces increased risks associated with fire loading and takes up valuable working or storage space. Chapter 16 provides an insight into legacy materials and issues around old equipment and facilities.

1.2.2 Category 2: Economical Barriers

When responding to our survey researchers have blamed a lack of finances and limited funding streams for poor safety performance. Proposals are often submitted without consideration of what safety implications will arise if the grant is awarded and the work begins. Ideally a risk assessment of the work will be completed so that the physical spaces and engineering control measures can be designed in and costs identified at the earliest stage. As not all applications result in funding, there is limited appetite to spend a considerable amount of time detailing required safety infrastructure or ongoing requirements for each proposal submitted. However, if the application is successful, additional funds may be needed to ensure the required safety controls can be implemented. This can be significant sums of money. The funding bodies may assume (or require) that the organisation will cover the safety costs of a project *via* central funds. For example, if a scientific proposal involves work with high powered lasers, the UK Artificial Optical Radiation Regulations 2010 require engineering controls to be in place to prevent exposure to hazardous lasers. The costs for a system to achieve this can be significant. These costs will depend on several factors including:

- What size is the space?
- What are the existing services?
- What new mechanical and engineering or electrical infrastructure will be required?
- What interlocks, containment levels, safety protocols or licences does the research involve and what requirements or restrictions does that impose?
- What furniture is required?
- What equipment (if any) is to be included?
- What level of professional fees will be required (*i.e.* how much design will be required from architects, engineers, specialists *etc.*)?

If the funding body cannot fund the safety requirements the department or researcher must. This can lead to delays if the design assessment has not been done in advance of the work being funded. This is where safety requirements will be perceived as "barriers for effective work" and the organisation itself as "not supportive" as the funded work will not be possible to begin until it is safe to do so. It also places the safety officer or adviser in an awkward position as they could be pressured to 'relax' the rules or are perceived as barriers to effective research.

It is unfortunate that these economical barriers are not getting more attention in the literature. An individual approach may be needed for each organisation so that a process for efficiently identifying potential costs for new research can be established.

1.2.3 Category 3: Organisational Barriers

Our survey indicated that researchers are concerned by the lack of clear guidance, insufficient supervision and training, lack of accountability and a burgeoning bureaucracy. The reliance on documentation is perceived as a way for the organisation to reduce regulatory risks rather than as tools to aid the scientist and reduce the risks to themselves. The varied nature of work in a research laboratory and decentralised management of universities can exacerbate these issues. This can result in multiple documents from different parts of the organisation for the management of the same risk. This can lead to confusion or delays, particularly as the increase in multi-disciplinary work between departments, schools and laboratories continues. The lack of ease of access to electronic health and safety management systems in the research sector is a major problem for researchers, managers and health and safety practitioners. If the end user cannot easily find the correct documents or forms to complete easily, the researcher can be tempted to ignore the requirement.

A common criticism of health and safety management is that it is reliant on documents, rather than practicality. Researchers need to be able to access high quality training that reflects their type of work and be directly involved with decision making when it comes to the implementation of new policies or procedures. Researcher consultation takes time but will result in better practical solutions to safety issues and higher rates of compliance. Some organisations have instigated student-led organisations,[11–13] involving students in the management of safety[14,15] and developing specific safety training programmes (especially with practical components).[16–21] As noted in the recent Nature Chemistry article, there is a pressing need for more research on how to devise effective safety training, how to measure its impact and achieve improved safety attitudes, not only in the undergraduate teaching labs but in the research labs as well.[1] This is addressed in our book in the chapters dedicated to safety management of new sites (Chapter 9), spin-outs (Chapter 15) and training of safe chemists (Chapter 6) and generally throughout the university (Chapter 5).

Our ability to teach and educate new students on approaches to risk and to have a sensible view of risk is a key area that needs further development. The undergraduates of today are the researchers of tomorrow, so if safety culture is to improve further, training at this level is required. There will be benefits to integrating risk and safety management into undergraduate degree programmes. This would need to be carefully designed to be relevant to the field of study and connected to work that the student is undertaking. It should also form part of the marks associated with the work. For example, an MRes course director at the Imperial College London Department of Life Sciences introduced training in fieldwork risks and completion of a fieldwork risk assessment into their course. Students are provided with training and information and asked to write the assessment for their proposed fieldwork. The assessment is then marked according to specified marking criteria and provides up to 5% of the marks towards the student's degree. This incentive towards the final marks resulted in significant improvements in the assessment quality and the students' feedback was very positive in terms of the learning outcomes and the skills gained for future employment. We are also seeing innovative approaches to teaching practical skills to students, for example, the 'Chemical Kitchen' is a part of the Imperial College London Pedagogy Transformation Project,[22] an interdisciplinary practical course that introduces students to the mindset and fundamental skills needed in a laboratory setting through a non-threatening parallel of cooking. It aims to teach practical laboratory skills, planning, creativity, safe working, precision, dexterity, making and recording observations, and the application of knowledge. Identifying opportunities for this type of learning and communicating safety knowledge to students in this manner increases the chances that key skills are learnt and embedded for future use. In this book, our authors provided insights into the teaching of students in Chapters 6 (Chemistry), 7 (Medicine) and 8 (outside the classroom).

Separately, a perceived lack of action on reported safety related infrastructure issues is highlighted in research institutions, particularly the university sector. Due to the complex nature of scientific buildings, an end user may not appreciate the complexity of fixing what appears a relatively simple issue. Feedback to end users on the current status of reported defects is critical to ensuring good communication between scientists and the estates team. Potentially delays in repairs to engineering controls (for example local exhaust ventilation systems) and therefore completion of experiments can increase risks if the experiments are conducted without the controls in place.

A successfully integrated safety programme has been reported that included institutional, administrative, and faculty efforts combined with an initiative from graduate and postdoctoral researchers.[21] However, there is a lack of successful documented examples on how these communication and collaborative networks can be achieved. Best practice examples in these areas would be a useful resource for research institutions to learn from.

Inevitably in the research lab there will be several rules for scientists to follow. In some cases, these rules are perceived as being prohibitive and overstated. For example, to comply with chemical storage requirements researchers will be required to store reagents in a dedicated location, not necessarily at their immediate work bench. If the reasons requiring the storage location are not made clear during training or induction the researcher may not follow the guidance, leading to a potentially hazardous situation.

Although discrete disciplines exist within a university setting, these lines are becoming blurred as more research funding focuses on interdisciplinary research and providing solutions to 'real world issues'. This can result in researchers from one discipline, where they have achieved success in educational and research terms, to venture into another domain where their knowledge of the risks associated with this new area may not be so competent. This is particularly an issue in 'hack spaces' or 'innovation spaces' where the crossing over of different disciplines is desired. Health and safety considerations, lab design and how these areas are managed are critical to ensuring a safe working environment. These issues are explored in detail in Chapter 14 'Innovation Spaces – the new campus risk paradigm'.

The safety practitioner has a critical role to play in coordinating safety management systems and providing advice. The challenge is to provide advice on a range of subjects to meet the needs of researchers in the modern laboratory. Multiple hazards and risks may be present in any one laboratory. It is unlikely that any one individual can provide competent advice at a technical level on a wide range of hazards, for example, chemicals, lasers, ionising radiation, biological agents, machinery *etc.* This requires a collaborative approach to health and safety management in these types of situation. Scientists and technical staff have the relevant technical expertise and the safety practitioner then assists to guide them through the key concepts of safety and risk management. Unfortunately, the feedback received in our survey showed that researchers sometimes receive inconsistent safety advice and safety professionals themselves don't always lead by example. Clear feedback on safety issues from those advising helps ensure researchers can continue working safely and with minimal delays. The health and safety practitioner must understand their own limits of knowledge and competence and seek advice where it is lacking, or insufficient and know when to refer to other competent professionals. Part of the complexity for higher education institutions and research institutes is the size and locations of their sites. Many universities are on several different campuses, or sharing campuses with other organisations, for example medical campuses and the NHS in the UK. This makes planning for emergency responses more complicated. Specific considerations should be made to ensure all those who may need help can access it quickly and efficiently while ensuring the emergency services themselves can be guided to the required location, which on campuses with multiple entry points and buildings is not always clear. Liaising with emergency services is the subject of Chapter 11 of this book.

1.2.4 Category 4: Behavioural Barriers

Apathetic attitudes, carelessness, ignorance, work culture, cultural background, poor communication, overconfidence, lack of common sense, "old" habits and laziness were often named among factors negatively affecting research safety. A response from the survey was: "People don't realize how bad something can be until it really happens". Researchers are often facing deadlines and may have to make a choice between productivity and safety. Much of the time the safety infringement will not result in an incident reinforcing the notion that safety is not required at that point in the experiment. Only when a combination of variables align will the incident occur. Behaviour in the lab setting must be led by the most experienced and senior managers. New members of a research group will generally replicate the behaviours they see around them. If that involves a lack of care when it comes to health and safety issues, this will in turn increase the risks of an incident occurring at some point in the future. Therefore, safety rules and procedures cannot be perceived as "unnecessary", "childish", and safety "micromanagement".

Solutions to these issues rely on the safety practitioners 'soft skills' and ability to create collaborative teams across areas that they have no line management responsibility for. Improving communication of safety matters,[23] "embedding" a safety professional in high hazard department,[24] and following established guidance and recommendations[25–28] also are essential to improving safety culture. We could find limited documented evidence of implementation of a safety behaviour programme in a research environment. This is an area worthy of further study.

In this book we have several chapters written by non-UK collaborators. Research science and higher education establishments will often have high levels of international staff and students. Researchers will often move between different institutions and countries, and their attitudes or approaches to safety culture and processes will be imported to their host institution. Communication of local safety culture and expectations is critical in a higher education or research environment. Chapter 3 of our book examines this issue from a European perspective.

1.2.5 Category 5: Industry-specific Barriers

Due to the nature of the research work environment and the preference for shared lab areas, several simultaneous unrelated experiments may be occurring in any one lab area. The hazards and risks associated with these will also be varied, as well as the controls required to keep risk down as low as reasonably practicable. It is impossible to build a specific lab for every process, especially as the processes often change as the research evolves. New reagents are constantly being tested, novel compounds developed, and experiments being amended due to previous results and published research. Researchers claim that "it is often the case that the safest way to perform an operation is not particularly convenient, especially where a minor increase in risk yields a significant increase in efficiency".

In the research environment, especially in universities, those in the lab and doing work may have the least amount of experience, for example PhD students and new postdoctoral researchers. They are often on short term contracts and under pressure to produce results to ensure their career can progress. The team leader will be a guiding role and will be available for advice but not generally in the lab with their staff and students. The tension between time required to complete study or work and safety rules creates pressure for those working at the lab bench to take short cuts when it comes to safety. The relatively high turnover of staff and students from these positions inevitably leads to research groups losing valuable experience and knowledge of safety matters. In our experience there is little evidence of teams having formal handovers of local safety information to new staff and students. The safety culture in any one research team is being undermined by the continual replacement of researchers. Even responsible researchers are struggling to pass on their knowledge when leaving to preserve the same level of safety awareness. Reducing the risks from this is difficult without having lab managers or technical staff to support the researchers and not only provide advice but also monitor safety procedures in the lab and ensure consistency of approach to lab rules that are in place over a long period of time. The constant turnover of lab staff and students also requires the institution to ensure its training and induction procedures are restated frequently to ensure new starters have the same information as those that came before them. Reiteration of safety training and providing clear and concise practical guidance will help to make work in the labs safe. However, nothing will work without the support of the supervisors and dedication of research and support staff. The organisation needs to be focused on safety and have a coherent plan and allocate enough resources to support the implementation and maintenance of safety. In our experiences the technical support teams within departments, schools or faculty provide an organisational memory and are often called upon to assist academics and teaching beyond their normal duties. They are often involved in training in equipment and procedures and have key skills that the research and teaching activities benefit from. The role of technical staff in maintaining good health and safety is the subject of Chapter 10.

One item that the university and research institutions should be leading on is how they can reduce their impact on the environment. Running science labs and supporting research and teaching is an energy hungry business and can have significant impacts on the environment if not suitably managed. Various national schemes are being implemented across the sector to provide incentives for improvements to sustainability and the environment. These issues are discussed in Chapter 17, 'Green and Sustainable while avoiding risks'.

An area that has received more attention in the last five years is stress and wellbeing within the higher education sector. Several initiatives have been launched for not only students but also staff, including access to confidential care lines, designing work programmes and teaching to reduce risks of overwork, raising awareness related to mental health issues and encouraging individuals to seek help where needed. Mental health and other training programmes for managers, tutors and others have been instigated and broadly they appear to be providing support where needed. A report by Universities UK in 2018[29] recommended a 'whole universities approach' meaning that all aspects of university life promote and support student and staff mental health. Many organisations have invested in training for mental health first aiders and awareness campaigns, however changing the core systemic issues to reduce risks of stress and overwork for staff will require in depth analysis, management structures and review of resources. Chapter 4 looks at stress and workplace wellbeing in a Higher Education (HE) and Research establishment in more detail and provides insights on how this is being managed in some institutions.

Our book ends with a chapter titled 'Organise not Agonise – Getting the Best from Audits and Inspections'. This chapter covers concerns, successes and causes of failures around auditing, starting with a focus on interfaces, where there is often uncertainty and lack of knowledge as to who is responsible or in control in a research or university environment. Such areas have been the cause of most health and safety management problems, and as such are generally the most fruitful for an auditor.

1.3 Closing Remarks

We hope the information contained in this book will be of interest to you and provide useful examples of how to approach certain issues within the research and higher education sector. Ideally, it will lead to greater discussion in these organisations at senior management level with Finance Officers, Human Resources and Safety Directors, about how to further improve safety performance. By putting these chapters together, we have learned a lot, and it is our wish that you do to.

Appendix 1

The survey provided some interesting views on how safety could be improved in the research setting. These are summarised here:

- Having a system where an individual requires retraining if they have been seen to use an unsafe method *x* number of times (*x* can be varied depending on how common the technique is and how unsafe they're being). Similar to a 'three strikes and you're out' system.
- Introducing a well-advertised (and adequate) safety upgrade funding scheme. This is infrastructure and can't be paid for, in the main, by research grants.
- Centralising and digitalising health and safety associated paperwork.
- Offering supervisors and managers more training into good management skills and how they need to balance managing experimental outcomes and workers mental and physical health. Often a culture prevails of pushing or demanding too much from their workforce, and this leads to mental health issues and lapses in health and safety.
- Ensuring that senior research investigators visit their laboratory space at least twice a week. I know of one group that has received only two visits from their supervisor/research lead in one year!
- Focussing on real risks, rather than perceived risk. A safety culture anchored in data and statistics would be very welcome.
- Getting supervisors to visit the labs more often and to accept the word/guidance of technical staff.
- Appointing informal safety officers (like fire wardens).
- Having "Keep your lab tidy" campaigns.
- Holding monthly updates on lab activity across the research group while paying close attention to any safety breaches.
- Implementing visible sanctions for non-compliance (such as closing labs).

Following on the last suggestion we also received a quote from one of the responders: "I would like to think health and safety could be improved by providing incentives but trying that I have achieved no tangible results in the past. Funny enough, it is by threatening disciplinary action that engages people more but for all the wrong reasons". We hope the experiences shared in this book will help you to avoid situations where sanctions and disciplinary actions have to be implemented.

References

1. A. D. Ménard and J. F. Trant, A review and critique of academic lab safety research, *Nat. Chem.*, 2019, **12**, 1–9.
2. R. Van Noorden, Safety survey reveals lab risks, *Nature*, 2013, **493**, 9–10.

3. I. Schröder, D. Y. Q. Huang, O. Ellis, J. H. Gibson and N. L. Wayne, Laboratory safety attitudes and practices: A comparison of academic, government, and industry researchers, *J. Chem. Health Saf.*, 2016, **23**, 12–23.
4. Wellcome Trust, *What Researchers Think About the Culture They Work In*, 2020, https://wellcome.org/reports/post-brexit-agreement-research-and-innovation-outcomes-simulated-negotiation-process, last accessed 23 June 2020.
5. S. Dekker *The Safety Anarchist: Relying on Human Expertise and Innovation, Reducing Bureaucracy and Compliance*, Routledge, 2018.
6. B. A. Charpentier, How will you collaborate in 2019? *Chem. Eng. News*, 2018, **96**, 1–3.
7. I. Schröder, D. Y. Q. Huang, O. Ellis, J. H. Gibson and N. L. Wayne, Laboratory safety attitudes and practices: A comparison of academic, government, and industry researchers, *J. Chem. Health Saf.*, 2016, **23**, 12–23.
8. P. Scherz, Risk, prudence and moral formation in the laboratory, *J. Moral Educ.*, 2018, **47**, 304–315.
9. USHA, *Leadership and Management of Health and Safety in Higher Education Institutions*, 2015, 1–20.
10. F. J. Peñas, A. Barona, A. Elías and M. Olazar, Implementation of industrial health and safety in chemical engineering teaching laboratories, *J. Chem. Health Saf.*, 2006, **13**, 19–23.
11. J. Ting, *et al.*, Building a stronger, sustainable safety culture at the University of Chicago, in *Abstracts of Papers, 257th ACS National Meeting & Exposition*, Orlando, FL, USA, 31 March–4 April 2019.
12. S. N. Anderson, V. P. N. Nziko, C.-W. T. Chang and A. C. Hengge, Enhancing safety in academic research laboratories: Creating a student based safety committee, in *Abstracts of Papers, 248th ACS National Meeting & Exposition*, San Francisco, CA, USA, 10–14 August 2014.
13. J. A. Martin, Secrets to success: Show up, do, in *Abstracts of Papers, 258th ACS National Meeting & Exposition*, San Diego, CA, USA, 25–29 August 2019, 2019.
14. N. Fredstrom, Building a safety mindset: An undergraduate's perspective on chemical safety in academia, in *Abstracts of Papers, 254th ACS National Meeting & Exposition*, Washington, DC, USA, 20–24 August 2017.
15. X. Wang, Peer to a mentor: Engaging graduate students in laboratory safety at northwestern, in *Abstracts of Papers, 257th ACS National Meeting & Exposition*, Orlando, FL, USA, 31 March–4 April 2019.
16. S. Tantayanon, Development of chemical safety program at Chulalongkorn University, in *Abstracts of Papers, 240th ACS National Meeting*, Boston, MA, USA, 22–26 August 2010.
17. C. Siburt, *et al.*, Empowering graduate students to lead a culture of safety: Developing a peereducation chemical safety training workshop, in *Abstracts of Papers, 252nd ACS National Meeting & Exposition*, Philadelphia, PA, USA, 21–25 August 2016.

18. R. J. Ceru and A. Azadnia, Hands- on approach to teaching a laboratory chemical safety course at Michigan State University, in *Abstracts of Papers, 241st ACS National Meeting & Exposition*, Anaheim, CA, USA, 27–31 March 2011.
19. S. Wiediger, Adding safety to the curriculum one research project after another, in *Abstracts of Papers, 255th ACS National Meeting & Exposition*, New Orleans, LA, USA, 18–22 March 2018.
20. K. A. Scanlon, K. E. Traum, S. E. Krall, D. Wilson and K. Johnstone, STAR-LITE: An innovative approach to delivering laboratory safety training to students, in *Abstracts of Papers, 235th ACS National Meeting*, New Orleans, LA, USA, 6–10 April 2008.
21. T. Woerner, S. Baldwin and P. McMillan, Safety in undergraduate chemistry: It takes the whole department, in *Abstracts of Papers, 252nd ACS National Meeting & Exposition*, Philadelphia, PA, USA, 21–25 August 2016.
22. R. Kneebone, A. Spivey and J. Youssef, Chemical Kitchen, 2019, https://www.imperial.ac.uk/about/leadership-and-strategy/provost/vice-provost-education/transforming-our-pedagogy/funded-projects-fons/chemical-kitchen/, last accessed 23 June 2020.
23. L. Redfern and M. Blayney, Safety minutes: Consistent way to promote and sustain the commitment to research safety, in *Abstracts of Papers, 258th ACS National Meeting & Exposition*, San Diego, CA, USA, 25–29 August 2019.
24. S. Sigmann, Collaborative efforts between faculty and embedded safety professionals to improve critical thinking skills of undergraduates, in *Abstracts of Papers, 254th ACS National Meeting & Exposition*, Washington, DC, USA, 20–24 August 2017.
25. N. R. Council, *Safe Science: Promoting a Culture of Safety in Academic Chemical Research*, 2014. DOI: 10.17226/18706.
26. D. C. Finsterty, *Safety in Academic Chemistry Laboratories*, American Chemical Society, 2017.
27. ACS, *Creating Safety Cultures in Academic Institutions: A Report of the Safety Culture Task Force of the ACS Committee on Chemical Safety*, 2012, https://www.acs.org/content/dam/acsorg/about/governance/committees/chemicalsafety/academic-safety-culture-report.pdf, Last accessed 23 June 2020.
28. C. Gaskell and C. Parkinson, *Responsible Research. Managing Health and Safety in Research: Guidance for the Not-for-profit Sector*, Institution of Occupational Safety and Health, 2012.
29. J. de Pury and A. Dicks, Stepchange: mentally healthy universities. https://www.universitiesuk.ac.uk/stepchange-mhu, last accessed 23 June 2020.

CHAPTER 2

Managing Legal Risks in an Uncertain Environment

PHILIP CROSBIE*

Health and Safety Solicitor, UK
*E-mail: p_crosbie@hotmail.com

2.1 Introduction

One of the key facets of good health and safety management is the assessment, and then mitigation and control, of health and safety hazards. Those responsible for assessing the hazards applicable to our activities are asked to consider the likelihood of that hazard arising, and the severity of the outcome should that hazard be realised. This is the core concept at the heart of safety management, and is captured within the broad legal duties under the Health and Safety at Work *etc.* Act 1974 ("HSWA").

Understanding the levels of risk then enables an organisation to properly allocate resources, time and effort to minimising those risks, or removing them altogether. The higher the risks, the more effort should be put into controlling them. No one can disagree with that approach and ultimately it is comforting to know that the regulators and the regulated are on the 'same page' in this regard.

Applying that task in the academic or research environment, when an assessor sits down to conduct that exercise, what is in their armoury? Risk assessors in all industries are assumed to have an understanding of both 'likelihood' and 'severity' ratings before identifying the appropriate measures of mitigation. However, in an environment where progress is based on

Challenges for Health and Safety in Higher Education and Research Organisations
Edited by Olga Kuzmina and Stefan Hoyle

Published by the Royal Society of Chemistry, www.rsc.org

innovation and pushing boundaries, there are often very few comparators or precedents to look to in order to understand the likely outcome.

Academics and researchers, therefore, must take a step into 'no man's land' to some extent. If institutions are restrained to only those activities where the outcomes are known, research and academic evolution will be severely limited. Yet who is brave enough to take a step into the unknown when those who propagate the health and safety message often refer to the potential prison sentences that await those that break the law?

2.2 Context

For the purposes of this chapter, I shall avoid reference to the moral and commercial realities of risks and their management, and shall focus on the legal regime. This is not to ignore the importance of making sure that our colleagues and friends are safe, but acknowledges that the legal regime typically acts independently of emotion.

Despite the stated avoidance, I do wish to highlight the success that safety management programmes can have when they focus on human reactions and social norms, rather than recounting the provisions of statute and regulation. In my experience, individuals will not be persuaded of the importance of health and safety by reference to the potential custodial sentences that may be imposed, or the fines that can be exerted on organisations that get it wrong. Often, any interest that follows such messaging is superficial and sporadic, and will wane as the threat of financial penalty or lack of liberty subsides. Readers may have seen similar situations where there is a regulatory focus on a particular topic within their portfolio; short-term interest that is directed at avoiding a liability, rather than resolving the issue.

A much more successful way forward is to consider the real reasons why people want to 'get health and safety right'. They want to ensure that those who work or study at an organisation are able to go home to their loved ones at the end of the day without coming to harm. If we forget this basic premise, and replace it with relatively empty threats of criminal sanction, then the battle to create a strong safety culture will be lost at the outset.

2.3 Legislative Background – Organisations

The broad obligations placed on organisations arise from three particular provisions contained within the HSWA. When we refer to an 'organisation', that will normally be a reference to a legal entity capable of employing others, such as a limited company, a charity, or a partnership.

2.3.1 Section 2 HSWA

Section 2 HSWA requires employers to take reasonably practicable steps to ensure the health, safety and welfare of employees. Whilst recognising a moral and commercial duty to protect employees, it is Section 2 that legally gives rise to the basic duty of care. Section 2 is broken down further to highlight elements of the duty that will be relatively obvious to the reader; the provision of safe equipment; maintenance of the workplace; suitable training and supervision; and appropriate welfare facilities. If there was an incident involving an employee, it is often a breach of Section 2 that forms the foundation of a regulatory investigation and subsequent potential enforcement.

Of interest at this stage is the reference within Section 2 to ensuring the health and welfare of employees, as well as their safety. In my experience, employers are well versed in protecting employee safety and have started to tackle issues of health somewhat. The management of welfare, however, is a topic that has received less attention until recently.

Over the last few years, wellbeing initiatives have become common across organisations. Staff have been offered free fruit, more flexible working arrangements, and other similar ideas, despite the fact that the duty towards staff welfare has existed since 1974, as with health and safety.

This chapter intends to consider all three aspects of the duty of care, in the context of an uncertain academic/research environment. It is simpler to see the safety risks associated with working in these circumstances, but the health and welfare risks are just as important. By way of example, the use of particular chemicals or the reactions from the same may have impacts on health that could not be predicted at the outset. From a welfare perspective, there are the obvious impacts on welfare where the conducting of testing or research requires long and unsociable hours to be worked. More widely in academia, however, there may also be uncertainties around funding and job security; these more remote factors have to be built into the overall risk management approach; they may create distractions that lead to physical safety risks. In essence, all three need to be considered in conjunction.

2.3.2 Section 3 HSWA

Section 3 HSWA encompasses similar concepts as the preceding section, but the target of the duty is the main differentiating factor. Section 3 creates a duty on employer organisations to ensure that non-employees are not exposed to risks to their health and safety. The wording of the duties within Sections 2 and 3 are different, yet the expectation on an employer is very similar.

In an academic or research environment, the reference to 'non-employees' within Section 3 casts the net extremely widely and brings about a consideration of a large number of individuals who may come into contact with academia or research. 'Non-employees' for this purpose captures anyone who may be affected by an organisation's undertaking, or who may come into contact with the organisation in some way.

Students (including post-graduate students) are an obvious example of those to which a Section 3 duty is owed. Students are not employees, but will be directly affected by the actions or inaction of an educational organisation.

Added to students are third party contractors, visitors, members of the public, and those employed by other organisations.

The way in which each of these categories of 'non-employee' will be protected in an uncertain environment will differ, for example:

- Students – The average student will be under the age of 25 and may require further consideration as to what activities they should be exposed to; it is accepted that these individuals will typically be less risk averse. This poses its own issues in an uncertain environment, as students may be more willing to accept the risk of the unknown than an organisation would. It is the task of the organisation, and its staff, to control and direct this enthusiasm, particularly where students engage in academic activities outside of the protection of the campus environment.
- Visitors and the public – Many education organisations are based on city centre campuses where the interaction with members of the public cannot be avoided. Visitors can pose their own challenges with regards to uncertainty; they are often unpredictable and, from experience, may not follow what would appear to be common sense. Why follow a clear pedestrian route when it is more direct to climb over a wall?
- Contractors and organisations – It must be remembered that just as organisations owe a duty under Section 3 to contractors and other organisations, those parties owe reciprocal duties. If neither party can predict an outcome or foresee the result, it would be advised to pool resources and knowledge to find the most appropriate way forward.

One of the primary aims of academia is to increase accessibility, whether that be through open days, secondments, or working with industry. Whilst admirable, it is easy to see how these priorities amplify the numbers of those to whom a duty under Section 3 is owed. This will only ever increase; as organisations look to attract more visitors to their buildings and greater attendance at events, as well as a growing expectation to attract students from more varied social backgrounds.

By way of example, some universities in the United Kingdom are well on their way to fulfilling the role of hotelier, festival organiser, or another similar role that decades ago would have been an entirely alien concept. The future development of the use and role of academic institutions and buildings presents another area of uncertainty looking forward. Those who are tasked with expanding the revenue streams of the organisation will no doubt face tensions with individuals charged with academic advancement. Can they co-exist safely?

2.3.3 Section 4 HSWA

One of the more modern concepts of academia or research involves inviting third parties to conduct research or other activities on an organisation's premises. Section 4 HSWA creates a duty on organisations that make places of work available to others. The duty requires organisations to ensure that premises can be used safely.

When thinking about managing risk in particular, the broad obligations contained within the HSWA and outlined above are refined by way of various regulatory obligations which provide more specific instruction on what compliance looks like.

The Management of Health and Safety at Work Regulations 1999 ("MHSWR"), Regulation 3 in particular, captures the basic obligation on employers to make suitable and sufficient assessment of risks to the health and safety of employees and non-employees, echoing the distinction created by Sections 2 and 3 HSWA. Further regulations exist and seek to govern a range of safety topics, including the control of asbestos, the management of working at height, the way in which hazardous substances shall be handled, and the obligations to provide suitable welfare facilities. I have not sought to detail those Regulations within this chapter as to do so would be a significant task, and would not take us much further in the debate. Compliance with the regulations is predicated on the fact that you know what risks you are dealing with. In an uncertain environment, we may not have such privilege.

2.4 Legislative Background – Individuals

Whilst we have recently seen prosecutions for organisations in the academic and research sectors, we have very rarely seen prosecutions of individuals working in the same environments. Whilst legal duties do exist and apply to individuals, a regulator will normally be reluctant to prosecute for breaches of the same in the absence of extreme negligence; preferring instead to focus on their employer. There are, however, important lessons to be drawn from the legal expectations placed on individuals and the lack of enforcement against individuals should not be seen as a lack of expectation.

2.4.1 Section 7 HSWA

Section 7 HSWA requires all employees to demonstrate reasonable care for their own health and safety, and that of others who may be affected by their actions or omissions. Employees should be reminded that all those who come to work are expected to demonstrate an element of common sense; following reasonable instructions, wearing the equipment provided by their employer, and reporting issues that they are aware of.

In an uncertain environment, organisations should take a level of comfort from the fact that employees are legally bound to exhibit a level of reasonable care. This will evidently extend to the consideration, and planning, for an uncertain environment. However, individual culpability will most likely be considered only where the organisation in which they are employed has given them the tools with which to succeed:

- An employee can only 'break the rules' if they are fully aware of what the rules are – employees should not be criticised for failing to have read and understood vast quantities of safety paperwork that is held on an impenetrable online platform, particularly where different documents may contradict one another with regard to safety expectations.

- An employee can only be considered to have 'misbehaved' if the organisation promotes good behaviour and condones those who do not comply – if everyone adopts poor practices and shortcuts, then one employee cannot be singled out for 'following the crowd'.
- An event can only be described as 'exceptional' if there are suitable methods to capture incidents and near misses and resolve the same.

In essence, an organisation will need to properly train and supervise its staff and then enforce the stated rules if it wants to subsequently benefit from saying an individual went on a frolic of their own. Individuals cannot, and in my experience will not, continue towards impending disaster in reliance on the fact that any legal breach will rest at the feet of their employer. The respective legal obligations simply do not come into the decision-making process. It is therefore incumbent on employers to create an environment in which the right decisions are the easiest.

2.4.2 Section 37 HSWA

Section 37 HSWA provides for a more enhanced expectation of senior employees or those responsible for 'putting people to work'. Those in senior posts within organisations may be caught by Section 37, which creates an offence for individuals who consent or connive in health and safety offending by their employer organisation, or contribute to organisational offending by way of their neglect.

Individuals who 'consent or connive' are those that are aware of a risk or danger, but fail to take the appropriate remedial action. When working in an uncertain environment, the lack of predictability will mean that 'consent and connivance' offences are unlikely as individuals will not hold the knowledge that they are then accused of failing to act upon. However, individuals will be aware of the fundamental health and safety requirements of the organisation, and will be expected to enforce the same.

The version of 'neglect' for these purposes is a reference to those who turn a blind eye or bury their head in the sand.[3] They do not wish to know about potential issues within an organisation, therefore Section 37 includes reference to 'neglect' to ensure that they are still held to account. For this reason, senior individuals should seek to inform themselves of the potential outcomes when dealing with an uncertain environment. They should equip themselves with an understanding of the way in which the risk assessment has been arrived at, and the assumptions and calculations that have been made. Whilst the senior individuals are unlikely to have any greater ability to foresee the actual outcome, they should be satisfied that planning for the most likely eventualities has been properly thought through.

2.5 The Health and Safety Journey...

Having considered the basic legal regime, we now turn to the key elements that make up any assessment of whether an organisation is meeting its legal duties.

2.5.1 What Risk?

In order to understand how organisations should approach the management of risk, it is important to firstly consider what is meant by that term 'risk'. In many health and safety discussions, the word 'risk' is too commonly abused, and there is a very real possibility of its misuse.

When referring to 'risk', it has been made clear in previous case law[1] that it must be something material; risks that are incidental to everyday life are not those where a legal duty to manage arises. The first step, therefore, when considering the approach to risk management, is to consider the risk portfolio of the organisation; asking why activity conducted within an institution or research body could be different from everyday life. An organisation should have a clear understanding of the risks within the institution that it is duty bound to control; the maintenance of a comprehensive risk register is often viewed as good practice.

This call to focus on 'material' risks allows those working in an uncertain environment to avoid the distractions that inevitably arise from academia and research. When dealing with the dangers of an uncertain environment, it is important that focus can properly be placed on where it can add the most value with regard to mitigation and control.

Such is the battle to reclaim the understanding of 'risk' that the regulator of health and safety, the Health and Safety Executive ("HSE"), has set up a Myth Busters Challenge Panel[2] in order to dispel misunderstandings around health and safety risks and to avoid individuals abusing the term as a way to avoid undertaking new activities. A robust approach to the 'risk' definition should not be seen as a blocker to research progress, but as a tool to ensure that it occurs in a safe way.

When working within an uncertain environment, therefore, organisations should ask what activities present material risks, *i.e.* what is it about the environment that moves it away from everyday life? I appreciate that in research environments this may generate a significant list, but it is better to capture those, and only those, than be distracted by matters that do not require further intervention.

Assessors should also be warned about the dangers of going too far in the drive to identify risks. If a risk is included within a risk assessment document, it amounts to the organisation accepting that it is a risk that should be controlled. In uncertain environments, individuals may be tempted to catastrophise events and include extremely remote or unlikely eventualities. To do so may create greater legal risks as the assessor collects in writing all the ways in which an experiment, for example, may go wrong, without any proper understanding of the likelihood of the same, or what can be done to mitigate that risk. Whilst predicting the unpredictable is an admirable talent, it is one that can lead to distraction from the risks that are much more likely to lead to harm.

2.5.2 What is 'Reasonably Practicable'?

If it is accepted that some risks fall within the remit of the organisation, and therefore require to be mitigated, organisations need to evidence that 'reasonably practicable' steps have been taken.

Whilst health and safety law is 'criminal' in nature and dealt with by the criminal courts, there is one significant difference to true crime. In normal criminal cases, the burden of proof is such that the accused is innocent until proven guilty. In a health and safety prosecution under the HSWA, the burden of proof is reversed such that it rests predominantly on the accused. In a prosecution for a breach of the duty under the HSWA, once a prosecutor has established a foreseeable risk that requires control, a dutyholder must show that it has taken reasonably practicable steps to control that risk in order to prove its innocence. If a dutyholder is unable to satisfy this requirement, it falls to be convicted.

As a result of the reversed burden of proof, organisations need to be confident in how they have approached an uncertain environment, and what they believe to be reasonably practicable to keep people safe. Unlike normal criminal law, organisations cannot simply wait for the prosecution to bring the case against them; organisations need to be seen to be on the front foot when tackling these issues.

Determining what is 'reasonably practicable' requires a balance to be struck between:

- the risks associated with the activity and its hazards
- the control measures necessary to mitigate those risks

The appropriateness of control measures will very much be dictated by matters such as their cost, the time it would take to implement them, and the complexity of the solutions available when the exercise is performed.

By creating this balancing exercise through legislation, guidance and case law,[4] it is acknowledged that dutyholders are not required to implement all physically possible control measures. There will be measures that can be taken today that will make activities safer, but they are not legally required on the basis of expense or the complexity of the solution.

Helpfully, the reference to 'reasonably practicable' supports those working in uncertain environments. The ultimate solution to an uncertain environment is to avoid any uncertainty at all, but that is not the requirement of the legal regime. The legal regime allows us to approach the unknown and explore the unforeseeable, provided that it is done within a setting that can react to what comes before it.

The balancing exercise required when deciding what is 'reasonably practicable' is just that; it is not static and must react to the knowledge and experience that exists within the organisation or can be taken from others facing similar issues.

This is captured within the legal regime itself. Regulation 3 MHSWR requires risk assessments to be reviewed regularly, and also whenever there is a suspicion that they are no longer valid. 'Regularly' is interpreted by many organisations as requiring an annual review, and in some cases this may be extended to every two or three years. The review is intended to capture new knowledge and understand whether anything about the previous year would suggest that the risk approach needs to change; crucial where working in an uncertain environment.

The second part of the duty, to review whenever it is suspected the assessment is no longer valid, requires a more involved exercise. In an uncertain environment, there may be many instances when the validity of the risk management approach needs to be confirmed. In my experience, in the most extreme circumstances, it would not be unreasonable to have weekly or monthly reviews of the assessment, to confirm it remains appropriate and to consider if there have been any recent events that would suggest a change is needed. This would be pertinent in the case of unique experiments and for the reasons described within this chapter, time to review the risk assessment should be built in to the overall supervision of the experiment, with an individual delegated the task, with authority to stop or question if issues are considered to arise. Incorporating such arrangements at the start of the experiment means that sufficient time is allocated and it is not rushed, or viewed as non-essential. The same advice would apply to field trips, where a review of the day's performance could be undertaken on the journey from the site of the field trip to the accommodation. In some cases, groups may also benefit from a third party 'umpire' who performs this review; someone who is not wedded to the success of the activity, or the continuation of the same, but who has the safety of its participants as the primary consideration.

It has to be right that risk assessments should be reviewed whenever there is a reason to suspect they are no longer valid. This duty captures the fact, as discussed above, that risk assessments are a constant balancing exercise, for which the risk assessment document simply serves as a snapshot in time. Many organisations will only use incidents and near misses as a reason to review risk assessments. It is my suggestion that this approach is outdated.

There are many other reasons, beyond an incident eventuating, that could be used to suggest a risk assessment is no longer valid. Taking the positive side of health and safety, it may be that the original risk assessor has been too negative in their assessment, and the activity has been undertaken for some time without any of the foreseen hazards arising. It would be acceptable to potentially reduce risk ratings, as well as increase them, as uncertain environments become more predictable over time. The control measures within a risk assessment may also change, particularly if new solutions or controls become more readily available to a dutyholder.

The risk assessment record should be seen as just that; a record of the balance that was struck at one moment in time. When the term 'risk assessment' is used, one may think of large piles of paper as we revert to thinking about the risk assessment document, and not the actual risk assessment activity. Much the same as an annual test on a vehicle; the risk assessment document is valid at the time of the test, but requires constant review to ensure it remains so.

2.5.3 Collecting the Knowledge that Affects the Balance – Silo Mentality

If it is accepted that in an uncertain environment organisations will not benefit significantly from reviewing past performance, then it must be ensured that contemporary experiences are captured in order to better understand the success or otherwise of safety management.

If previous incidents are not available, organisations would be well-placed to have a proper system for capturing near misses and incidents so that the same can be used as learning points. Most organisations will have suitable near miss and accident reporting processes in place, some of which will have benefited from significant investment in terms of money and time. Features of some of the best reporting systems I have seen include:

- Accessibility – Organisations are now exploring how to make incident reporting easier and more intuitive. Mobile applications have been developed in the hope that easier reporting will encourage more reporting.
- Feedback – A clear feedback loop so that those reporting incidents feel that they are being listened to and their actions are acknowledged. If individuals view accident reporting as akin to a 'black hole' then they will soon be dissuaded from engaging with a valuable resource.

However, one of the issues I see on occasions is a failure to properly share knowledge and learnings from incidents and near misses across an organisation and, more widely, across sectors. There is a very real possibility that what may be an uncertain environment for one part of the organisation, could be very familiar to another.

Too often a silo mentality, be that between labs, departments or faculties, creates a very real risk that one part of an organisation is 'on notice' of a risk, and another is entirely oblivious. Organisations should give thought to how messages pertinent to a wider audience are shared. Some successful examples I have seen include:

- Safety bulletins sent out across entire organisations for the most serious incidents. Best practice when drafting such notices is to try and make them as generic and widely applicable as possible. This can include avoiding reference to the specifics of an incident, but focusing more on the root causes (see further below). The root causes are likely to have resonance with a wider audience than any specific reference to a limited set of circumstances.

- Cross-department or cross-campus safety groups dedicated to sharing the learning from each other. One of the interesting challenges posed in the academic environment is having a range of departments with differing risk profiles all within the same organisation. Whilst this can create a view from the 'low risk' departments that they can do less when managing risk, I have found that having all teams within the same room generates valuable and insightful discussion.
- Discussing safety issues at all meetings, including those that may not have safety as the focus. Safety discussions, particularly in the presence of management, can be powerful vehicles for change. Organisations that prohibit manager attendance for fear of hampering honest feedback are the ones we should fear the most. However, the place of safety within an agenda is almost as important as its presence at all. Too often, organisations will place safety 'at the top of the agenda'. Too often, safety becomes a bolt-on, or a hurdle, that must be dealt with before the true 'value' of the meeting can be brought out. Safety should be seen to be on the agenda, but should be just as flexible as other topics that feature; in some meetings, there will be more important themes. Safety should form part of the agenda because individuals want it to, rather than because they feel obliged to.
- Programmes where members of different departments visit unfamiliar parts of the organisation, providing a 'fresh pair of eyes' and bringing about different viewpoints and challenges. There is a real risk that individuals working in the same location for a long period of time may become 'risk blind' or even complacent to the risks that they are dealing with and the practices in force within the location.
- I have seen significant success stories where organisations share their risk assessments for all within the organisation to view, rather than keeping them within their own teams or departments. The first benefit of this approach is that an organisation will have a centralised database of risk assessment documents which can be used to monitor their review and to ensure that all individuals within the organisation have access to the most recent documents. The second benefit is that those tackling similar issues may not have to 're-invent the wheel'; they can refer to work previously performed for guidance or suggestions. An even more successful approach can be achieved where a search function is permitted within a risk assessment database, allowing individuals to identify risk factors or topics that may not be immediately evident from the title of the risk assessment document.

The law is clear: the duties held by an organisation do not distinguish between different departments or teams; the employing entity is viewed as a whole. What is known by one part of an organisation should be known by every other. What is predictable to one part of the organisation will arguably form the knowledge of every other.

Broadening out the silo concept, it is worthwhile considering the risks of silos within the industry or profession. It is often said that there is no commercial competition in safety, and therefore nothing to lose by sharing knowledge, good practice, and experiences. When judging what is reasonably practicable, regulators will look to the competitors of dutyholders to understand what others are able to do with similar resources and risk profiles. It is therefore important for safety professionals and organisations to look out beyond their own organisation to understand what others are doing when faced with similar challenges.

Given what else is said in this chapter, it is important that looking outwards is not limited to what is achievable within academia; educational bodies will be viewed in a similar way to private industry, and lessons may be learned from anywhere. I would encourage all of those involved in managing risks to try to identify appropriate comparator organisations and industries and then seek to understand what lessons can be learned from them.

Understanding the prevalence and severity of incidents outside of their organisation allows dutyholders to test whether their assessment of the level of risk is as accurate as possible. Breaking that silo mentality also helps dutyholders understand whether the control measures employed are on par with others on the same level.

2.5.4 Does the Reference to 'Reasonably Practicable' Help Us?

As discussed previously, the reference to a 'reasonably practicable' test provides some comfort to organisations; they are not expected to create a risk-free environment, and it is accepted that individuals may be exposed to suitably controlled risk without fear of criminal sanction.

In the learning environment, perhaps the reference to 'reasonable practicability' helps us further. I would suggest that there is an acceptance that mistakes will occur, provided that they are made within an environment that learns from those mistakes and reacts appropriately. Compared to industries such as manufacturing where processes are long-standing and risks are relatively easier to acknowledge, surely there should be some allowance for academic institutions and research facilities where the unexpected is more likely to occur, and outcomes may not be so easy to predict?

What is reasonably practicable within a stable environment must be more predictable than in an experiment or research scenario. Perhaps a better parallel to draw in the manufacturing sector is the situation encountered during unplanned maintenance; *i.e.* something has gone wrong, it is not clear what, and it will be necessary to try a number of 'fixes' before the correct solution may be found.

There have been many health and safety prosecutions following injuries and fatalities to those involved in maintenance 'problem solving'. However, the majority of the prosecutions arose from a failure to undertake the basics, and not from a failure to deal with an uncertain environment. The courts regularly see cases involving documented processes that are not enforced in practice or incidents resulting from shortcuts that have become 'the norm'. The much more foreseeable problems will often be those that cause the greatest risk; a failure to isolate, not wearing the appropriate protective equipment, or inadequate supervision.

The fact that we work in an uncertain environment, therefore, does not excuse failing to do everything reasonably practicable, but does provide the context to the duty that rests upon our organisations.

2.5.5 Convincing the Jury

Whilst we have set out the legal test above, and discussed the difficulties associated with an uncertain environment, there is yet a further layer of unpredictability associated with any subsequent enforcement of health and safety law.

In the worst-case scenario, where a health and safety prosecution is brought and the now defendant decides to fight the allegations, the trial will most likely take place in front of a jury. Members of the public selected from the electoral roll will have to listen to the evidence and legal argument and determine whether the organisation or individual has met the duties imposed upon them.

Jury members properly come to the process with their own viewpoints and life experiences. However, it would not be unreasonable for a jury member to expect a high standard of care from an organisation such as a university. Educational institutions are trusted with the young people of our society and there will be an expectation that they are kept safe. In my opinion, explaining the uncertain environment which we face to a jury, particularly where there has been a student incident, will be a difficult task indeed. For that reason, where the environment is uncertain, educational bodies should consider all of the practical points raised within this chapter so that they may go some way to convincing a court that, as a minimum, the fundamental safety issues were dealt with.

Compare that to a research institution, particularly one related to healthcare, and jury members may permit a level of forgiveness if they can see the overall public benefit in the work that is being performed.

Ultimately, juries can be uncertain environments in themselves and we never know what has swayed any particular decision. The focus of organisations and individuals, therefore, should be avoiding ever being before one.

2.6 Individual Responsibility and Engagement

Echoing what has been said previously regarding the duty on organisations, the Section 7 HSWA duty is understood in terms of 'reasonable care'. What is 'reasonable' in any given scenario will depend on the environment that is being worked in and the experience and knowledge of the individual concerned.

In an uncertain environment, it would not be acceptable to allege a Section 7 failure against an individual for an event occurring that was unforeseeable and unpredictable. However, as individuals progress in their employment, they will gain experience and knowledge and their actions or inaction will be judged based on the experience that they hold. It is therefore vital that individuals are engaged on the topic of safety because they may have invaluable experience and understanding that can help dutyholders to better understand whether they are taking the right approach.

Some good examples I have seen include:

- Staff consultation committees – Getting employees together to discuss their collective experiences and challenges can often be a useful way of ensuring that learning and good practice is shared across an organisation. It is important that these meetings are well-structured and that there is a communication flow between this committee and the senior leadership groups within an organisation; in my experience, without this, such meetings can often become a 'talking shop' where grievances are aired, but no action results.
- Asking staff to review risk assessments – Those undertaking the activities are best placed to comment on whether a particular safety management approach is going to be effective. I always encourage engagement of employees in the risk assessment process. It also ensures that subsequent enforcement of the requirements of the assessment is easier, as it is not viewed as being an imposition from others.
- Engaging with staff on business decisions – I have seen many examples where protective equipment has been purchased by an organisation and rolled out to staff with no engagement or discussion around whether the equipment is suitable. Asking staff to wear bulky equipment when they need to perform delicate operations will only lead to frustration; and most likely them not wearing the equipment.

2.7 Not Just a One-off, it's Got to be Bigger than that

In an uncertain environment, harmful incidents can often be unique, sporadic and without warning; hence the difficulty of preparing for and managing the same.

However, when we consider legal breaches that have come before the courts, they often share similar traits; they typically follow long-standing or systemic failures. Often when representing those involved in serious workplace accidents, I see that a significant focus of the attention is placed on the incident itself. An organisation will often become wrapped up in the detail of the incident timing, the stories of the individuals involved, and the theories of causation.

Many organisations fail to appreciate that the incident is simply the trigger for regulatory investigation; it is not often the focus as the inquiry progresses.

When a regulator conducts an investigation, they are often not looking at one particular incident, they appreciate the need to consider wider failings to understand whether legal duties have been breached. Is this incident a symptom of more significant issues within the dutyholder? The law covered previously does not charge an organisation for a particular incident; it charges an organisation with failing to ensure health and safety 'in the round'.

Whilst, therefore, one-off incidents may attract the attention of a regulator, and the potential for legal risk, an organisation that can satisfy the regulator that the incident was in the context of a sophisticated and effective health and safety management system will be best placed to avoid regulatory enforcement.

2.8 Incident Investigation

An organisation that wants to manage its legal risks should consider the 'bigger picture' revealed by an incident. One very important way in which an organisation can achieve this is by way of its investigation process.

One of the criticisms levied by regulators at dutyholders conducting investigations is that they too often jump on the immediate cause of an incident, rather than considering potential root causes.

The guidance to organisations therefore is to take a 'step back' following any incident and think about what particular circumstances have given rise to this potentially isolated event. The following immediate causes could give rise to more fundamental concerns:

- If rules have not been followed, then dutyholders should consider whether the rules have been properly understood, or if management properly enforced the rules, or perhaps shortcuts have been developed over time.
- If there has been a failure by an individual to follow their training, this may suggest that the training provided was unsuitable, delivered in a way that compromised understanding, or simply needs refreshing.
- If an incident involves work equipment, organisations should investigate the maintenance of that equipment and whether it remains a suitable tool for the job required. Do individuals understand how to use the equipment safely, or does training need to be refreshed?

A failure in the incident investigation process to properly uncover the causes may also be indicative of a failure to have competent investigators available. Organisations should consider how they equip their investigation teams to 'get to the bottom' of incidents. One common problem I encounter is that investigation teams do not include those at a senior level. This can sometimes create a challenge for the investigator, who is having to question the judgement of senior individuals within the organisation. An investigator facing that situation may avoid doing so altogether for fear of other ramifications, thus compromising the effectiveness of the inquiry. Investigation teams should be empowered with a clear mandate to investigate the entire organisation, and senior managers who attempt to be elusive should be tackled.

There are also concerns, particularly in academia, around the number of competent investigators and how that compares with the number of incidents requiring investigation. If left solely to health and safety practitioners, properly investigating incidents becomes a significant burden, hence why more operational personnel should be involved; those individuals need to own safety, and that includes considering when things go wrong. Proper reporting systems should also be used to group particular incidents in order to identify trends; grouping incidents will permit an investigation to focus on a more limited number of incident themes.

When dealing with the content and format of the investigation output, there is a fine line to be drawn between capturing the right level of detail and being too prescriptive and therefore at risk of creating a 'tick box' mentality. Many organisations provide a template investigation form, which is commendable as it ensures that the necessary details are not overlooked. However, the same document should also allow sufficient lateral thinking that means investigators can properly record their thoughts and findings. This is particularly important in an uncertain environment where it is needed to react, and be flexible, to new findings and new risks, rather than trying to 'crowbar' an investigation into 'drop boxes' of fixed outcomes.

Crucial to a sound incident investigation process is the ability to identify trends and patterns, especially when working in an uncertain environment. There needs to be someone within the organisation who is charged with the responsibility of collating investigation reports and considering whether they suggest the emergence of new risks or unexpected trends. This may not be immediately obvious from studying graphs or 'incident cause' charts that many automated systems generate. There should ideally be an additional holistic review that considers investigation outcomes 'in the round' so that someone is mapping the risk emergence journey across the entire organisation.

A further challenge in this regard is the concept of corporate memory. As referred to above, one of the difficulties associated with the uncertain environment is a lack of history on which to base risk management decisions. What if there was a history, but that existed in the minds of employees who may be susceptible to long periods of illness or unexpectedly leave their employment? Organisations should give serious thought to how the corporate memory of an organisation is captured, preserved and then communicated to those who subsequently join. It would be entirely unsatisfactory, and certainly frustrating, if the knowledge of an organisation is wiped clean every time employees move on. This is a challenge that many struggle to grip, particularly in industries where there is a high turnover of staff.

2.9 Focus on the Basics in the Learning Environment

If we are forgiven in a learning environment for those risks that are unexpected or unforeseen, it is unforgiveable for an organisation to miss the basics. There are some fundamentals of good health and safety that should be seen as non-negotiable for those working in an unpredictable environment:

2.9.1 Competent Individuals

Individuals should have training so as to allow them to act in the safest way possible. Basic training relating to lab safety and the use of equipment are fundamental. In addition to having received that training, however, organisations should consider how individuals are competent in understanding and managing risks.

For those responsible for managing safety, I would suggest that competency is a very different concept to that of being trained. I have experienced organisations which excel at collecting training records evidencing when individuals have attended courses. There is less evidence to show how the organisations assess the skills, knowledge and experience of individuals to determine their competency. Organisations should keep the matter of 'competence' under constant review and consider whether there is anything that occurs during their relationship with employees that would call into question their competence; near misses and accidents would be good examples.

More positively, I would encourage organisations and safety professionals to consider the value received from their training programme. Often upon enquiry, I have been informed that the training programme 'has always been like this' without any real understanding as to whether the training still achieves its intended outcome. Organisations should think about whether there are better ways to train people.

The breadth of technology available to organisations has created new avenues for training that are more effective, and arguably more interesting to the trainee. When was the last time that you asked an individual who attended a training course whether it was relevant to their role or useful to them? Those working in academia face influxes of new personnel annually, all of whom need to be provided with an awareness of good health and safety practice that is tailored to the environment they are about to enter. Employers need to think carefully about how this is delivered in a way that is impactful and not simply rely on what has always been done before.

2.9.2 Suitable Equipment

The safety equipment provided to those working in an uncertain environment should be suited to that environment; it should be easy to use and should be comfortable to wear. Individuals should be encouraged to engage with the selection of safety equipment so that the equipment best suits their needs, and it becomes easier for organisations to enforce wearing the same.

Organisations should ensure that equipment, particularly personal protective equipment, is compatible and that multiple pieces of equipment can be worn and used in collaboration. A particular issue to bear in mind in the educational environment is to ensure that any safety equipment provided fits the user. This will be pertinent where younger students may be asked to wear larger lab coats for example. There have been tragic incidents when young apprentices have caught ill-fitting sleeves in moving machinery. Interaction of protective equipment should also be considered; the wearing of protective goggles over spectacles is something that is often missed.

2.9.3 Housekeeping

From experience, it is often the case that where there is lack of ownership within a particular location or area, there will be a lack of care when dealing with that area. Often the 'forgotten' areas of labs or classrooms are seen as dumping grounds for disused equipment or storage areas for lesser used materials. I would encourage all organisations to allocate responsibility for all areas of a premises, so that there are no such gaps.

2.9.4 Proper Supervision

A crucial part of academia relies on more senior and experienced personnel assisting in the planning and supervising of activities. Whilst those providing supervision may not be any more enlightened in the potential outcome of an experiment or research activity, they are a key presence to ensure the basic rules are enforced.

Organisations should think carefully about how supervision is effected within their teams. The following summarises some key lessons learned:

- Supervision is best provided by those who are familiar with the activities under supervision, but not so familiar that they have developed their own shortcuts or become risk blind themselves. This is why refresher training can add real benefits, highlighting and correcting bad habits that have been developed over time.
- Those providing supervision should be of such character to challenge poor practices or bad behaviour where necessary; not something that necessarily comes easily to everyone in the workplace.
- Supervision should be available at all times, particularly high-risk times such as outside of normal hours or during particularly sensitive times of an experiment.
- Supervision should be dynamic; too often senior individuals view walk-throughs of the working area as supervision. Supervision needs to be

much less predictable and much more flexible. A short conversation with one individual can be just as effective as patrolling an area for the same length of time.

In organisations with the right culture, it is helpful to see upwards supervision, *i.e.* individuals challenging their superiors in relation to matters such as safety equipment and practice. However, I appreciate the difficulties in creating a culture where individuals feel able to challenge those further up the management chain. This is particularly so where the individuals to be challenged have control over careers or research prospects and those with concerns are rightly worried about looking like a troublemaker. It may take a brave person to challenge those in higher management positions but once that barrier is broken down, it creates a much more open and honest culture, where challenge is seen as a way to learn and improve, rather than blame and chastise.

2.9.5 Comprehensive Documentation

One of the real risks in an uncertain environment is for an organisation to have produced reams and reams of safety documentation that is neither consulted nor helpful when dealing with the everyday reality of risk management. Too often, organisations take false comfort from the availability of documentation as being some form of protection in the event of an incident.

Experience shows however that lengthy documentation can present its own risks:

- Lengthy documents are more difficult to disseminate across the workforce with any confidence that they have been read and understood. Organisations in academia and research will hopefully acknowledge that there is unlikely to be time within the average day for an individual to sit down and review large amounts of safety documentation in isolation, without that time being specifically allocated for that purpose and then also protected from interruption.
- Lengthy documents are much more likely to contain a promise or rule that is not enforced in practice. As documents progress, there is often a temptation for authors to make aspirational statements of intent, rather than accurately capture the 'reasonably practicable' steps that can be accepted by an organisation. Where documents and practice do not match a gap is created that could be exposed by a regulator in any enforcement action.
- It is very easy for lengthy documents to confuse and fail to highlight the key elements of good safety management. Distilling safety messages into direct and focused documents has a number of benefits, although organisations need to recognise that time needs to be dedicated to this;

creating a focused document can often require just as much effort as drafting a more comprehensive set of instructions, as authors need to apply their minds to removing that which is genuinely unhelpful.

Documentation should cover the activities and risks encountered by an organisation. However, there has to be acceptance that documents will not be able to provide for every eventuality and there should be some flexibility built into documentation to allow for those following them to appreciate that a level of appropriately competent autonomy is allowed. By way of example, the risk of lone working is often prescriptively controlled within safety documentation, despite the fact that individuals may work alone in a vast range of scenarios, with numerous possibilities for risk to eventuate. Documents should provide the framework for safe lone working, but allow for it to be interpreted to fit all scenarios.

Experience shows that documentation may be drafted by safety teams within an organisation with the expectation that the requirements of the same are implemented across the entire spectrum of activities undertaken. Where organisations have a significant portfolio of risks, ranging from lab work to desk work, there is a real challenge to draft documentation that is effective for all. In such circumstances, an element of flexibility may avoid having to enforce a rigid approach that is unachievable for some, and not enough for others. By way of example, organisations could avoid setting prescriptive rules, and instead prefer an 'outcome' approach; focusing on the end result to be achieved, allowing those with differing risk profiles some degree of flexibility as to how to achieve it. Referring back to lone working, a lone working policy that always requires a 'buddy' to be present is unlikely to be enforceable, as there will be occasions when lone working is sensible and can be done safely. In lab environments however, lone working would need to be much more closely controlled.

2.9.6 Clear Documentation

Having a suite of clear documentation with defined aims and objectives is much more effective when working in an uncertain environment. Common mistakes that I often see include:

- Documentation that conflicts, such as two policy documents that appear to cover the same risks, without acknowledging each other. There should be great care taken when drafting widely applicable policies such as those relating to risk assessment, use of equipment and supervision, as they will often trespass into more specific policy and process documents that deal with similar themes. There should be a clear hierarchy.
- Documentation that does not describe individual roles properly. By way of example, the reference to 'line manager' is made in a great many documents. However, in some organisations, individuals would struggle to identify who that actually is.

- Documentation that asserts control that is not achievable in reality. If a document applies to individuals beyond employees, such as contractors, visitors or students, then organisations should give thought to how it is brought to their attention, and whether it can actually be policed. In many cases, policies are said to apply to a number of people when in reality those individuals are unlikely to have seen the policy, and they certainly do not feel bound by it.
- Authors of safety documentation appear sometimes to be tempted to over-complicate process or language in the vain hope of making a process appear more sophisticated. I always advise those involved in the drafting process to set out what they want readers to understand, and not be tempted to elaborate further.

Taking an example in signage across the country; there has been a drive to reduce wording and make signage as universally understandable as possible. The same approach should be taken for safety documentation. A document that includes diagrams and pictures can be much more useful, and effective, than lengthy prose.

2.9.7 Reporting and Role Profiles

Experience tells us that understanding a reporting structure within an organisation, particularly academia, can be fraught with difficulty. However, in an uncertain environment, it is important that everyone knows who they report to and who is ultimately responsible for their development and safety. Dutyholders are encouraged to capture reporting structures within visual charts. Those that have difficulty in capturing their structure risk being accused of having a system that is too complicated.

Dutyholders should also think about job descriptions or role profiles. Whilst Section 7 HSWA imposes a duty on every employee to take reasonable care for their own health and safety, organisations should be clearer about what this means in their organisation. What is expected of those individual employees, how will they be held accountable, and how do they know if they have met the expectations of them. Job descriptions can quickly become outdated, or superseded by operational changes, but it remains important that in an uncertain environment, there is at least some certainty about the expectations of the workforce.

2.10 If We Need to Know about Potential Gaps, How Do Leaders Encourage Reporting?

For many years, health and safety has had a very negative perception. It has been viewed as the reason not to do things, or the excuse for not pushing boundaries or limits. Those involved in health and safety management are often frustrated by this perception.

Health and safety should be viewed as a positive facet of everything an academic or research institution wants to achieve. Part of this battle is dependent on the way that safety is discussed within an organisation. Think about the following examples which may dictate how safety is viewed in your workplace:

- When performing health and safety audits, is the focus on finding non-compliance and faults, or is there also celebration of the successes and the good safety management behaviours?

  I have experience of reviewing health and safety audit reports which tend to be a collection of bad practice and problems. Even if the bad practice is very much in the minority, there is little, if any, celebration of all the positive observations and things that have gone well during the audit. All of the actions that then arise from an audit relate to the need to 'fix' issues, rather than celebrate and seek to replicate good behaviours.

- As well as recording incidents and near misses, is there a way in which good practice and behaviours can be captured, reported and celebrated?

  Some organisations use a similar process for capturing good behaviours as they do for bad, using the same 'near miss' reporting process to record 'safety observations' where there have been successes. A powerful next step is for there to be some form of prize or commendation for a selection of good practices. This initiative does not require significant investment or expense, but it can be invaluable in resetting the impression of health and safety.

- Do senior staff regularly praise individuals for good health and safety traits, at least as much as they criticise for health and safety failures?

  When was the last time someone in your organisation received a compliment for their own health and safety behaviours?

2.11 What Constitutes an Effective Management Regime?

In determining whether the organisation is getting risk management right, it is not sufficient to wait for an incident or near miss to suggest that it is not. The impact or damage caused by such an incident could be far too severe.

Many organisations use accident statistics in order to understand their own performance and effectiveness when managing health and safety. To do so may be further behind the curve than the law expects, where there has been a real focus in the industry on leading indicators rather than lagging indicators; looking at where the next risk is going to arise, rather than dwelling on what has already occurred.

In order to understand whether a safety management system is effective, I would encourage organisations to identify the trends and behaviours that it would wish to see as symptoms of good health and safety management. By understanding these, and using them as a way to 'mark' the organisation, everyone will be better equipped to assess the progress of the organisation and what more can be done.

Experience has shown that there are some useful indicators that can be used in order to measure effectiveness:

- The number of positive safety interactions – This can be a real indicator of the culture of an organisation and the way in which people view health and safety and the importance of having pro-active and positive discussions about it.
- Management Walkarounds – An organisation should understand whether managers and supervisors are on the 'shop floor'. There should also be monitoring of where managers visit; do they always visit locations that are 'on the way' elsewhere, or do they make a genuine effort to visit unfamiliar areas? Asking staff from other departments or teams to visit their contemporaries in other organisations can be invaluable.
- Document Delivery – Understanding if staff are properly accessing and reviewing documents; where it is possible to monitor access, to see who has read documents and for how long. If it is not possible to monitor online access, then having discussions with those on the 'shop floor' about documents following their dissemination will serve as a good indicator of whether they have reached their intended target.
- Spending on Safety Initiatives – Is money being spent wisely, is it bringing good results, and what else is available? A safer environment is not necessarily guaranteed by additional spending, but an organisation should understand how much resource is being targeted towards health and safety, as a proportion of spend on other facets of the organisation.

2.12 Creating the Right Culture – Knowing When to Stop in a More Commercialised Environment

In my experience, the vast majority of health and safety incidents involve well-intentioned individuals who genuinely believe they are doing their best for their employer. Employees will often view an unplanned situation as a unique learning opportunity, rather than something that presents an unknown element of risk. Research assistants taking shortcuts will often be mindful of the benefits of conducting experiments as quickly as possible, rather than the risk a shortcut could pose. No individual will come to work with an intention to harm themselves or a reckless abandon for the safety of others within the organisation.

Organisations, therefore, need to create an environment where individuals can see the benefit of and, more importantly, feel empowered to follow the rules and requirements of the organisation. Often in an uncertain environment, this revolves around feeling empowered and able to stop proceedings when things do not 'go to plan'.

I have experience of speaking with those working in academia who recount the pressures of deadlines, budgets and teaching demands. Organisations who want to properly manage the legal risks of their activities need to ensure that all of these demands do not encourage or tempt individuals to undertake unsafe actions or cut corners.

Various serious health and safety incidents have shown that harmful outcomes could have been avoided if the individuals involved had stopped and thought about whether they should still proceed; or if there was a need to stop and rethink. Place yourselves in the shoes of those individuals and think about why they might not want to stop and how we can change this:

- Do they feel unreasonable pressure from management to complete their work?
- Do they fear blame or negative comments if they want to stop?
- Are they concerned about the financial implications if the work does not go ahead?

Regulators are fully aware of the commercialisation of academia, and educational institutions in general, and they will want to ensure that individuals are not being put under unacceptable pressure. The commercial impact (ignoring the personal and legal elements) of a serious incident within an organisation can be catastrophic; activities may be ceased indefinitely, individuals could be interviewed by regulators and under suspicion themselves, and 'normal' activities may not resume for months or years. This outcome, however, is often so far removed from day to day decisions that individuals may need to be reminded that taking a few more moments to 'think again' is a better option.

2.13 What about Activities Abroad?

Taking operations outside of the UK adds a further layer of uncertainty as the environment may be unknown or more difficult to control.

Organisations need to take a structured approach to activities conducted abroad to ensure that those risks which can be controlled are properly considered, and as mentioned previously, individuals dealing with the risks arising feel empowered to make the right decisions.

There are examples of where organisations potentially attempt to go too far. Trying to enforce UK standards in countries that do not share the same legal requirements can present significant difficulties.

Organisations should do what they can to understand where risks may arise:

- Websites such as that provided by the Foreign & Commonwealth Office[5] assist with a broad range of information on known risks and threats, and also provide useful guidance on local customs so that visitors can properly prepare themselves, and be best able to keep themselves safe whilst on visits abroad.
- In many cases, others in the organisation will have undertaken similar trips, and this knowledge should be tapped into and shared with those planning the current trip.
- Those responsible for the trip abroad should think about a 'Plan B' – Sometimes the pressure to complete 'Plan A' can put people at risk; particularly if students or research projects are reliant on a particular visit. Organisations should think about alternatives if things don't go to plan.
- Finally, those responsible for the trip should be provided with details of how to summon help when abroad. Is an advice line available to call? If phone numbers of those left behind in the UK are to be used in an emergency, then consider the impact of time zone differences. In more extreme environments, the lack of phone signal may also present challenges that need to be resolved.

2.14 The Difficulties with Looking after Mental Health

Throughout this chapter we have been considering the risks posed by working in an uncertain environment; it is difficult to predict outcomes, and there are likely to be a variety of potential difficulties arising from the same scenario. If those are the challenges, then mental health presents the perfect example.

Whilst managing physical safety presents obstacles, the task is relatively easier as a physical hazard will typically present the same potential harm to any person who may be affected by it. A fall from height will typically result in a similar potential for injury, as will the result of an unexpected explosion. Mental health is entirely different; two individuals under similar pressures and with similar demands on their time and resources may react in completely different ways.

The management of mental health and wellbeing is truly an uncertain environment. When thinking about managing the legal risks and keeping people healthy, all of the points and practices discussed above are equally relevant:

- Individuals should be provided with clear communications and documents so that they understand what is required of them in their role.

- As with changes to the physical environment affecting safety measures, organisations should consider how organisational change may impact the mental health of the workforce.
- There needs to be a culture of honesty and openness about mental health and mental illness. It is imperative that leaders within organisations speak positively about the desire to protect mental health in the same way as physical health. The mental health of individuals should be discussed in peer reviews and performance discussions.
- Those seeking to effectively manage mental health need to identify their own key performance indicators. Absence statistics are a useful indicator, but an effective organisation should be monitoring engagement with the mental health services, as well as the rate at which managers are building discussions about mental health into their wider contact with the workforce.
- When investigating cases of illness, organisations should be led by the need to identify the root causes and consider whether elements of an organisation's operation could have contributed to poor mental health. Organisations have been seen to invest significant resources into remedying the symptoms of poor mental health, whilst doing less to consider the causes.
- Finally, organisations should empower members of their workforce to come forward if they feel as if they are suffering, or believe others are. As with incident reporting, individuals should know how to get help if they need it.

2.15 Final Remarks

Uncertainty is not something that the health and safety regime copes with well. Those responsible for managing safety would much prefer to have predictable outcomes of activities that can be used to scope the 'reasonably practicable' control measures that can be adopted.

However, in an environment that thrives on innovation and pioneering, we need to find a regime that can cope with the unexpected and the unforeseeable. Thankfully, the legal regime in this regard does assist; the law does not require organisations to create risk-free environments, but simply to have a culture and workforce that is able to adapt and to make the right decisions to achieve as safe an environment as possible.

Throughout this chapter, we have considered the ways in which an organisation dealing with the uncertain can be certain about one thing; that the right decision will be taken when the time comes. This does not require an attitude change, as those working in academia and research are well-intentioned; but it does require a framework which encourages and supports those who wish to 'take some time' or 'think again'.

My experience as a lawyer is typically an involvement when things have 'gone wrong'. All too often it is easy to see how an incident could have been avoided and, in hindsight, often the solution is obvious. It is that hindsight that I hope can be brought into your foresight having read this chapter.

References

1. Regina v Porter [2008] ICR 1259.
2. Busting the Health and Safety Myths [Internet] Health and Safety Executive, *Health and Safety Myths*. Available from www.hse.gov.uk/myth/, last accessed 05 March 2020.
3. R v P and another [2007] ALL ER (D) 173 (July).
4. Edwards v National Coal Board [1949] 1 All ER 743.
5. *Foreign & Commonwealth Office Website [Internet]*. Available from https://www.gov.uk/government/organisations/foreign-commonwealth-office, last accessed on 05 March 2020.

CHAPTER 3

The EU and Safety Management in Higher Education

DICK HOENEVELD*

Delft University of Technology, Faculty of Technology, Policy and Management, Delft, Netherlands
*E-mail: d.hoeneveld@tudelft.nl

3.1 Introduction

This chapter focusses on the special character of occupational safety and health (OSH) management in higher education, OSH trends, options for improvement and the role of European regulation.

I will argue why OSH management in higher education requires a different approach to OSH management in industry. OSH in higher education is not only special in character, but it is also challenged by special trends that influence the direction of its OSH management. Because industry dominates OSH management development, OSH in higher education partially needs to develop its own methods. Most solutions to cope with OSH challenges are already embedded in higher education organisations, we just have to find and harvest them. Options for improvement of OSH management in higher education will be discussed in more detail. To complete the chapter I will make a step towards OSH legislation in the European Union, its influence on national legislating in the UK after Brexit and the contribution to OSH in higher education of the European Agency for Health and Safety at Work.

Finally, I will look forward towards OSH information flows in the future.

Challenges for Health and Safety in Higher Education and Research Organisations
Edited by Olga Kuzmina and Stefan Hoyle

Published by the Royal Society of Chemistry, www.rsc.org

3.2 OSH Management in Higher Education

3.2.1 Summary

Higher education institutes operate under conditions that are very different from industry and this influences their OSH management. Amongst other things, higher education distinguishes itself by a high rate of staff (and student) turnover, the multicultural setting, the wide range of safety risks, the constantly evolving activities as well as varying multidisciplinary research and above all, the concentration of its thinking power in the research groups: the operational units. Most of these distinctions can be found in industry too, but never in such a combined setting and this characteristic defines the needs for dedicated OSH management in higher education.

Within higher education, a number of OSH trends can be recognised that need to be addressed by safety management to maintain good levels of safety. Some of these trends are directly connected to central research agendas such as the greening of technology, but other trends originate in the organisational structure of higher education, such as the high administrative burden for academics.

In section I am going to analyse the following:

- the need for a flexible and middle management orientated approach in OSH management;
- the social impact of technology progress;
- the rise of new concepts such as the evidence-based approach for OSH tools;
- the need for (more) international cooperation.

I will also highlight some promising safety management approaches, like cooperation between safety science and OSH practitioners. Such initiatives could help to control novel risks that are emerging more and more in higher education and could overcome value conflicts between researcher autonomy and business control. Such conflicts are embedded in the structure of higher education institutes.

N.B.: A value conflict is a difference of opinion based on deep-rooted moral beliefs. Because such conflicts are belief-based, they cannot be resolved with facts and require mutual understanding between the stakeholders involved.

Looking at the dynamics of today's safety management in higher education, I will demonstrate that options for improvement are available. Yet, it should not be ignored that some challenges are becoming too complex to handle for a single institute. Improved interuniversity cooperation on safety management might become a necessity in the near future.

3.2.2 Four Characteristics of Higher Education

Safety management systems can be defined as organisational systems that steer the direction of an organisation towards safer operation by influencing its sociotechnical system (technology, worker's competences, structures, policy, culture and regulations). However, the most appropriate steering system may vary depending on the type of organisation. As all organisations are in some way unique, good OSH management should be able to deliver custom made solutions. When looking at the characteristics of higher education, it's not hard to classify OSH management in higher education as being in a league of its own.

Dominant characteristics that define special OSH requirements in higher education are:

(a) the informal power of the front-line workers;
(b) the relatively poor controllability of safety;
(c) the high degree of internationalisation;
(d) the high staff and student turnover rate.

Of course one can think of several other characteristics that are relevant in higher education and any selection is arbitrary, but these four characteristics generally dominate in higher education. Let us discuss them in more detail:

3.2.2.1 *Informal Power of the Front-line Workers*

If you have worked in higher education for some years, you will have noticed the supremacy of the academic staff, the lecturers, researchers, PIs and professors. From hereon I will address this group as 'academics'. Executive boards, deans and heads of departments may have formal power in our institutes but in reality, the operational management of the organisation is dominated by the academics: they possess 'informal power'. This is certainly the case for the icons in higher education, the potential Nobel Prize winners and the ones with outstanding science careers, but even the management influence of common academics is much higher than anyone from the front-line level in industry.

Academics in higher education do not only control the output of higher education, they are by far their biggest asset. And if institutes no longer meet their requirements, established academics can, and sometimes do, move to other institutes together with their staff. This puts academics in a very powerful position. Furthermore, the sheer volume of the academic's combined thinking power exceeds the resources of the top management. From an organisational point of view, academics coordinate institutes of higher education. The famous management researcher Henry Mintzberg[1] has identified such organisations as 'professional bureaucracies' and academics in higher education as the 'operational core'. In a professional bureaucracy, it is useless to force OSH regulation top down upon the work-floor. In the end, the operational core controls the organisation's output. If top down regulation is not accepted by academics, they will ignore it, oppose it and outsmart it, with all the mighty power they have. In this regard, the literature of Mathieu Weggeman[2] is interesting, in this he gives numerous, often hilarious examples of how academics (as part of our middle management) deal with counterproductive management systems and the creativity by which they deploy all their resources to frustrate such systems. To illustrate this, I include an example of Weggeman[2] about creative academics defeating an administrative monster:

Long live the random generator!

"After submitting and resubmitting, clarifying, defending and again submitting numerous letters, a research consortium finally succeeded in securing an externally financed European Union project that provided work for three years. They were prepared to jump through all the necessary bureaucratic hoops because they were intrinsically motivated by the complexity of the project. But the celebrations soon gave way to the hangover; all project teams concerned had to log their time in units of 6 minutes. The subsidising organisation had so many administrative staff, accountants, auditors (actually accountants, checking the other accountants) that they needed an enormous amount of data to keep them busy. "Time logging in units of 6 minutes for a project that lasts three years? That implies a huge loss of time! They had, of course, forgotten to include the time for filling in the forms in the project application. So of course, we're not going to do that," said the researchers. And so, they first invested in writing a computer program that would produce fake but credible timesheets. At the start of the project each project member had to provide some personal information (sub-projects he/she was working on, his/her allocation factors in the project, the most likely holiday periods and days off, whether you were an early starter and an early leaver, or the other way around, someone who took long lunches, *etc.*) and then the random generator took care of the rest. One day you went to lunch at half past twelve, the next at five to twelve. One day you left at a quarter past four and the next day you worked until seven. The program processed things in such a way that at the end of the month the accumulated time logging data was just slightly higher than the figure allowed for by the subsidising organisation, based on all their other projects. Every four weeks the project members received four completed week forms from the server with the request to check whether the data was realistic. If that was not the case for certain items you could make corrections online and in real-time and a new print-out was made. After that you signed off four times and you're good to go! Most of the time, all the data was fine. What's more, you could, if you wanted, request a time sheet from the system for the third week in October next year.

Of course, the bureaucrats were fooled, but that was not the aim of the exercise. It was meant to provide maximum time for and focus on the execution of the project for which all researchers were highly motivated. The other day I met one of them and she told me enthusiastically that they had sold the time logging program to another organisation that had run into the same problem!"[2]

This excerpt illustrates that academic workers will find ways around the bureaucratic burden and additional admin tasks do not guarantee the job is made as prescribed.

3.2.2.2 Limited Control of Safety

It is not uncommon for insiders and outsiders to consider higher education a much safer working environment than industry, and a shallow look at the accident statistics seem to support this feeling, but statistics can be deceptive. So, let us look at the case in more detail.

Without doubt the numbers of deaths and severe injuries in industry are more striking than in higher education, but safety management is not only about severe work accidents but also about long term exposure to risks and the ability to control risks. Are these well managed in higher education or have we just been lucky?

A survey done in the late 1990s at the Department of Chemistry of the Delft University of Technology could give some insight. The purpose of that survey was to stimulate internal discussion about safety by ranking the department on the same safety scale as chemical industry at the time, including Shell, BP and AkzoNobel. A Shell manager who ranked chemical plants worldwide, was hired to audit the department in the same manner. Each laboratory cluster was visited by the manager and the university OSH manager – by that time that was me – and we interviewed technicians, PhDs, researchers and students present in the labs. Many dangerous situations were revealed, such as overnight hydrogen leakages, beverages and food stored next to reagents, hydrofluoric acid incidents, toxic fumes, implosions of large glass vessels *etc.* Each of our findings was discussed with the lab manager involved and then ranked using the Shell risk matrix. All findings were prioritised and plotted against the safety profiles of the chemical industry. Much to the surprise of the department, the Shell manager and myself, the risks at the Chemistry Department appeared to be sky-high and did not even fit on the industry diagram. In industry, such a score would have resulted in several deaths per week and immediate closure of the plant.

We started to look for answers. Why has this potentially lethal situation not led to casualties already? Our conclusion was that although the number of unsafe situations was huge and wide-ranging, the scale of the research experiments was extremely small. So, accidents did happen all day long, but thanks to the small scale, any incidents had only small impacts and were therefore considered business as usual. Yet, the major reason for the relative low accident rate was the continuous ad hoc safety intervention by technicians, lab managers and supervising staff; basically, all tenured staff. On a daily basis, they had to shut off gas valves or heaters, adjust pump rates, remove flammable liquids, switch off mains and even pull students away from danger. Tenured staff did not so much prevent accidents from happening, but they controlled the impact and were very effective in that. The findings of our survey were discussed extensively, not only within the department, faculty and university but also in several Dutch interuniversity safety platforms. The outcome of the discussion was that the Delft chemistry situation was not uncommon for most universities in the Netherlands, possibly for universities abroad as well. We concluded that lethal accidents would occur sooner or later if we were unable to shift from repression towards prevention. The urgency of this shift towards prevention became even more eminent after the slow but continuous decline of technical staff numbers which was observed in most universities. In this context, it is worth noting the number of deaths that have been reported in US universities over recent years. Apart from the 2009 accident at UCLA that is discussed later in this chapter, also Yale (2011), Texas Tech (2010), University of Florida (2010), Florida State University (2015) and University of Hawaii (2016) were confronted with fatal and near fatal accidents. For details, I refer to the APLU report.[3] At first sight, this US accident profile in higher education appeared different from the European one and one theory was that the higher numbers of tenured technical staff in European higher education might have a quenching effect on the impact of accidents. However, such a conclusion needs in-depth study, which has not been done yet.

3.2.2.3 Multicultural Environment

Of the workers in OECD countries (Organisation for Economic Co-operation and Development), 12.5% are of foreign origin, while 33% of the PhD students in the USA and 47% of the MSc students in Science and Engineering come from abroad; 70% of the PhD students of Delft University of Technology (Netherlands) are of foreign origin, which is an increase of 100% from 2003. These are some illustrations of the influx of foreign knowledge workers and students into higher education. The numbers have increased over recent decades, leading to more and more laboratories and classrooms being populated by people with different cultural backgrounds. This drive will continue as higher education and research universities seek to employ the most capable and bright in their fields. Competition is truly international in the university sector.

As most of us might have experienced in diversity programmes in many institutes, change in heterogeneity adds to improved resilience and thinking power of research groups, so cultural diversity is an asset. Literature[4] states, however, that multicultural teams can be more unsafe than mono-cultural ones, but why? Safety management is about understanding each other, verbally but also non-verbally. When organisations internationalise, new concepts of communication and new (safety) perceptions are introduced by the fresh international staff, challenging existing management approaches. The 'old' safety approaches were developed over the years by trial and error and might not fit the new population and the dynamics of the current risks, technologies and novel methods anymore. In reaction to this, phrases sometimes heard in the field are: 'Foreign students do not have the practical skills', or 'We have a general feeling that foreign students are involved in accidents more frequently', or 'He always says "yes", even when he doesn't understand at all' and 'Our safety management fails in the new situation'. These are all expressions of friction between existing safety approaches and a new multicultural environment. Therefore, amidst the ever-increasing globalisation of worker migration, cultural differences pose a genuine challenge to OSH managers. Worker groups are more and more comprised of individuals from a range of different cultural backgrounds, and strategies which have consistently worked for OSH managers in the past may become less suitable over time. In the new situation, cultural intelligence is needed, a form of human intelligence that has been described as an individual's knowledge, skills, abilities and capability to effectively manage themselves and others in cross-cultural situations and environments.[5] Cultural intelligence is focused on the abilities to grasp, reason, and behave effectively in culturally diverse situations. Individual OSH managers and middle managers in general, who are equipped with cultural intelligence tend to have abilities across a range of cultural settings and can aid less skilled managers to overcome these challenges. The aim is to work out a safety management approach that supports multicultural teams.

3.2.2.4 High Staff and Student Renewal Rate

The high staff and student renewal rates are inherent aspects of higher education. The average stay of PhD students in one organisation is 4 to 5 years, masters students and postdocs stay for shorter times and even tenured staff tend to be more mobile than before. So, safety awareness and safety management programmes have to be repeated again and again to a constantly renewing population. In this process, the middle managers of research groups have a central role. Middle managers in general are less mobile then other academics in higher education, they form the most stable management layer amongst academics and by that they become the collective memory of the institute, the carriers of the organisational culture. Though organisational culture is sustained by the people within an organisation, it is known that such a culture often continues, long after the persons who established that culture have left.[6] Middle management plays an important role in preserving this cultural memory. Despite the importance of a cultural memory, positive updates and changes must still be considered. A constant influx of newcomers and import of new approaches and solutions to safety challenges, can lead to a better safety culture if managed correctly. Adaption to a changing environment is a good thing but staying in control needs to be a conscious process. It is advisable for OSH management to support the middle management in handling changes in the safety culture to ensure improvements are integrated and threats do not endure.

3.2.3 Four Trends in Safety Management

According to the European Agency for Safety and Health at Work (EU-OSHA), work and workplaces are constantly changing as a result of the introduction of new technologies, new substances and new work processes.[7] The structure of the workforce changes, organisations develop and the work itself changes.

This change may give rise to new risks and challenges to workers' health and safety. In order to ensure safe and healthy workplaces in the future, trends in safety must be anticipated and addressed.

In 2019 – on behalf of the Dutch Ministry for Infrastructure and Water management – I was involved in a survey on values and safety management regimes in the Dutch research institute setting.[8] The survey identified four major safety trends that are currently dominating safety management in higher education. These four trends are:

(a) greening of technologies and research products
(b) automation in research processes and products
(c) privatisation in safety research and safety standard development
(d) increased administrative burden on researchers affecting the efficiency and creativity of the scientific process

Below I will address each of these trends in more detail:

3.2.3.1 The Greening of Technologies

We tend to associate the word 'green' with safety, but what is expected to be good for the environment is not necessarily good for the safety and health of humans. To give an example, the massive shift from fossil fuel powered vehicles to battery powered ones, introduces a new fire risk; not only on the public roads, but also in test facilities of higher education and in offices when charging hoverboards or bicycle batteries. And let us not forget the exploitation of third world workers in the lithium mining industry and the rise of political tensions in lithium-rich regions. New 'green' technologies or working processes can lead to new hazards, which call for new combinations of skills to deal with them: the 'old' OSH management cannot simply be transferred to them.

The greening of technologies is a major trend in current scientific research, emphasising conflicts between recognised safety approaches and the introduction of new types of risks and potential failures. Furthermore, the trend introduces potential conflicts between human safety and environmental safety, not only in the research phase, but also in the later stages of innovation (start-up, scale-up and full production). This mechanism has increased over recent years, driven by societal appeals for the 'greening' of existing technologies. In order to change existing technologies such as the fossil fuel economy into more sustainable ones (greening), it is more and more likely that we will enter a world of novel risks. Large grants, massive programmes (EU Green Deal) and a strong societal momentum pushes the greening of technology to the limits of safety.

Greening of technology and the application of novel innovative solutions requires particular attention in the light of the 'safety paradox'.[9] The safety paradox states that achievement of high safety levels generally results in lower attention to safety. The key problem in the greening of technology is that the achieved high safety levels correspond with the 'old' technology and not with the new green technology. False perception of safety does not accentuate the need of constant adaptation to new technology, materials, structures, processes, and systems. Economic and political pressure could lead to OSH matters being overlooked and OSH managers as well as researchers need to be aware of this development and possible implications.

3.2.3.2 The Automation Trend

In general, the automation trend can be characterised as an enhancement, or even replacement, of human performed activities and duties with software-based solutions. This is a rather broad definition that nevertheless captures a single key development critical to safety management: the re-distribution of human factor risks. The crucial consideration here is that while automation can be seen as the reduction of human risk with the help of automated systems, in fact it pushes these risks onto other stakeholders, for instance the manufacturers and maintenance people of the automation equipment. For laboratory safety, automation might lead to situations where computers decide about acceptability of risks activities, leading to a generation of laboratory workers who are risk averse or don't know how to assess risk. At Delft University of Technology, automatic risk assessment tools were deliberately banned from the corporate safety management system (Lab Servant[10]) in order to stimulate safety reflection and deliberation by individual researchers.

Developments in digital technologies are rapid and include artificial intelligence (AI), machine learning and robotics. The social impact of such digital technology on safety management is still not fully recognised.

OSH risks of automation are generally easy to control in the research phase, where research workers can install physical barriers between themselves and the automation devices such as robots or self-driving vehicles. However, in the later phases of development when inventions are tried out in open areas or public spaces the OSH risks increase. When innovating products for society, researchers should anticipate risks in later phases of the innovation life cycle, like the scale up phase, the production phase, the user stage and the waste phase. This overarching research approach is known as safety-by-design (SbD). In SbD, OSH practitioners can be ideal sparring partners for researchers to reveal hidden safety risks, provided communications between them and researchers are transparent and based on shared moral values. OSH management should be aware that development of robots and AI for society is becoming an important issue in higher education; consequently, the number of test facilities is expected to grow and consequently OSH risks will increase.

A special challenge in automation is the social impact of technological progress. This arises from disruptive growth of technological improvements such as social media, sensor technology and artificial intelligence. To give an example of the impact of technological improvements on people and people's behaviour: it is easier for our academics to browse for safety advice on the internet or to have a chat with a colleague abroad, than to make an appointment with the department safety officer. Nothing wrong with pro-activity, but internet answers or amical advice might not always be a reliable source. In OSH matters, a consultation with the local safety officer is always advisable. But technological improvements can also be of help to safety management. To give an example: equipping all laboratory workers with a sensor app on their smartphone, having them sampling exposure data and sharing that data with the OSH manager, might give far better insight into the safety situation than a daylong measurement by the occupational hygienist. Of course, privacy should be safeguarded in data-sampling and data-sharing, as a legal requirement and to avoid opposition from laboratory workers. The challenge for OSH management here is to understand the social impact of new technology and to use it in favour of safety. Later in this chapter we will discuss whether this challenge is within the capacity of autonomous institutes in higher education.

3.2.3.3 The Privatisation Trend

The openness of safety research and its transparency is undermined by privatisation of safety research and safety standards. Privatisation is not a standalone or novel development, self-regulation in privately funded research has been going on since the 1990s, but at this moment it is entering a completely new level, where the sharing of essential safety information in the research environment comes into conflict with financial interests of third parties. To give examples: to protect their intellectual property, suppliers normally do not share safety information of equipment components with laboratory workers. Essential safety information, such as safety standards or calibration, is developed by private companies and considered a classified part of their business model. On one hand the privatisation process can be considered a welcome development that promotes a wider set of stakeholders engaging in safety research and practices. On the other hand, privatisation creates new value conflicts because in many contexts there are good reasons to consider safety as an imponderable value that cannot be expressed in monetary terms and therefore requires public control and should not be left to the market. Furthermore, the value of the openness of research, instrumental for safety strategies may come into conflict with the considerations of intellectual property and market interests. In the context of academic research, privatisation of safety standards can sometimes be regarded as an approach driven by the value of efficiency. However, on the larger scale, privatisation of the safety standards and associated efficiency can come into conflict with the values of transparency and openness of research. We should be aware that the sharing of safety information in commercial enterprises is not straight forward.

3.2.3.4 The Administrative Burden Trend

The 2008 US Faculty Burden Survey[11] revealed that faculty members who serve as Principle Investigators (PI) spend up to 42% of their federally funded research time on administrative tasks. Overall, 84% of the PIs reported that the administrative burdens associated with their research have increased in recent years. The US findings were confirmed by studies of the Association of Universities in the Netherlands[12] and similar studies in Germany.[13]

As was investigated in the US, the increase in administrative burden comes partially from internal regulation and partially from (inter)national regulation relating to biohazards, chemical weapons convention, drugs and drug precursors, genetic modification technologies and the increasing attention that is being paid to these areas by the enforcement agencies. This adds strains on academic activities and leads to poor perception of 'safety' issues with the academics.

The administrative burden in higher education is causing a value conflict between the administrative tasks of academics and the efficiency of their research.

Considering the slow but continuous increase in the number of safety regulations, one could argue if adding more safety regimes to an already overloaded system is in anyway beneficial. It could lead to increased numbers of researchers trying to avoid tick-box exercises (see also Weggeman[2]). In the Dutch survey on Safety-By-Design as part of a Basic Responsible Research & Innovation Framework,[8] we showed that in an average university, up to fifteen different safety management regimes can be present. It is not uncommon for a researcher to have to deal with ten regimes during one experiment. This number corresponds with the analyses of Decker[11] who identified twelve regimes to be considered as administrative burdens for researchers in the US.

To give an example from Delft University: we found that five regimes (a. Institute protocols, b. Training of personnel and students, c. Data information security compliance, d. Safety planning and monitoring, and e. Equipment and supply purchases) were managed by different support staff units and subsequently each of the regimes needed their own set of information of which up to 50% can be considered double data entry for the researchers involved.

The major problem here is that as a result of different safety regime owners, safety regimes regularly do not share information between each other, forcing academics into routine administration and double data entry, resulting in unnecessary restraint on the research process in general and research creativity specifically. This reflects the multi risk nature of research as opposed to industry, where the number of risks is smaller and consequences larger. The bottom line of such developments in higher education would be reduced safety levels. Altogether this situation creates a value conflict between safety management and researcher productivity and creativity.

The conclusion for safety management regimes is that they unintentionally can create value conflicts between safety management and academics. Yet, harmonising of safety management regimes is not complicated and should lead to less administrative burden for academics, improved research creativity and more efficient safety management.

As an example: in the Netherlands, we concluded that the solution for harmonisation of regimes was to programme them all within the same software platform and grouped around the same database, for instance BIO assessments, risk assessments and GDPR assessments. Information that is already present in the database is automatically retrieved by other regimes. Such reuse of information can lead to better insight in the overall safety situation and can enrich individual regimes with information that was previously not available.

3.2.4 OSH Management Options for Higher Education

So far, we have seen that OSH management in higher education is different from OSH in industry because of the special characteristics of higher education. And we have seen that the OSH management will be confronted by emerging trends in safety. If we look at this from a more holistic point of view, what kind of OSH landscape do we see?

Overall, we see a world of change in higher education, novel risks, technology push and limited structural controllability of safety. From the perspective of safety science, we expect the sector to be in transition from planning-based OSH management towards dynamic, reaction-based OSH management. This concept is represented below in more detail.

If we want to stay successful and safe, academics and middle management need to play more visible roles in this new OSH management system. To address emerging challenges and the values of human and environmental safety, we have to find a balance between the current rigid safety management and new dynamic and ever-evolving safety management. In their book "On the challenges for safety management", Hale and Baram[14] already recognised that changes in safety management were never so rapid and radical. One can question if safety management principles, that were designed to fit established organisations of the last century with well-defined rules and rigid planning strategies, are still able to cope with the dynamics of present day problems. 'What worked in the past, may not help us today', as Marco Steinberg puts it.[15] Steinberg is active in what is considered social and strategic design, where everything revolves around the question of what an organisation wants to be and stand for and how it wants to get there. Strategic design is about how a design approach can bring solutions to problems. What are those problems, what do they say about an organisation? What are the underlying values? From a strategic design perspective, safety management might need to shift from detailed planning to more flexible management and design orientated approaches. Tools and methods for design based management have been developed by industrial designers (short cycle programmes, prototyping, decentralised organisations, stakeholder participation). Design based management is still a new concept for policy makers, but it is slowly being acknowledged by higher education, industry and governments. See the lecture by Marco Steinberg: Design as a Strategic Tool for Systemic Change, available on YouTube.[15]

Obviously, national regulators also need to adapt to this changing environment and focus more on the dynamic approach to risks, particularly in research and development. Furthermore, they should be aware that much of the administrative burdens originate from the complex regulatory landscape they create in national rules and regulations.

In my view, the major challenge for OSH in higher education is grouped around dynamic safety and the need for organisational flexibility. We saw the increase of novel risks – partially pushed by the greening of technologies – and we saw the limited control of safety in higher education. Limited control of safety and increase of novel risks are examples of dynamic risks. Dynamic risks tend to be unpredictable and require swift action. Safety management planning is not really the best approach when confronted with dynamic risks. Safety management planning, especially in large organisations, follows the annual cycle of budgeting and planning, which have long intervals in which dynamic risks may have changed several times. Furthermore, decision makers in large organisations are generally positioned high in the organisation and they have a large span of control, so communication channels tend to be slow and formal. Hence, OSH management based on central planning is not really fit to control today's dynamic risks. Where safety management planning is not an option, a system of quick and adaptive reaction is an alternative. Paradoxically, such a reaction based organisation has to be planned as well, but this is more a matter of organising safety instead of central planning. Quick reacting organisations tend to be organised in small groups, near to the work floor so they are able to detect derogations in safety. In higher education, one could consider research groups as a quick reaction group and middle managers being the core of that group. We have seen in Section 3.2.2 while discussing the poor control of safety in higher education, that tenured staff (technicians and academics) are an important layer of defence against major accidents, even though they generally are not part of the formal OSH organisation.

Academics and middle managers stand out as key players in future OSH management in higher education; they are the key players who can adapt quickly and intelligently to emerging trends and can steer the organisation from within. We also saw that cultural intelligence of this specific management layer is of great importance to cope with the challenges of multicultural miscommunication on the work floor. And finally, we noticed that the middle management layer is the carrier of the collective memory of the institute, the organisational safety culture. This quality explains why the common approach of pure top-down (safety) management is rarely effective in higher education. In higher education, middle management needs to be involved in OSH management decisions which has to be organised. Here I must stress that a drastic switch towards full OSH management by middle management would be a mistake. In the first place, middle management in higher education – who are mostly academics themselves – is generally not interested in OSH as a profession. And secondly, because of strategic goals and budget allocations, OSH management in higher education institutes requires at least some top-down coordination and central strategy. So, the proper approach should be hybrid OSH management that combines top-down with bottom-up approaches. Hybrid management is a much better solution to the OSH challenges in higher education. Future-proof OSH management in higher education also requires better training of students, PhDs and Postdocs to assess risk in a dynamic manner. Risk and risk judgements need to be a formal course and part of degrees including a final mark that counts towards their result or is a professional requirement to become a fellow or academic. Only then, will the majority of individuals working in the lab be able to adapt to safety changes and make rational decisions on the risks they are taking. As for the competence of OSH professionals, a shift from operational safety advice towards direct support of middle management may be a future skill that needs to be developed.

In the alliance of middle management and OSH professionals, the latter should play the role of experts, stepping in when safety challenges become too complicated. OSH professionals should be equipped with knowledge from safety scientists and from their OSH network with colleagues in other institutes. In this, OSH professionals move away from direct intervention and concentrate on advice and support of middle management. On a strategic level, OSH professionals also have an advisory role towards higher management and leadership in higher education, aimed at strategical OSH goals of the organisation.

3.2.4.1 Evidence Based Approach

Although not frequent, serious accidents do occur in higher education. In 2009, UCLA was shocked by a terrible accident in one of the laboratories in the molecular science building. A spill of *t*-butyl lithium – a pyrophoric reagent – over the body of a research assistant led to severe burns and subsequently her death eighteen days later. In the aftermath of the accident, including state prosecution of the principal investigator (PI) involved, accident investigation by the authorities concluded that not wearing a lab coat was the most significant factor in the severity of the burns. UCLA management was puzzled by the findings, as personal protective equipment (PPE) was available in all laboratories and safety protocols actively distributed. But OSH management misjudged the relaxed local safety culture and did not recognise signals appropriately. When reviewing the OSH management system, UCLA Health & Safety Office revealed a serious gap between OSH theory and OSH practice. Over-all, UCLA academics considered OSH requirements not to be practical and as a result, OSH responsibilities and a culture of safety were not internalised by principal investigators (PI) and their staff. There were great doubts amongst academics whether safety protocols (special operational procedures/SOPs) were actually evidence based. This scepticism proved to be right. A review of the Health & Safety Office showed that only a few of the available OSH approaches were evidence based. Most tools, including commercial ones, were based on assumptions or poor understanding of OSH research. There are no records of similar reviews outside of the US, but several safety scientists expect that worldwide, a large number of OSH tools have a weak theoretical basis.

As for UCLA, the accident was a huge wake-up call. And where many institutions might have had an initial response and then gone back to business as usual, UCLA had a sense that someone needed to step up in fostering a more proactive approach. Together with the other Californian universities, UCLA founded the University of California Centre for Laboratory Safety (UCCLS), with a central task to stimulate the development of evidence based OSH tools within the US and also worldwide. In the vision of the UCCLS, support staff should not impose tools upon academics if they are unable to prove that the tools are really effective.[16] From a moral point of view: if academic work in universities is evidence based, should support staff in higher education not have an obligation to do the same? Since 2012, evidence based OSH tools have been developed with seed money from UCCLS and the evidence based approach for OSH in higher education has slowly been accepted by other universities in North America as well as in Europe.[17] In this respect, I would like to stress that there is much potential in closer cooperation between OSH professionals and safety scientists. Safety science is not an extensive discipline in higher education, most of the renowned safety scientists are working in our institutes. It would be only logical if OSH professionals would cooperate more closely with safety scientists in higher education. This does not only mean that OSH management will be able to continuously innovate OSH approaches according to scientific insights, but it also means that academics appreciate evidence based OSH approaches. Clearly, such efforts are of great assistance to maintain a good balance between OSH management and the scientific process. Closer cooperation between safety professionals and safety scientists, including behavioural and social scientists, can be an asset for several institutes.

Nevertheless, misunderstanding of safety science research and insufficient insight in the interaction of OSH tools in socio-technical systems is still a problem that exists in today's world of OSH. OSH managers should be aware of that weakness when introducing new safety tools in their institutes such as risk assessment methods, accident analyses systems and safety culture programmes.

3.2.4.2 International OSH Networks for Higher Education

As we have seen so far, OSH management in higher education is challenged by many technological and social developments, faster and on a larger scale then ever seen before. From a health and safety perspective, higher education faces challenges that are quite different from ten to fifteen years ago. To highlight a few:

- Internationalisation and the global reach of higher education has led to a significant increase in students and workers with different cultural backgrounds attending higher education. This requires the host institution to ensure the health and safety management system and culture is conveyed so that it can be understood across a wide range of backgrounds.
- The social impact of technology progress is becoming a decisive factor on (safety) management: people spread information through social networks much faster and more efficiently than through corporate communication. Google has become the dominant information source, researchers expect an online experience to manage safety that is seamless and useable on multiple formats, includes research targets that change all the time, sponsors pop ups, covers new risks as they emerge, and policies that are increasingly incident driven. Interuniversity mobility amongst researchers is high and inevitably they transfer safety habits from one institute to another.
- Technical staff used to be the backbone of safety management in laboratories; a continuous reduction in technical staff is observed while numbers of inexperienced lab-workers are rising.
- The advent of new technologies and novel risks (synthetic biology, graphene research and nanotechnologies) and the need to understand the risks associated with these issues where data is not present or limited.
- The increase in the (inter)national regulatory burden relating to biohazards, chemical weapons convention, drugs and drug precursors and the increasing attention being paid to these areas by the enforcement agencies puts an added strain on academics and safety support services.

These examples illustrate that national and international developments confront higher education with safety management challenges that are generally too complex for a single institute to handle. There is a rising awareness amongst OSH practitioners nationally and internationally that the answer lies in cooperation, sharing knowledge, tools and data, and shared development of evidence based tools. Recent examples of such cooperation are the nation-wide cooperation programme of the UCCLS, the biyearly Dresden international conference on safety in higher education and the OSH experts exchange programme of the Dutch universities.

Over the years, many institutes have developed their own safety tools. The costs and efforts of developing, maintaining and keeping the content up to date has shown to be a serious burden for single institutes. Considering the fact that the OSH challenges of most institutes are comparable and therefore require similar tools, the call for cooperation on OSH is becoming louder. Furthermore, today's scientists have a high global mobility and should not face yet another new approach when they start work at a new institute.

Newcomers would surely benefit from aligning such tools. From an efficiency perspective, one could even argue that closer cooperation between institutes could free money for science. So, if we really want to create safety impact at our campuses, cooperation is an answer. Such cooperation can be on a personal level and between people that share an interest in OSH challenges. But it can also be organised in a more structured way. In Europe for instance, such a cooperation could be hosted by the Conference of European Schools for Advanced Engineering Education and Research (CESAER), a non-profit association of leading universities of science and technology in Europe. The association unites over 50 leading universities of science and technology in 25 countries. Amongst others, the aim of CESAER's mission is to learn from one another, share information and best practice in the areas of higher education, research, innovation and university governance. OSH management fits very well in this ambition.

3.3 OSH Regulation in the European Union

3.3.1 Summary

Safety management in higher education, as we discussed earlier, does not function in isolation, but within the regulatory framework of society: sector based guidelines, national law and international agreements. For most European countries, the OSH regulation of the European Union (EU) is leading even when countries are not formally a part of the EU, such as Iceland, Norway and Switzerland. In this chapter, I will argue why central elements of EU OSH regulation are expected to maintain their influence in the UK after Brexit is completed.

It should be noted that the EU legislative system gives substantial freedom to tailor OSH regulation into local laws, rules and cultures. For a better understanding, I will explain the structure of EU OSH regulation and the role of the EU OSHA agency in more detail. I will highlight some developments in European regulation as well as differences between groups of member states, regarding OSH regulation, especially the diversion between higher education institutions in the northern and southern European countries in matters of OSH and social security legislation.

3.3.2 EU Regulatory Framework

Safety is a concept with a strong moral appeal to people. Although safety perception can vary greatly amongst social levels, age, gender and cultural background, most of us have an opinion about safety and consider protection against safety risks one of the most important responsibilities of authorities and employers. As a result, safety is generally well regulated by concerns, autonomous states and international communities such as the ILO and EU.

OSH regulation in the European Union (EU) is largely defined by international law, partly by the EU itself and partly by the International Labour Organisation (ILO). The ILO (related to the United Nations) regulates the safety and security of workers worldwide. ILO agreements have been signed by most countries around the globe, so ILO regulation can be considered as universal workers' rights on which other legislation such as EU laws are based. EU legislation is the basis of national legislation in member states. Although up to 90% of European OSH regulation is derived from international bodies (ILO/EU), they only determine the outlines and it is national laws that regulate OSH in individual states.

Far more than anything else, the EU is an economic union and not a social one. As a result, the organisation of social security and – to a somewhat lesser extent – the regulation of occupational safety and health, is almost completely in the hands of the individual EU member states. Overall EU OSH legislation is enacted in the "OSH Framework Directive" (89/391/EEC). This framework sets a standard for EU member states as well as for European countries that are affiliated to the EU. One could safely say that thirty years of OSH harmonisation led to an embedded European OSH strategy that is likely to last in individual European nations, whatever the future of the EU will bc.

What is the EU OSH strategy all about? The strategy is based on three pillars:

1. OSH prevention through Risk Assessment (89/391 Art. 6).
2. OSH competence by the appointment of a safety advisor/representative within companies (89/391 Art. 7).
3. Work accidents registration (89/391 Art. 9).

EU directives do not come into force straight away. To come into effect, directives have to be implemented in national law first. By giving preference to national implementation, EU regulation is intended to merge with national laws already in place, preserving coherency within the national legislative system of member states. As a result, slight differences can be observed in the way the three OSH pillars (RA, prevention officer, accident registration) are implemented in individual EU countries.

3.3.3 Soft Law

We can distinguish different types of regulation, one of them being the distinction between hard law and soft law. Hard law is the most common kind of regulation, it is about strict regulation, the kind we are all familiar with. It is the outcome of a lengthy legislative process, created by specialists, supervised by parliament, governed by legal bodies, and enforced by authorities. Soft law on the other hand, is the outcome of consultation and negotiation between private stakeholders – within the boundaries of hard law – leading to norms, certification, guidelines and agreements. Soft law is voluntary but not optional once parties have agreed to accept the soft law. Soft law is not enforced as such, but neglecting soft law responsibilities could result in losing licences, clients, suppliers, public trust *etc.*

There is a tendency in most western countries towards a larger role for soft law in society.[18] Hard law has great difficulty in keeping up with the high speed of technological changes, because of its thorough legislative process and because it is time consuming and highly bureaucratic. Soft law, however, is much more flexible and faster to develop. We are currently in a process where hard law is slowly reformulated from normative and detailed regulation into more overall regulation, giving room for soft law to take over details. Because soft law regulation processes are market processes, OSH managers in higher education have opportunities to participate in expert groups of norm committees or advisory bodies for safety guideline development and have influence on the future of OSH regulation. Because soft law regulation processes are market processes, OSH managers in higher education have opportunities to participate in expert groups of norm committees or advisory bodies for guideline development and have influence on the future of OSH regulation. As an illustration, I refer to the nanosafety guidelines for higher education, where government inspired OSH experts from higher education developed their own standards in the field of nanosafety.

3.3.4 North–South

For OSH practitioners in higher education who are generally in contact with institutes elsewhere in Europe, it can be valuable to have a rough idea about a north–south diversion in OSH regulation. In general, one can observe a Northern European and a Southern European model in OSH regulation. Roughly, the Southern European model relies on government control and government involvement in OSH regulatory bodies. Northern European countries on the other hand rely much more on a good dialogue between employer and worker organisations to reach good OSH levels. The northern model expects government to be more a facilitator than a controller. One should be aware that the north model is not better than the south model, just that there is a difference in approach. The north-south observation has remarkable parallels with the cross-cultural research of Hofstede,[19] who did empirical research on cultural differences between nations. Hofstede observed that Southern European countries tend to favour an implicit organisation model that resembles a pyramid structure with the (central) government controlling all activities. According to Hofstede, northern European countries favour a model that facilitates local negotiation between stakeholders, symbolised by Hofstede as a market. Good examples of cultural differences are given by Alasdair White[20] when writing: "...Are the Brexit negotiations a matter of mutual incomprehension between the individualistic attitudes of the United Kingdom and the collective attitude of the European Union? Was the Greek Sovereign Debt Crisis of 2009–2015 a problem of economics or a clash of cultures between individualistic northern states exemplified by Germany and then collectivist, risk-averse Greeks?"

Again, there is no right culture, just different approaches that are based on deeply rooted local ways of solving problems.

Also, in the social security systems of the European countries, a north – south diversion can be observed. A 2019 Dutch survey[21] on the involvement of external OSH experts in risk assessment, showed differences that follow the boundaries between two kinds of national social security systems. The researchers found that based on the funding of the social security systems, the EU can be divided in two models: the Bismarck model and the Beveridge model.[21] In the Bismarck model, social security is funded out of contributions from employers and worker unions. In the Beveridge model social security is funded by government. Here too, a general division can be made between north and south; the Beveridge model is primarily used in northern European countries and the Bismarck model generally in southern European countries. However, this north-south division is less clear as Germany, Belgium and Denmark can be grouped under the Bismarck system (south). Countries that are be grouped under the Bismarck model almost all use a comprehensive compulsory social security system for compensation of costs related to work accidents and occupational diseases. 'Beveridge countries', however, have in most cases no specific compulsory social security system; in case of work accidents, employees rely on their health care insurance. In the Beveridge model, the cause of the damage is of little importance.

What does the north–south division mean for OSH in higher education? For one thing, there is substantial variation in the financial compensation for damage from accidents and occupational diseases, within the EU. Sometimes commuting to and from work is covered as well as work accidents, in other cases students are insured similar to employees or the compensation of damage may vary depending on the cause of accidents. This might be interesting information for higher education in case staff and students are moving to other institutes within Europe.

The north – south diversion is also interesting for accident data analyses, because it influences the quality and quantity of data sets. Under the Bismarck model, travel accidents are included in occupational accidents. Therefore, southern European countries (including Germany, Denmark and Belgium) are likely to gather different accident statistic data from northern European countries. Furthermore, the more comfortable social insurances under the Bismarck model stimulate reporting of incidents, leading to higher reporting numbers. And as a spin off, occupational accidents within the Bismarck model are systematically filed on a national level by the insurance bodies.

Safety data analysts should be aware of the north – south effect, when pooling accident data from different European countries. Overviews could give a wrong picture of safety situations within other institutes for higher education, if the north – south effect is not recognised.

3.3.5 EU-OSHA

In response to the high number of workplace accidents in Europe (yearly 4 million of which 8000 fatal) in the early 1990s,[22] the European Commission declared 1992 the European Year of Safety and Health at Work and founded the European Agency for Safety and Health at Work (EU-OSHA) in 1994.[23] EU-OSHA is an agency of the European Union with the task of collecting, analysing and disseminating information on occupational safety and health. The agency has its headquarters in Bilbao, Spain, with a staff of OSH, communication and administrative specialists. The main task of the agency is to provide OSH policymakers and OSH professionals in the EU with evidence based reports, in-depth publications on specific OSH topics and the development of practically useful safety management tools.

EU-OSHA is largely a decentralised organisation that works through a network of national focal points. Member states, a number of other European countries (EEA, EFTA), the western Balkans and Turkey are participants in the EU-OSHA network. The network covers three distinct areas: 1. analysis and research, 2. prevention, and 3. campaigning and awareness raising. The Healthy Workplaces Campaigns, including a bi-yearly safety award competition, are part of the agency's flagship activities. Next to the healthy workplaces campaign, three EU OSHA programmes are especially interesting: European Risk Observatory (ERO), European Survey of Enterprises on New and Emerging Risks (ESENER) and Foresight.

ERO: In 2005, EU-OSHA established the European Risk Observatory (ERO)[7] that provides an overview of safety and health at work in Europe, describes the trends and underlying factors, and anticipates changes in work and their likely impacts on occupational safety and health. ERO is especially focussed on the influence of new technology and shifting economic and social conditions, which causes workplaces, work practices and processes to change constantly. These new situations bring with them new risks and challenges for workers and employers, which in turn demand political, administrative and technical approaches that ensure high levels of safety and health at work.

ESENER:[24] In 2009, EU-OSHA carried out the first ESENER survey on new and emerging risks (European Survey of Enterprises on New and Emerging Risks). Within ESENER, nearly 50 000 interviews were held with managers and health and safety representatives with the aim of getting an insight into how European companies were managing health and safety issues. In 2019 the third ESENER (ESENER-3) was carried out and the first results are available on the EU-OSHA website.[24]

Foresight: The Foresight project[25] involves looking beyond those workplace risks that are currently emerging, to try to anticipate change in the longer term. It involves developing scenarios in certain subject areas, looking at what might happen given certain technological developments or changes in society, and the implications for OSH, which decision makers can act on. The first and still ongoing Foresight project is looking at 'green' jobs; those which involve technology or practices which are designed to reduce environmental impact.

How can EU-OSHA contribute to safety management in higher education? The most noticeable value comes from analyses and reports, which can be used as a compass for future OSH developments and also as guidelines to develop new OSH approaches in higher education, for instance on risks of nanomaterials, the 'greening' of technology and digitalisation. Through its wide network, EU-OSHA is not only able to identify OSH trends at an early stage, but it also has the means for in-depth research and development of evidence based OSH tools. All EU-OSHA material is published for free on the EU-OSHA website. For a full overview see: https://osha.europa.eu.

3.4 Looking Forward

I would like to end by briefly touching on some options for the improvement of safety management in the future.

In this chapter on safety management in higher education, I have argued that OSH management in higher education is confronted by new technological and organisational challenges that need to be addressed by empowering middle management. Simultaneously, OSH professionals should shift from an operational orientated role to a more coaching orientated role. This should be a gradual shift, proportional to the needs of the institute and not a revolution.

However, shifting towards middle management centred OSH, will lead to OSH information flows that are operational orientated and not to keeping central management informed. Access for central management to OSH information will probably be reduced. Such lack of OSH information for the higher management levels weakens strategic OSH, which is needed by the higher leadership to take balanced decisions. The solution could be in making digital information available that is already present in the different corporate systems, such as ordering of chemicals, reporting of failures, maintenance schedules, permits, inspections, inventories and training schemes *etc.* If all this information can be pooled, processed and visualised, it could become an excellent OSH indicator without much additional administrative burden for the work floor. To build such a holistic information system would be a challenge to ICT of course, but an achievable one; it has been done by other institutes. Amongst them are CERN in Geneva, who developed about ten years ago, a software building safety inspection module within their corporate Engineering and Equipment Data Management Service (EDMS)[26] support tool, part of the Large Hadron Collider project. The module did not only assist in revealing weak points in the safety management system for middle and top management (repeating failures, inert actions, building performance), but it also related failures in maintenance track records. With the approval of CERN, the inspection module stood as a model for a web based version developed by Delft University and the webtool is now in use by a number of Dutch universities (Lab Servant[10])· This is a good example of clever tool development as well as international cooperation.

References

1. H. Mintzberg, *Structure in Fives, Designing Effective Organisations*, Prentice-Hall Int, London, 1993.
2. M. Weggeman, *Managing Professionals? Don't! How to Step Back to Go Forward; a Continental European Perspective*, Warden Press, 2014.
3. APLU Council on Research Task Force on Laboratory Safety, *A Guide to Implementing a Safety Culture in Our Universities. CoR Paper 1*, Association of Public and Land-grant Universities, Washington, DC, 2016.
4. G. Benton, Multicultural crews and the culture of globalization, in *International Association of Maritime Universities (IAMU) 6th Annual General Assembly and Conference*, 2005.
5. P. Yorio, J. Edwards and D. Hoeneveld, *Safety and Health Competence: A Guide for Cultures of Prevention*, Chapter 6: Cultural Intelligence, a construct to improve occupational safety and health in the face of globalization and worker mobility across national borders, Taylor & Francis, Abingdon, 2020.
6. E. H. Schein, *Organisational Culture and Leadership*, Jossey-Bass, San Francisco, 2004.
7. EU-OSHA, *European Risk Observatory*[Internet]. Available from https://osha.europa.eu/en/about-eu-osha/what-we-do/european-risk-observatory. Last accessed on 3 March 2020.
8. P. Van Gelder, P. Vermaas, D. Hoeneveld and G. Ishmaev, *Safety-by-design as Part of a Basic Responsible Research & Innovation Framework, A study on mapping of values in the context of SbD and surveying safety management regimes in a research institute setting*, Dutch Ministry of Infrastructure and Watermanagement (in Dutch only), 2019.
9. J. Reason, Safety Paradoxes and Safety Culture, *Inj. Control Saf. Promot.*, 2010, 7(1), 3–14.
10. Lab Servant, a safety management webtool for higher education: https://www.labservant.nl.
11. R. S. Decker, *Federal Demonstration Partnership Faculty Burden Survey*, Washington, DC, 2008.
12. Association of Universities in the Netherlands, Pamphlet: United for academic education, 2016. Availabe from https://www.vsnu.nl/files/documents/Publications/Pamphlet%20-%20United%20for%20academic%20education.pdf. Last accessed on 3 March 2020.
13. *4th University Conference: Safe and Healthy University*, Dresden 28–30 September 2015 https://www.dguv.de/iag/veranstaltungen/hochschultagung/2015/index-2.jsp, last accessed on 3 March 2020.
14. A. Hale and M. Baram, *Safety Management, The Challenge of Change*, Pergamon, 1998.
15. M. Steinberg, Design as a strategic tool for systemic change, Youtube video added by Tretaroka Društvo, 2014. Available from https://www.youtube.com/watch?v=696KFvBT4Xg, last accessed on 3 March 2020.

16. J. H. Gibson and N. L. Wayne, *J. Chem. Health Saf.*, 2013, **20**, 4–17. Proceedings of the 2012 University of California Center for Laboratory Safety Workshop.
17. E. Czornyj, D. Newcomer, I. Schröder, N. L. Wayne and C. A. Merlic, *J. Chem. Health Saf.*, 2018, **25**(4), 36–49. Proceedings of the 2016 Workshop Safety By Design – Improving safety in research laboratories.
18. E. Stokes, Double movements in the regulation of new technologies: the case of nanotechnology. Regulatory Transformations: Rethinking Economy-Society Interactions, *Law Soc. Rev.*, 2017, **51**(2), 451–453.
19. G. Hofstede, *Culture's Consequences: Comparing Values, Behaviors, Institutions, and Organisations Across Nations*, 2nd [revised and updated] edn, Sage Publications, 2001.
20. A. White, *When Cultures Collide: An Evidence-based Review of How National Culture Affects the Way We Perceive Our World*, Brussels, White & MacLean, 2017, https://www.researchgate.net/publication/320863629_When_Cultures_Collide_An_evidence_based_review_of_how_national_culture_affects_the_way_we_perceive_our_world, last accessed on 3 March 2020.
21. J. van Peteghem and V. de Broeck, Research on the implementation and execution of Articles 6, 7 and 9 of the Framework Directive 89/391 in 10 European countries (in Dutch only), 2019.
22. *Causes and Circumstances of Accidents at Work in the EU*, European Commission, Directorate-General for Employment, Social Affairs and Equal Opportunities F4 unit EC, 2008.
23. 1992 – The European Year of Safety, Hygiene and Health at Work [Internet] European Commission, Press release from 27 June 1991. Available from https://ec.europa.eu/commission/presscorner/detail/en/IP_91_608. Last accessed on 3 March 2020.
24. EU-OSHA ESSENER, Third ESENER (ESENER-3) 2019. Available from EU-OSHA website: https://osha.europa.eu/en/publications/l_en?text=&sort_by=field_publication_date, last accessed on 3 March 2020.
25. EU-OSHA Foresight. Available from EU-OSHA website: https://osha.europa.eu/en/publications/foresight-new-and-emerging-occupational-safety-and-health-risks-associated, last accessed on 3 March 2020.
26. CERN EDMS. Available from CERN website: https://home.cern/news/announcement/cern/introducing-new-edms, last accessed on 3 March 2020.

CHAPTER 4

Stress and Workplace Wellbeing in a Higher Education and Research Organisation

MONICA KANWAR*

Director Health, Safety & Sustainability, University of Hertfordshire, College Lane, Hatfield, Herts AL10 9AB, UK
*E-mail: m.kanwar@herts.ac.uk

4.1 Safety Culture, How is it Different in Research and Higher Educational Organisations?

There has been a lot of work on safety culture in businesses over the years but less so in research and Higher Educational Institutes (HEIs). These organisations tend to have a more complex management structure than traditional commercial companies. The challenges come from a different approach to work, complex employment relationships and groups working in more isolated structures. Academic school and research organisational structures have very strong embedded identities and complex subcultures based on very specialised expertise and knowledge. Policy and political landscapes can change rapidly which results in the development of an environment that can feel insecure. Other factors include the employment status of staff and researchers who may be on fixed or short term contracts due to funding allocations. There has been, in recent years, a greater focus on targets through the regulatory changes and criteria amendments to submissions

Challenges for Health and Safety in Higher Education and Research Organisations
Edited by Olga Kuzmina and Stefan Hoyle

Published by the Royal Society of Chemistry, www.rsc.org

such as Research Excellence Framework (REF), Teaching Excellence Framework (TEF), and formation of the Office for Students (OFS). This adds to the changing expectations and workload of academic colleagues, that are 'seen' to be out of their immediate control.[1]

No single approach to health and safety fits all Higher Education (HE) environments, and it is difficult to provide a silver bullet. This chapter looks at possible influences of culture in an HE environment and how the culture of an organisation influences the stress and wellbeing of individuals. Finally, it looks at practical resources that could assist locally to enable steps to be taken to support the HE environment.

Developing a culture and managing stress in an organisation, or team, requires continual attention with many different influences across the tiers of management in HE. The next section will take a brief look at the current understanding of safety culture, stress and wellbeing and what this means in an HE environment.

4.2 Safety Culture

The expression of safety culture historically has come from official inquiries such as the 1987 Kings Cross (London) underground station fire and the 1988 Piper Alpha North Sea oil rig platform explosion[2] and it has become part of health and safety nomenclature. The expression has more recently evolved to include a broader approach to workplace attitudes, behaviours, and personal responsibilities as well as organisational improvements such as training and people development, which can influence the identification of hazards and control measures to minimise risk. The current approach is that safety culture is a more inclusive concept of organisational culture.[3] Safety culture is not something that can be easily defined, however there are many definitions relating to organisation and individual values, attitudes, beliefs, risk perception and behaviour, as they relate to a proactive approach to employee safety.[2] The most used definition has been produced by the Advisory Committee on the Safety of Nuclear Installations[4]:

> "The safety culture of an organisation is the product of individual and group values, attitudes, perceptions, competencies, and patterns of behaviour that determine the commitment to, and the style and proficiency of, an organisation's health and safety management. Organisations with a positive safety culture are characterised by communications founded on mutual trust, by shared perceptions of the importance of safety and by confidence in the efficacy of preventive measures."

The terminology also becomes more entangled when safety culture is linked to and confused with safety climate.[5] Safety culture is something an organisation 'is' or 'has', whereas safety climate is a 'snapshot' of employee perceptions on the priority given to safety compared to other organisational outcomes. The Health and Safety Executive, HSE, states that safety culture is influenced by several factors (listed below) and that management is the key influencer of an organisation's safety culture:[4]

1. Management's commitment to safety.

It is well established and understood that positive attitudes towards safety at senior management level are vital to a positive safety culture and it cannot be assumed these will cascade down the organisation.[5] It has been suggested[6] the development of a positive safety culture should be within the context of the organisational culture as a whole and therefore embedded throughout.

2. Employee involvement in the safety management system.

This relates to safety engagement, consultation and communication elements of the safety system. It involves bringing together management and safety representatives, so it becomes an extension of management commitment to safety and a measure of the communication flows on safety matters across the organisation and having a 'shared' interest.

3. Training and competence in safety related behaviour.

This relates to the understanding of safety across all levels of an organisation. An important aspect is to support the development of skills, knowledge and experience. Building a common understanding of risk management and the role of people who manage risks is also important in HE.

4. Communication between management and workforce on safety matters.

The importance of having effective communication structures in place based on shared beliefs and mutual trust to achieve collective goals.[7] It is about an organisation articulating its clear objectives on health and safety to everyone.

5. Compliance with procedures and regulations.

The importance of meeting the requirements of procedures and regulations. This requires the development of local monitoring strategies to be able to see how safety procedures are working and are being embedded. Development of effective local level monitoring; ranging from reactive measures (to accidents) or proactive measures (behaviour observation tools), is an important aspect. This is linked to managing risk at every level of the organisation.

6. Organisational learning regarding safety issues.

An important indicator of the maturity of a safety culture is how lessons are learnt from near misses, accidents, or other incidents in order to set an approach that looks at the processes rather than the individual. This approach enables learning outcomes to be embedded across appropriate levels of the organisation. This is underpinned by effective organisational monitoring strategies to measure safety performance. An organisation can learn a lot by looking back to move forward and then taking the opportunity to regularly review, reboot and refresh where change is a regular factor.

Culture is a concept of shared corporate values within an organisation which influences the attitudes and behaviours of its members (academic, research, professional and technical staff). Safety is part of that and is seen as influencing the attitudes and beliefs of members in terms of health and safety performance.[8]

4.3 Stress at Work

HSE in their stress management guidance[9] define stress as 'the adverse reaction people have to excessive pressures or other types of demand placed on them'.

Employees feel stressed when they can't cope with pressures, for example employees can get stressed if they feel they don't have the skills or time to meet tight deadlines. Translated into the HE context, this can relate to deadlines from research, assignment marking, funding applications *etc*. Providing planning, training and support can reduce pressure and bring stress levels down. Sometimes the HE ecosystem with its flexible boundaries adds an element of complexity that makes it more difficult to put in structured planning, training and support.

Stress affects people differently. Factors like skills and experience, age or disability may all affect whether an employee can cope with the day to day aspects of work. In an HE environment the complexity of deadline pressures, the transition from student to young researcher with more 'independent' working and the training of staff who have a role to support, supervise or manage team members may be seen as less of a local departmental priority than development in teaching or research. This can therefore elevate the associated stress within an HE environment.

There are six key factors which can affect stress levels at work as stated in the HSE stress management standards.[9] These are outlined below with the aim of looking at them through an HE perspective:

- *Demands*: in HE demands come from deadlines from research grants, meeting students' needs such as assignment deadlines, research funding expectations, looking after a team, meeting the needs of external regulators, *etc.* Additional demands have emerged from increasing the marketisation of higher education, surveillance through the many feedback questionnaires[1] showed that the pace of change, complex employment models and decreased autonomy are having an influence on mental health throughout the sector. Pressure Vessels[1] detailed how development of new workload allocation models has led to a change in workload demands.

- *Control*: how much say someone has over the way they do their work. Management by metrics (papers published, Research Excellence Framework 'star' ratings, citations, journal impact factors, grants awarded, Teaching Excellence Framework, National Student Survey scores and so on) often establish deadlines outside an individual's control.
- *Support*: the encouragement, sponsorship and resources provided by the organisation, line management and colleagues. Within the HE environment this could be through departments such as Human Resources, Health and Safety, Equality Inclusivity and Diversity, Occupational Health and Training and Development teams, it may also be wellbeing support such as Employee Assistance Programmes (EAP). In an HE environment there are many informal and formal support networks. This can also be through immediate colleagues, stretching across the organisation, but also through networks that have been built from across relevant teaching, research or technical sector associations. This can be very strong in HE environments, the network of support that cuts across institutional boundaries cannot be underestimated, and it is a strength that is particularly developed in the HE environment.
- *Relationships*: promoting positive working to avoid conflict and deal with unacceptable behaviour, is particularly important where there is a greater transient population. Developing and maintaining relationships with people, are common causes of stress. It may also relate to how staff are treated and therefore linked to the safety culture and, by default, to the culture of the organisation. Is the environment highly demanding, target driven, high paced or is it value driven? While these are not mutually exclusive, how we behave can impact others.
- *Role*: whether people understand their role within the organisation and the organisation ensures that they do not have a conflicting role. In an HE environment a person may have multiple roles (*e.g.* mentor, manager, teaching associate), this can introduce competing demands of each role. These roles can also change over time which can become problematic when managing expectations on conflicting demands and priorities in a matrix management approach.
- *Change*: how well organisational change (large or small) is managed and communicated. This has been a major factor in HE environments including changes to tuition fees, as well as new drivers from the Research Excellence Framework and the Teaching Excellence Framework. Students' expectations have also changed following the increase in tuition fees as they are increasingly 'seen' as customers.

All the above factors have an impact in the HE sector including on its employees. These are important triggers for workplace stress. Colleagues in HE have also been seen to have a very high sense of personal responsibility for students which is very similar to the ethos in other educational environments, and should be considered in any approach that is taken to managing stress in different categories of staff in an HE environment. Studies suggest that the role of social support and relationships in the context of health and wellbeing are directly and causally related to health in general and are positively associated with physical health. There is now evidence that social support from a direct supervisor is important and, in its absence, was associated with poor work and health outcomes.[10]

4.4 Wellbeing

Wellbeing has now become a commonly used term, however sometimes the term is interchanged to refer to stress and mental health. Wellbeing is generally regarded as a broader term than mental health.

The World Health Organisation states that "health" is a state of complete physical, mental and social wellbeing and not merely the absence of disease or infirmity.[11]

The role of a health and safety professional is important to support any wellbeing approach. The Health Foundation Parliamentary Group in 2016 defined wellbeing as a sense of meaning and purpose.[12] Wellbeing is used as a 'buzzword', where we all expect more from life. The Education Support Partnership report in 2018 suggested that the word 'balance' helped to define a common view on what wellbeing meant to a group of academics that were questioned.[13] This word resonates well in an HE environment where a duty of care to each other is very strong, however it can sometimes be difficult to avoid the interchangeability between work and home and breaking the boundaries between them.

A useful visual approach is to think of wellbeing as an 'upstream' intervention which provides an opportunity to prevent psychological risks occurring. As with stress, the challenges around an individual's wellbeing are that it can be influenced by non-work factors which can be improved or made worse by work. It is also difficult to measure short term benefits as it has a genuine long-life perspective.[14] It can be impacted by many factors in the same way as stress, for example transitions through roles from student, to becoming a member of staff through to becoming a new researcher or being promoted to a management position.

It is important to decide on the values and expectations being placed on individuals and to determine the balance of responsibility of action between the state, employers, families and individuals as stated in the Foresight report in 2008.[15] This report also devised an approach to wellbeing called the 'Five ways to wellbeing' based on the nutritional approach to consuming five fruit and vegetables a day, see Table 4.1. This model has been a way to

Table 4.1 Resources and tools to help with improving stress, wellbeing in the workplace.

Tool or resource	Summary	Link	Possible application at: **Individual Team Organisational**
Health and Safety Executive Stress Management Line Manager Competency index	This tool allows a simple self- assessment of managers based on their own interpretation of how they behave.	https://www.hse.gov.uk/stress/mcit.htm	Aimed at managers but could be used as part of organisational training for managers on stress management and awareness sessions for all staff.
Healthy University tool kit	A Healthy University aspires to create a learning environment and organisational culture that enhances health, wellbeing and sustainability. The Healthy University tool kit is designed to support HEIs that wish to adopt and/or embed a whole system Healthy University approach.	https://healthyuniversities.ac.uk/	To be used at organisational level to develop a holistic approach to a healthy campus.
Five Ways to Wellbeing: Connect, Be Active, Take Notice, Keep Learning, Give	The Five Ways to Wellbeing are a set of evidence-based actions which promote people's wellbeing. While they don't claim to be the biggest determinants of wellbeing, they offer a set of simple things individuals can do in their everyday lives.	https://www.gov.uk/government/collections/mental-capital-and-wellbeing	Can be used at all levels and is a really simple tool to use. Enables conversations to take place and actions to be set and owned.
Thriving at Work Report	This review provides a set of Mental Health Standards for Employers to provide a framework to set out action plan to implement.	https://www.gov.uk/government/publications/thriving-at-work-a-review-of-mental-health-and-employers	To be used by the employer as an organisational tool.
University Mental Health Charter	This charter provides a framework for universities to adopt the 'whole university' approach to mental health, and support them to become places that promote mental health and wellbeing of all members of the community.	https://www.studentminds.org.uk/charter.html	To be used by the University as framework.

facilitate an understanding in a clear and simple concept. It can work at different levels from an individual through to an organisational approach and inform policy. It has been seen to be a simple tool to enable conversations around wellbeing, See Table 4.1 for details.

4.5 Why is it Important to Consider Wellbeing and Stress at Work in HE?

The HE environment has changed dramatically over the last five years. These changes include:

An increased reliance on digital technologies and increased dependence on online tools which can bring their own digital literacy challenges. The speed of change in digital technologies over the past five years has been significant. The availability of data, the 'always on' culture and the portability of devices have all seen positive impacts on lives but can also have negative impacts on individuals. Such negative impacts include trying to keep on top of changes to systems and technologies through to managing expectations of both the employer and student to responding to communication in an environment that expects 24/7 availability. The boundaries between home and work life can sometimes feel blurred impeding the ability to relax from work.[1]

There has been an increased use of data from surveys such as the National Student Survey, staff surveys and other internal data gathering. It is important that metrics should be used responsibly and the reasons behind the application of the data should be shared and understood as stated in the report produced by Higher Education Policy Institute (HEPI) in 2018.[16]

The primary focus for HE has been around student wellbeing, rather than on staff wellbeing· The whole university approach by the Health Universities Network is now taking centre stage on a way forward to integrating staff and student wellbeing and is referred to in literature.[17,23,24]

Although measures for staff are not as well established as student wellbeing measures, there is now more data relating to staff wellbeing in HE and there has been an increased interest in this area. The data is being produced by the University Colleges and Employers Association (UCEA) and there are some more established engagement and wellbeing networks that share best practice across the sector, one of which is the Healthy University network.[17]

Unfortunately, the data is incomplete with underreporting of sickness absence by academic staff, this should be considered when comparisons are made against wider economy benchmarks. In a recent UCEA survey 68.5% of respondents reported that academic absence was either moderately or significantly underreported according to UCEA, 2018–2019 sickness absence report.[18]

As in previous years, the proportion of long term (over twenty consecutive days) to short term sickness absence is disproportionately higher for academic staff, senior managers, and those in lower grade non-office based roles than for professional office based staff.[18] The levels for academic staff are likely to be linked to the underreporting of short term sickness absence as longer term absences are more likely to get reported than short term ones.

Consistent with previous years, the main cause of sickness absence in terms of days lost was mental health – accounting for 25.6% of days lost. Half of mental health-related absence is attributable to stress, with the total percentage of sickness absence related to stress ranging from 3.5% for senior management to 18.5% for academic staff.

Mental health is the top reason for long term sickness absence with 77.6% citing this as the main reason and a further 15.3% citing this as the secondary reason. The top reason for short term sickness absence was 'cold, cough and flu' with 78.6% of HEIs reporting it as their top reason for short term sickness absence while stomach, digestive or gastrointestinal problems were in second place, followed by mental health. There is a steady upward trend in the proportion of long term sickness absence due to mental health which has risen from 17.5% since the first UCEA survey in 2011–2012 to 25.6% in 2018–2019.[18]

More information and analysis from a variety of networks is building a picture through reports for different categories of staff. The importance of technical staff in the role of supporting students and the importance of providing 'invisible' pastoral care was seen as part of the support provided according to a recent report by the Science Council in 2019.[19] Looking at measures for wellbeing will enable improved conversations around this important subject as referred to in HEPI in 2019.[16] More work on the analysis of data for staff and students together would be helpful as both groups are strengthened by higher levels of wellbeing.[16]

Development of tools to support staff wellbeing and looking at stress management in HE is important and can really make a difference in HE, translating through to what must be the final objective of having happier staff, which in return results in happier students and a healthier and safer working environment.[20] Currently the absence of a universal measure for staff wellbeing leaves a gap in the understanding of wellbeing in an HE environment.[16]

Taking a whole university approach could be helpful to fulfil organisational engagement objectives but also individual objectives that could then result in a more innovative and creative environment. A continual improvement approach used in safety management systems provides an integrated approach to both. Given that the HE environment is so dynamic this approach links well to bringing safety management and safety culture to act as an enabler to support teams and HE institutional strategic objectives.

The relationship between safety culture, wellbeing and stress is complex, they are interwoven between workplace, societal, health, and economic factors. An HE environment is unlike other workplaces in that the population can stretch across many countries, cultures, short term contracts, long hours, short deadlines for teaching, marking, research, and very flexible management structures. On the other hand, informal networks that stretch across organisations and institutes build strong communities of practice which can be seen to counterbalance some of the elements that can trigger workplace stress.

The culture of an organisation can set the 'tone' on how stress is managed, and it can build a community around wellbeing. This can become the foundation of its culture through becoming a value driven approach. However, this approach takes constant attention and is not necessarily easily measured. Engaged employees stay engaged because they have the autonomy to impact upon their work in a way that it becomes not only more resourceful, but also more challenging. This insight is important for practice because it indicates that engaged employees need active jobs in order to stay engaged. The competitive advantage of engaged employees is that they do not rely solely on management to achieve their goals. Instead, they are able and willing to initiate the required changes themselves.[21]

The common link to culture, stress management and wellbeing is good line management skills. The importance of healthy relationships, shared values and an empowering culture is critical. Open and transparent communication with the ability to have fair measures to monitor and review are important. The National Institute of Clinical Excellence published a standard on healthy workplaces in 2017.[22] This guidance highlights the importance of the line manager's role as a link to an engaged employee.

What is important in developing an approach to workplace wellbeing; below are some key principles:

- good communication at all levels
- good balance between work and home life
- the ability to have some control over the way a person does their work
- job security
- maintaining a relevant and skilled workforce
- having good workplace relationships
- having a healthy mix of challenge and support

4.6 Resources Available to Help with Supporting Staff in HE

In Table 4.1, I have identified resources that are currently in use and are not specific to the sector apart from 'The University Mental Health Charter',[23] this is a very new charter and little is currently known about the impact of its implementation.

The specific resources that have been identified are incorporated within the principles of the University Mental Health Charter. The Charter is a route map to build a Whole University approach to Wellbeing as outlined by the Universities UK (UUK) Step change framework.[24]

The Thriving to Work[25] report published the Mental Health Core Standards in 2017 is included as part of the whole university approach as set out by UUK.[24] It sets out a framework for an organisation to develop a preventative strategy to support and manage psychological risk in the workplace. The core standards were published to enable employers to draw up an action plan to help them manage mental health in the workplace. This includes training, development of staff to have better conversations, supporting the development of line management skills and monitoring sickness absence data.

An important aspect of any approach is to have a process to actively monitor progress and to check how interventions are performing. In much the same way as looking at how controls are working in a physical safety environment it is important to have an approach that can be monitored and measured in a psychological environment. If this is not possible this could result in systematic failure or result in short comings.[26] What this looks like is based on what the interventions are within any workplace wellbeing strategy or action plan. Indicators such as sickness absence data or information on engagement, such as number of attendances to wellbeing sessions, provides a very important starting point. Those indicators are linked to enable an organisation to maintain and improve a system, in much the same way as health and safety. The application of a continual improvement cycle provides a flexible tool for implementation.

What can help with any approach is to be able to have engagement across all levels of the organisation. Using active conversations at different levels to engage colleagues is important and links with a coaching led approach, which is also now being used more widely in health and safety management. It is important to maintain openness, mutual respect and participation, and to enable and empower teams to allow wellbeing approaches to develop so they can take local safety ownership. Wellbeing is one of those areas that works best through a whole university approach as there are so many stakeholders who can have a positive impact. In much the same way as safety management, the importance of working together to get the best outcome tends to be the most sustainable approach and enable continual improvement.

Creating a wellbeing culture in the same way as an organisation would aim to develop a safety culture is to move away from focusing on 'reactive' measures, such as sickness absence, to 'proactive' measures such as early intervention plans through improved skill development and early risk management. As with safety it is a step change to an approach, but it aligns to the approach in safety where improvement can take place through monitoring non-injury data. Applying targeted wellbeing interventions will assist in focusing on 'upstream' support to enable prevention.

Table 4.1 has brought together important authorities and their resources. These have been listed with a summary of the approach and is intended to highlight the level within the organisation at which they can be used. The Table also highlights the development needs to allow the implementation of the different resources.

4.7 Case Study of a Wellbeing Approach – an Engagement Led Approach to Wellbeing

Below I describe an example of an approach taken at one institution which has been one of the small steps of change through an engagement led approach to workplace wellbeing. The case study focusses on the development and journey of this institute over a ten-year period from 2009 to 2019.

4.7.1 First Phase

The planning of a coordinated approach to wellbeing was undertaken through a cross department engagement model. Internal stakeholders across the University input into the development of a University wide strategic approach. There was no additional funding attached to this work and it was really through the passion, drive and strength of engagement across the organisation that this took place. Importantly there was senior management commitment and endorsement for this framework to take placc.

Formalisation by the senior management team took place once a framework was developed. The framework was used as the strategic driver over a period to bring together internal stakeholders to drive workplace wellbeing forward. The early years were about small steps and driving engagement. The stakeholders were from across the institution both academic schools, such as Sports Science and Psychology, and professional departments, such as Occupational Health, Student Support, Sports and Health and Safety. The group was chaired by the Director of HR who was one of the senior managers in the organisation. This framework linked into the work that was being undertaken at a sector level.[27]

4.7.2 Impacts of Early Years of Wellbeing (2009–2014)

This group and approach took some positive steps on engagement through:

- development of a Wellbeing Framework – See Figure 4.1.
- initiatives such as Wellbeing Wednesdays
- operational meetings with stakeholders to drive forward on initiatives
- annual Wellbeing Fair
- health MOTs
- development of a logo
- direct reporting to senior managers

There were marginal gains but there were no clear measures. One of the benefits noticed was excellent engagement and the ability to work across an institution with many stakeholders.

Figure 4.1 University of Hertfordshire Wellbeing Framework.

4.7.3 Second Phase 2015–2018

There was a change of focus with a development of the strategic plan at an organisation level. The approach changed to one of bringing 'health' back into health and safety This required the expertise of the health and safety professionals but also the occupational health professionals to build on the already strong institutional internal stakeholder network. This led to the approach of 'happy staff happy students'. There was a link to the strategic plan of the organisation but also a link to the national HSE strategic plan 2016.[28]

This phase focussed on:

- Enhancing the Occupational Health Service to include a more comprehensive health promotion programme. This included early intervention to support staff.
- Engagement with the HSE GB Working Well strategic approach.
- Developing measures to monitor progress.
- First campaign focused on an institutional wide health (ergonomic) risk – see Table 4.2 and 4.3.
- Annual conferences across the institution on health and wellbeing.
- New active staff and student programme developed by the University Sports department.
- Engagement with more 'national' health and wellbeing days being promoted.

This approach worked well to increase engagement across the institution. There was a focus on measuring achievement, reporting on progress to a variety of committees and using available data to further engage with colleagues across the institution. There was a top-level framework but a lot of what was happening was being driven through specific local interventions.

Table 4.2 Campaign plan.

Objective	To improve uptake of ergonomic online courses and raise awareness of musculoskeletal disorders
Length of period	12 months campaign
Set targets	Use measures • Data on uptake of activities • Data on uptake of online courses • Data on attendance at awareness sessions
Reporting	• Reporting to each department on a regular basis to encourage engagement • Reporting to senior managers on uptake to encourage engagement • Reporting to safety committees
Impact	• Submission of GB work well best practice and accepted as a case study • Over a 12 month period staff were more aware of MSDs and how to look after their backs

Table 4.3 Campaign outcomes over a 12 month period.

Activity	Data	Impact
Online course update providing an incentive of a free coffee	Increase in uptake on online courses related to ergonomics	Awareness of musculo-skeletal disorders
Lunch time seminars on ergonomic awareness	Attendance data	Greater awareness
Physical activities	118 activities organised Over 500 attendances to a range of activities	Awareness of activities on campus
Get Britain Standing Day promotion	244 hits to an in-house video produced to encourage people to move away from their desk	Video is still being used regularly

Data from these interventions such as talks on wellbeing or free swim sessions provided existing evidence where available, otherwise national statistics were used to drive institutional improvement.

The increased uptake and interest in available services and activities formed the basis to make a case to have a wellbeing champion network and a fixed term adviser to coordinate the development of the network and drive further engagement. They were also used to look at further developing engagement measures and new tools to help monitor progress with the aim of developing the institution's first annual report on Health and Wellbeing.

4.7.4 Impact Over Ten Years

Ten years on from the small initial steps, the University's approach to workplace wellbeing is now part of business as usual and, as with safety, it is about continual improvement. The next phase will continue to drive improvement with a focus on two main health risks in ergonomic and psychological risk.

The impact over the ten years (see Figure 4.2) has been significant:

- Campaign was included in the HSE GB working well commitments.
- Nominated internal awards for the initial group of wellbeing stakeholders.
- Shortlisted for two safety and health national awards and winning one.
- Achieving Vice Chancellor's award for the Wellbeing Adviser in 2019.

4.8 Finally

We are now in a new phase of life, at the start of this chapter the COVID-19 pandemic did not exist. However, in the UK, now having experienced many weeks of lockdown, there has been a significant change to people's lives at home and at work. The HE environment as we knew it in early 2020 is now a very different place. The rapid change that took place to deliver a high-quality online presence for students and the adaptability of researchers and technical staff to support the needs of health care providers, has shown the incredible versatility of the HE sector. Staff and students have been supported in different ways, such as virtual teaching support and online counselling which have been shown to be effective. Thinking about workplace wellbeing, stress and safety culture with the advent of physical distancing is

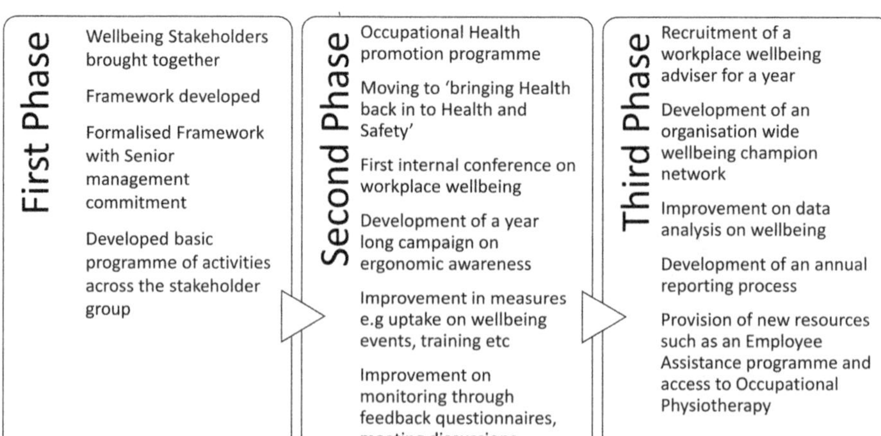

Figure 4.2 Ten-year journey on workplace wellbeing.

very different. Societal perceptions have changed, and workplace wellbeing has moved to the forefront of many organisations. We have had to adapt fast; change has been part of everyday life and our relationships have taken on new challenges through online working and our complete reliance on digital technology to see us through lockdown. The needs of staff have been very different, and our ways of monitoring have needed to adapt very quickly.

Now more than ever, organisations should be bringing safety and wellbeing front and centre of any organisational strategy. Adapting to a 'new normal' and resetting and rebooting organisational cultures is necessary for HEIs to be ready to support our staff and students. Balance as quoted in Education Support Partnership report, 2018[13] could not be more important and how we maintain community support through lockdown will set organisations apart. It remains true that looking after our staff at this time will mean we are in a better place to look after our students. It is the foundation for a productive HE environment. Looking after our staff will help organisations rebuild post COVID-19 pandemic. Using the simple tools and the approach outlined in this chapter may help teams, departments and HEs in achieving this.

Bibliography

1. L. Moorish, *Pressure Vessals: The Epidemic of Poor Mental Health Among Higher Education Staff*, HEPI, Oxford, 2019.
2. T. Lee and K. Harrison, Assessing safety culture in nuclear power stations, *Saf. Sci.*, 2000, **34**, 61–97.
3. A. I. Glendon and N. A. Stanton, Perspectives on safety culture, *Saf. Sci.*, 2000, **34**, 193–214.
4. *Advisory Committee on the Safety of nuclear Installations, Study Group on Human Factors*, HSE books, 1993.
5. R. M. Choudhry, D. Fang and S. Mohomed, The nature of safety culture: A survey of the state -of-the-art, *Saf. Sci.*, 2007, **45**, 993–1012.
6. S. Clark, Perceptions of organizational safety: implications for the development of safety culture, *J. Organization Behav.*, 1999, **20**, 186–198.
7. A. I. Glendon and E. F. McKenn, *Human Safety and Risk Management*, London, Chapman and Hall, 1995.
8. D. Cooper, Towards a model of safety culture, *Saf. Sci.*, 2000, **36**, 111–136.
9. *Managing the Causes of Workplace Stress – a Step-by-step Approach Using the Management Standards, HSG 218*, HSE, 2007.
10. O. Hammig, Health and well-being at work: The key role of supervisor, *SSM Popul. Health*, 2017, **3**, 393–402.
11. World Health Organisation (WHO), Constitution of the World Health Organisation, 1948.
12. All-Party Palimentary Health Group, *A Healthier Life for All: The Case for Cross Government Action*, The Health Foundation, London, 2016.
13. D. Guiney, T. O. ' Brien and Dr Dennis, *Staff Wellbeing in Higher Education*, Education Support Partnership, London, 2018.

14. Department of Health, *Wellbeing – Why It Mattes to Health – Compendium Factsheet*, Department of Health, 2014.
15. Government Office for Science, *Mental Capita and Wellbeing: Making The-most of Ourselves in the 21st Century*, HMSO, 2008.
16. Higher Education Policy Institute Policy Note 13, *Measuring Well-Being in Higher Education*, HEPI, 2019.
17. M. Dorris, A. Farrier, S. Doherty, M. Holt, R. Monks and S. Powell, The UK Health University self-review tool:Whole system impact, *Health Promot. Int.*, 2018, 448–457.
18. University & Colleges Employers Association. *Sickness Absence in Higher Education 2018-19*, UCEA, 2019.
19. *Technicians: Providing Frontline and Vital Support for Student Mental Health and Wellbeing*, Science Council, 2019.
20. P. Litchfield, C. Cooper, C. Hancock and P. Watt, Work and Wellbeing in the 21st Century, *Int. J. Environ. Res. Public Health*, 2016, **13**(11), 1065.
21. A. Bakker, E. Demerouti and D. Xanthopoulou, *How Do Engaged Employees Stay Engaged*, 2012, www.cienciaytrabajo.cl.
22. National Institute of Health and Care Excellence, *Healthy Workplaces: Improving Employee Mental and Physical Health and Wellbeing Quality Standard*, NICE, 2017.
23. G. Hughes and L. Spanner, *The University Mental Health Charter*, Student Minds, 2019.
24. *StepChange: Mentally Healthy Universities*, UUK, London, 2020.
25. P. Farmer and D. Stevenson, *Thriving at Work: The Independent Review of Mental Health and Employers*, 2017.
26. A. Ghahramani, Diagnosis of poor safety culture as a major shortcoming in OHSAS 18001- certified companies, *Ind. Health*, 2017, **55**, 138–148.
27. *Improving Performance through Wellbeing & Engagement- Essential Tools for a Changing HE Environment*, HEFCE, London, 2011.
28. *Helping Great Britain Work Well*, HSE, 2016.

CHAPTER 5

Educate Don't Indoctrinate: Planning and Delivering Training and Support for Workplaces with High Staff Turnover

D. M. HEALY*[a] AND P. O'LEARY[b]

[a]Brunel University London, London, UK; [b]National University of Ireland Galway, Galway, Ireland, UK
*E-mail: derek.healy@brunel.ac.uk

5.1 Introduction

In order to understand, appreciate and progress our training methodologies to best serve those we seek to train, we must first understand what it is that we are trying to achieve, or conversely what we are trying to avoid. We want to discuss the premise of this chapter before we go forward – the educate *vs.* indoctrinate of it all. This bit is a little academic, but we think that it is important to clarify the difference between the two before we move into the details of planning and delivering your training, so please bear with us.

Challenges for Health and Safety in Higher Education and Research Organisations
Edited by Olga Kuzmina and Stefan Hoyle

Published by the Royal Society of Chemistry, www.rsc.org

Merriam-Webster defines education and indoctrination as follows:
Educate:

> "to train by formal instruction and supervised practice especially in a skill, trade, or profession"

> "to develop mentally, morally, or aesthetically especially by instruction"

Indoctrinate:

> "to imbue with a usually partisan or sectarian opinion, point of view, or principle"

> "to instruct especially in fundamentals or rudiments"

If you take a little time to read a paper by Ioana *et al.*[1] you will find that they discuss the subtle differences between these two approaches to learning, explaining that this subtlety makes it extremely difficult to clearly differentiate between the two. Of all the papers, pages *etc.* we've looked at we find their approach to explaining the difference between the two to be the easiest to digest. They propose that the term education must be synonymous with development and growth (from an intellectual perspective) and that fundamentally to educate means to help another to develop the ability to think critically with regards to the information that they are presented with. Education therefore allows an individual to be able to assimilate and process information in order to be able to apply what is learned to a number of different and often unique situations.

Quite distinct from this, they suggest that indoctrination requires that the individual being taught accepts the information that they are presented with, without questioning, taking it, as it were, at face value or as gospel.

From our perspective, the difference in the two approaches will mean the difference between your staff or students being able to memorise and recite information and their ability to apply the information learned in a useful and productive manner.

So, you now know the difference between education and indoctrination – what next?

How do you go about ensuring that the training that you deliver will be suitable and effective for your work environment, and that the approach will match the needs of your employer and your audience?

We've taken the liberty of breaking the approach down into palatable steps to help us to work through the process together.

5.2 Step 1 – Is More Training What you Need?

We came across an interesting article by David Machles[2] in which he discusses the importance of understanding your organisational needs, the unique challenges associated with your institution, and your own internal strengths and weaknesses as a health and safety department, in order to be able to deliver useful and effective training. He suggests that you first need to understand your environment (read as your institution), to appreciate the positives and negatives of what you deliver, before you can move forward – and we thoroughly agree. Introducing new training when there are more fundamental issues within your organisation may do more harm than good, and waste valuable resources, and may even impact further on employee/ employer relations. We've all seen it – where problems with communication, workloads, access to resources have meant that the safety aspects of work are side-lined, or worse overlooked completely. This can be even more pronounced in an academic institution where you need to be able to balance the needs of your staff (academic researchers, technicians and an array of professional services staff that form part of your workforce) and your students.

Quite often we run the risk of throwing more training at a problem and hoping that it fixes it. Sometimes we also need to consider that there may be fundamental issues with the safety culture in our organisations that need to be addressed first. Perhaps safety equipment and PPE aren't easily accessible or well maintained. Perhaps the method of delivery of training or the content isn't what's needed – we need to appreciate that our methods may be outdated or out of sync with the needs of our audience. Difficult as this may be to take on board, we must do so in order to develop and grow. With such a diverse workforce you will need to consider having a multifaceted approach to training.

To get a better understanding of the particular challenges at your own organisation, it can be helpful to ask those receiving the training for some feedback – either through online surveys (we have found that the anonymity often promotes openness), focus groups, questionnaires or interviews. These forums can be very beneficial and can sometimes highlight gaps in training that you may not be aware of. Perhaps spend some time shadowing people from different areas to try and gain a better understanding of the day-to-day routine. If you do decide to take this approach there are a couple of invaluable pieces of advice we feel you should follow: (1) make sure that people know the purpose of your presence – it's not to spy or find fault, it's to gain a better understanding of the environment and the process; (2) follow the protocols and safety advice for the particular department – don't perpetuate the "do as I say, not as I do" approach to management and safety.

In our personal experience we have often found that the best resources are those people doing the work on a day-to-day basis – building relationships, and having regular chats and catch-ups with different members of the university can greatly help you to better understand the day-to-day challenges that they are facing, and in turn can help you to better facilitate their safety training needs. Don't neglect this personal touch, all too often we forget that we are dealing with people.

The other challenge here is that some of the issues that you identify may be more complex and might require the involvement of other departments, for example, human resources or senior management, to help tackle them. As health and safety professionals we need to remember that communication is key and a good relationship with all areas of your organisation will benefit everyone in the long term.

Once you have a better idea of what the problems are and where they are, you can start to put together a plan to address them.

5.3 Step 2 – What is it you are Trying to Achieve?

It's important that, after you've done the groundwork, you take what you've learned and try to come up with effective solutions. Spending time and energy conducting surveys and focus groups and getting feedback from people will be a complete waste of time if you don't use that information and try to address the issues highlighted. Having given our opinions on numerous feedback surveys at the various places where we've worked or studied over the years (yes, we are those people who feel it necessary to contribute!), one of the most frustrating aspects is the lack of follow up – please don't fall foul of this trap because it will do more harm than good to the new and improved relationship you're trying to build with your audience. Always make sure that you at least try to address all of the issues that people have taken the time to highlight for you – and, where you can, offer productive solutions or advice – or if you're onto a winner – both!

Ok, so you've done your gap analysis, you've talked to your people and you have an idea of what the challenges you must overcome are – so now what?

In order to make an effective plan you need to know exactly what it is you are trying to achieve within your organisation and what resources are available to you to do this. There's not much point in planning for augmented reality training sessions and online learning if your budget won't stretch beyond a training room and some complimentary biscuits. Be realistic in what you can do and set achievable goals.

Are you taking a tick-box approach to compliance? Just enough to meet the statutory requirements? Are you looking to refresh existing skill sets through refresher training? Or are you looking for a dynamic approach incorporating both of these aspects, but also with room to upskill and further develop your employees?

Think carefully about what it is you want to achieve. If you're unsure, start with asking yourself this simple (yet somewhat pretentious) question:

What is your *raison d'être*? What is the purpose of a health and safety department?

Fundamentally our role is to protect. The health and safety team have a duty of care under the Health and Safety at Work Act *etc.* 1974 to ensure that our staff, our students, our contractors and the general public are kept safe while they continue doing whatever it is that our individual institutions do – research, teaching, science, engineering, fashion, design, business, law and everything that falls in between.

According to an article in Safety and Health Practitioner[3] (SHP), the modern health and safety professional needs to be able to wear multiple hats (Figure 5.1). We are all well aware of the fact that we are expected to be excellent communicators; legal experts; tech-savvy; business and finance orientated, that we need to be coach, mentor, guru and PPE savants who are well versed in all aspects of safety – biological, chemical, construction, ergonomics, and breathe – the list goes on. Needless to say, we have a lot to live up to.

With all of these pressures it's easy to lose sight of the essential reason we exist – to help people to stay safe while at work by ensuring that they are equipped to do their jobs safely. So, start there.

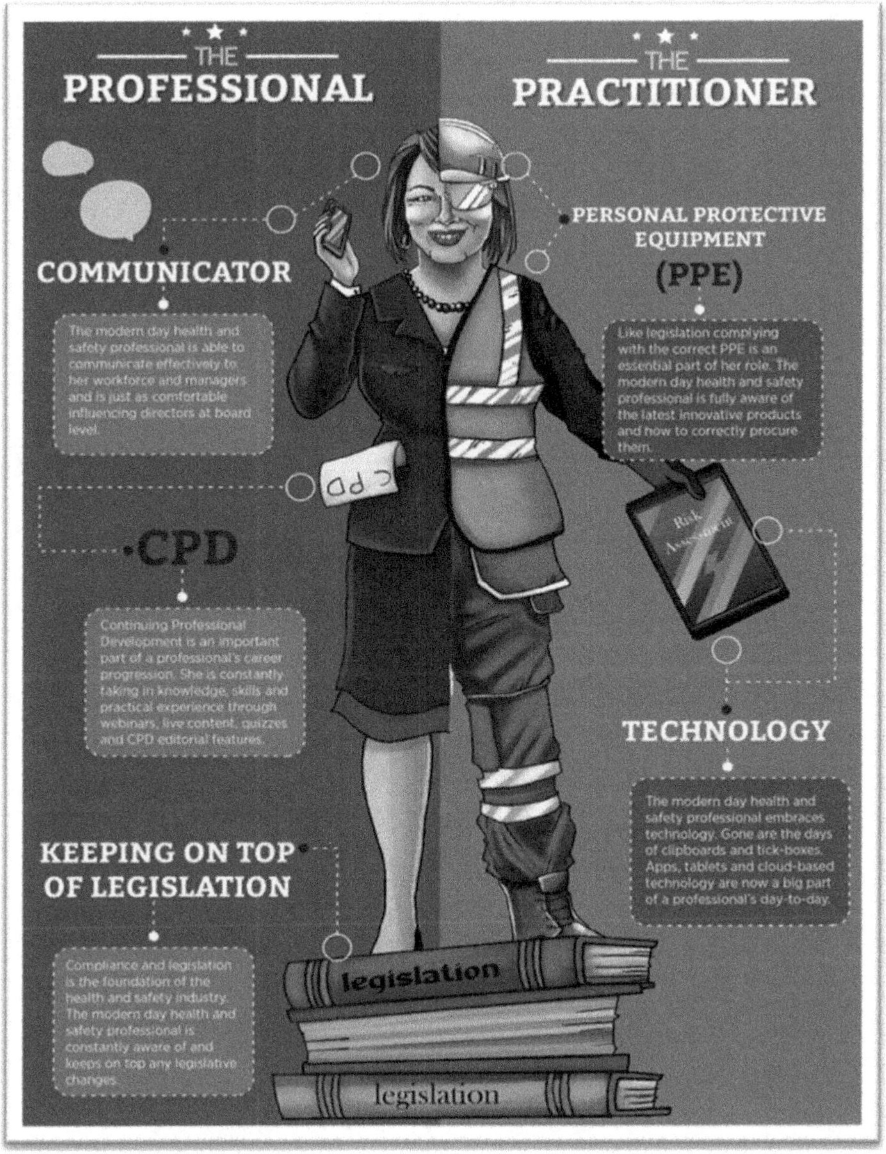

Figure 5.1 The many hats a health and safety practitioner has to wear. Reproduced with permission from ref. 3, Safety and Health Practitioner. https://www.shponline.co.uk/infographic/interactive-the-modern-health-and-safety-professional/.

5.4 Step 3 – Who Needs Training? and What are Some of the Challenges?

In this section we're going to discuss who needs training and the types of challenges that you might want to consider.

5.4.1 Who Needs Training?

The easiest way to approach this is to compartmentalise your target audience. Take a look at Figure 5.2 – we have used a university model and broken it down into some very simplified headings to get us started: Academic staff; professional services, technical staff and undergraduates. We appreciate that there can and will be overlap in each of these areas, but this simple approach allows you to start to understand who your audience is and what they might need.

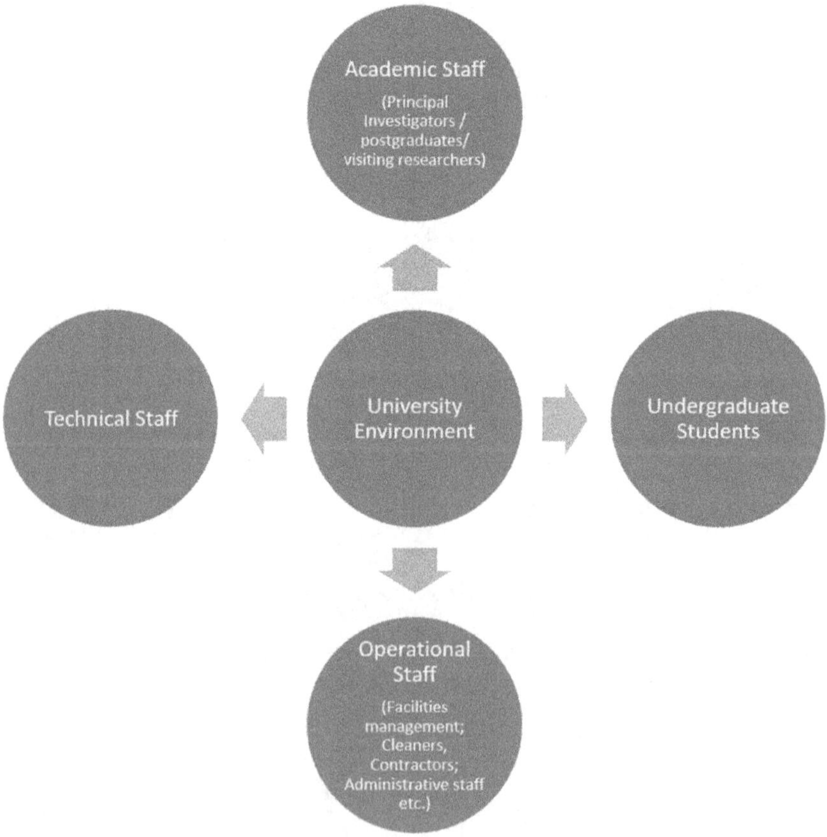

Figure 5.2 A simplified overview of the different staff and students requiring training within a university structure.

Let's start with our "academics". Quite often we are looking at individuals who are early career researchers such as postdocs and fellows and/or the Principal Investigators (PIs) on one or multiple research grants. These individuals will often have research, teaching, administrative and pastoral obligations for undergraduates as part of their day-to-day roles. They will also, more often than not, be responsible for other budding academics – either masters students or PhD students (henceforth grouped together as postgraduate students) – all of whom will be working together to achieve the research aims of the project. From experience, you will also need to account for the fact that academic/research staff don't always work to the standard "nine to five" working day and are often in at different times throughout a term, and sometimes not at all for long periods. Add into the mix here the extra complication of visiting researchers, these people pose an added dimension of risk – quite often they are only present for short periods of time and depending on how your training is structured you may not be able to address their needs beyond the basic safety induction. If you're lucky enough that your visiting researchers are on-site for longer periods of time and subject to the same rules as your "regular" staff, then you can consider them as you would any other academic members of staff.

We must also consider our technical staff and all of the duties that they may be undertaking. Technicians act as vital support for higher education institutions, both in terms of research and teaching. Quite often our technical staff need to juggle the general laboratory issues (chemical stocks, risk assessments, service maintenance and management of the laboratory/workshop spaces on a day-to-day basis) along with teaching elements including experimental set ups, safety training and experiment demonstration for undergraduates, and any other number of tasks within the different departments. It is the technical staff who are often heavily relied on to undertake practical aspects of managing access and use of the laboratory and workshop tools and equipment. Depending on the type of set up (sciences, engineering, arts, design *etc.*) the technical staff may need a broad range of additional support in terms of training and continuous professional development.

Undergraduate students need to be considered as non-experts in general – a blank canvas if you will – though obviously as they progress their background knowledge grows. In lots of ways these are actually the easier target audience for training. The undergraduate will need to be aware of the safety requirements and practices within a university environment. This includes laboratory and workshop safety as well as general safety (fire, sources of first aid, accident reporting *etc.*).

Professional services staff cover a very broad spectrum indeed. Whether in facilities and maintenance, cleaning staff, staff dealing with student accommodation and residences, commercial outlets (shops or coffee shops *etc.* that might be college/university run), administrative staff, student services, human resources, marketing – to name but a few – all with very different roles and often very different needs. (Here you may also need to think about not only your obligations to those staff directly employed by your institution, but also any contractors. It's important to know that these staff have had appropriate induction and are aware of the health and safety process at your institution.)

5.4.2 What are Some of the Challenges?

Now ask yourself, what are the challenges associated with delivering training to this diverse group of individuals? In our combined experience the challenges we have often experienced are outlined below and will hopefully give an overview of some of the common issues we all face as health and safety practitioners:

- Not enough time – one of the complaints we regularly hear from staff is that their time is limited and that they can't fit in training. "I have way too much on to attend a training session" is a common refrain.
- High turnover of people – in a constantly changing and evolving environment it can be difficult to maintain basic levels of training and competence. This, coupled with often little or no succession planning or handover procedures, means that people with key skills are regularly lost.
- Experience – varied levels of experience mean that your training programmes must meet different levels of need across your workplace. This can be quite a challenge to assess and coordinate continuously.
- Differences in background and culture – we must always be conscious that we are privileged to work in environments that are quite often very diverse, with different attitudes to training and safety, but this can pose its own challenges.
- Communication – one of the key aspects to training is communication, we need to remain aware of the fact that not all staff (or students) will have the same proficiency in English and must try, where possible, to address this.
- Different needs – we must also consider those who have a disability, and who may need adjustments or assistance in order to ensure that they can access the training provision.
- Targeting appropriate training – training delivered should build on and enhance existing skill sets and assist people in ensuring that they are fully competent to do the work that is expected of them. It can often be challenging to know what is required and when, and to ensure upskilling beyond the initial burst of training.
- Aside from the above issues surrounding training design and delivery there are also practical challenges associated with maintaining up to date records of training, particularly for large, complex organisations where staff have access to many different sources of training aside from the centralised delivery.

This list is by no means exhaustive and you may have items you wish to add, but in general we think most health and safety practitioners would agree with the points raised above.

Not attempting to address these challenges will mean that you are faced with the same issues arising repeatedly and you will spend your valuable time firefighting instead of delivering the training needed for competency to be maintained. We're not suggesting that there is an easy fix to this – I wish we could lay out a "how to" that suited everyone, but experience has taught us that the best approach is a multifaceted approach.

5.5 Step 4 – Develop and Deliver Your Training

As we've mentioned earlier in the chapter, it is quite possible that you have come to this point and just decided that some basic health and safety training (whatever that means?) will suffice to get you through – nothing more is needed. If that's your plan, then congratulations – you're done – feel free to stop reading any further and move on to the next chapter.

If you're still with us, then I guess you want something more.

Following suitable investigation and analysis, you've now decided that it is definitely a quality training plan that you need, and you also now have a pretty good idea about who it is you need it for. You've done the hard work and considered the potential challenges and pitfalls, but you now actually have to do it – you have to sit and make the plan.

One of my (DH) favourite poems during my studies was "The Old Sailor" by A. A. Milne[4] – in it he refers to a sailor who becomes shipwrecked on an island and then proceeds to try and make a plan to survive. The sailor knows what he needs, shelter, food, drinking water and a multitude of other things, but he never quite manages to get anything done because he never knows where to start. In the end the poor guy does nothing at all – because he is so overwhelmed – and just waits for someone to rescue him (know the feeling?).

Basically, in that little aside, we're trying to address that sense of dread we all feel when we have a big job that needs to be done. It's so easy to just sit and plan *ad infinitum* and never actually do or achieve anything. The wonderful art of procrastination.

But that's not what you're here for.

The easiest way to sum up the rest of this section is with the old saying

> "How do you eat an Elephant?
>
> - One bite at a time" (origin unknown)

If you don't get what I mean by this, it's that in order to be able to tackle what seems like a big challenge, you need to tackle it in small chunks. I think it was Frances of Assisi who summed it up best:

> "Start by doing what's necessary; then do what's possible; and suddenly you are doing the impossible."

Ok then – let's do that, let's start with what's necessary.

If you cast your mind back a couple of sections you will remember that we talked about why it is we are doing what it is we are doing and we mentioned the Health and Safety at Work Act *etc.* 1974 (for our readers in Ireland the Safety, Health and Welfare at Work Act 2005 with subsequent amendments will provide you with similar guidance). With regards to training, this act means that as an employer you must "provide whatever information, instruction, training and supervision as is necessary to ensure, so far as is reasonably practicable, the health and safety at work of your employees" Health and Safety Executive[5] (HSE). This point is further expanded by the Management of Health and Safety at Work Regulations 1999, which clearly identifies particular situations where training in health and safety is especially important, *e.g.* "when people start work, on exposure to new or increased risks and where existing skills may have become rusty or need updating" (HSE[5]).

Keeping this in mind, you (and your organisation) need to decide what exactly your fundamental training will include, and how you are going to ensure that all of your employees will be aware of this and have completed it. You will also need to consider how that information, or at least the relevant parts of it, are conveyed to your undergraduate students. You need to look at the basic health and safety message that your organisation is trying to impart and decide on the best way to do this.

In general, all higher education providers will have an overarching health and safety policy that outlines the organisation's approach to health and safety and what you are committing to as an employer, as well as what is expected of your employees. Our advice would be to ensure that this policy addresses training. The health and safety policy should reaffirm the necessity for compliance training for all, and the need for specific training depending on your role. If your policy doesn't have this, then start there. Review, re-write and re-approve with these additions so that you have a very clear message to all staff and students on the importance of training. If people think that training is an add-on or an optional extra, then more often than not, you're already in trouble. How can you address the challenges of time allocated to safety matters and communication if you don't have clear stipulation for this in your policy documents?

It's also important to note that, coupled with the information provided below, it is essential to have a local induction system embedded in the process. This will reinforce the safety message for new staff and also remind those doing the training about what the safety measures involved in the work are. Induction can consist of one to one meetings between the staff member and their line manager – some institutions have a checklist for the manager to work through to ensure that all aspects are covered. A sample of what this might contain can be seen in Table 5.1.

From personal experience, and anecdotal evidence from others across the sector, training delivery can be somewhat simplified by committing to providing core compulsory modules for all members of staff, along with job specific training to be completed by those members of staff who require it. Outside of this you can offer a repertoire of additional optional career enhancing training (Figure 5.3).

Table 5.1 An example of an induction checklist.

Task	Further information	Completed by	Date
Roles and responsibilities	Introduce the new member of staff to the team and explain their roles and responsibilities.		
Emergency procedures, first aid, accident and incident reporting	Show the new member of staff where the emergency exits are. Explain the procedures for accident and incident reporting and current first aid provisions.		
Compulsory compliance training	Explain which courses need to be taken and assist the new member of staff with enrolling on them.		
Job specific training	Explain which courses need to be taken and assist the new member of staff with enrolling on them.		
Risk assessments	Ensure that the new member of staff is aware of the risk assessment for the area. Suggest that they enrol on a risk assessment training course.		
Employee signature		Date	
Manager signature		Date	

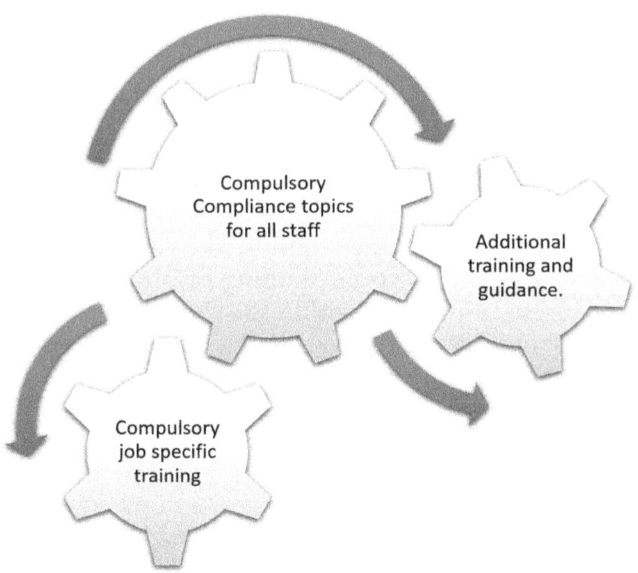

Figure 5.3 A simplified overview of training programmes. Compulsory compliance training for all staff addressing core health and safety issues that are relevant to all. Compulsory job specific training that is role dependent. Additional training that is available to all but not compulsory.

5.5.1 Compulsory Training Development

Examples of compulsory training include health and safety modules that all staff (regardless of their role or position) must undertake within three to six months (for example) of joining the organisation. This can be delivered in any number of formats including (but not limited to) workshops and/or online modules (we'll discuss the options of different types of delivery a little later on). Within these core modules you can introduce staff to the organisational approach to health and safety, clearly outlining the expectations both employer and employee should have moving forward. Some key topics we have seen included in the sessions are:

- Roles and responsibilities – a very clear message is given relating to the expectations and requirements for all staff
- Risk management
- Accident and incident reporting
- Fire safety
- First aid
- Manual handling
- Working with display screen equipment (DSE)

All Staff, regardless of their role or position, are introduced to these concepts and their purpose is explained along with guidance and further information, as well as further training options available both internally and externally to the organisation.

5.5.2 Compulsory Training Delivery

The key aims here are simplicity – a clear, coherent message that everyone can follow and accessibility – you want all of your staff to be able to access this training. We have experienced a number of approaches to this and the most successful (in our opinion) was a multi-pronged approach.

The aim with this type of first-tier training is to make people aware of how health and safety is managed within their workplace, to inform them about their responsibilities to their employer and to themselves, and to inform them of how their employer is delivering on their responsibility to the employee.

Option 1: Workshops designed and delivered by members of staff (from your health and safety department, or any training or staff development departments that you might have).

Delivering set sessions at the beginning of the year that all new starters can sign up to, with extra sessions delivered throughout the year.

This type of delivery does have some success but is completely dependent on the size and complexity of your organisation. The challenges we discussed earlier will still apply – staff unable to attend due to time constraints, or perhaps having to wait well beyond an acceptable period before they can book a workshop that fits in with their work. We've all come across this situation numerous times, people who've worked at an institution for one or two years and are still waiting to get onto an induction course because of time pressures, or because they had to "hit the ground running".

Option 2: Online delivery.

Designing a module that can be accessed from anywhere, at any time, has its benefits – it means that people can no longer use the excuse of not being able to get to a training session, as they can access it from their PCs or mobile devices. Possible downsides to this approach are (1) lack of interaction – people sitting alone running through slides without anyone to talk to about it or ask questions. Also not all material can be taught online and sometimes hands-on training is required; (2) it can be a bit of a tick-box exercise if it's not done properly – people skipping through the slides to the final "You have completed this module" screen. Though this can be compensated for if you throw a little pop quiz in at the end with an expected pass rate or questions on understanding throughout the course material. Not always popular, but effective; (3) without proper monitoring and review these platforms can become stale and outdated, and depending upon your provider they may prove costly or difficult to update.

Option 3: A combination of methodologies.

One of our favourite options is a combination of these methodologies – or "Compliance Cocktails" as they are called at Brunel University London – a mixture of workshops delivered throughout the year and combined with online learning to ensure that all staff, whether they are new to the institution or looking to refresh their core compliance skills can pick and choose the sessions that suit their learning needs – whether that is through attending an interactive workshop or completing online training.

Along with the delivery method, content here is key. Don't expect people to come away from these sessions as experts – what you want to do is to get them to go away from these training sessions thinking and questioning – I know what I need to do now, but where can I find out more information? Am I doing all I need to do? I should speak to the health and safety team or my management.

Take a look at the current content of your compliance training portfolio and ask yourself – Is this content engaging and informative?

5.5.3 Job Specific Training Development

The key aspect to consider here is what kind of work is undertaken across your organisation, and the type of information that would be needed to ensure that this work is undertaken safely. The approach you take is completely up to you. It may be as simple as those working with hazardous chemicals must undertake a training course on Control of Substances Hazardous to Health (COSHH), and if you don't work with hazardous substances then this course becomes optional (so as not to exclude anyone from training opportunities). There may be someone who works with human tissues or relevant material at your organisation – if so then you may require them to attend a training course on human tissue work.

5.5.4 Job Specific Training Delivery

To be able to provide, or facilitate, job specific training it's important that you have a good understanding of exactly what happens within your organisation. This doesn't mean that you have to be an expert in every aspect of the work and research that is ongoing, but you do have to know what is happening so that you can help to identify any specific training needs that may be required. It's easy enough to look at chemistry laboratories and think "ah – these people need a COSHH course to enhance their safety performance", but what about people working with abrasive wheels – do they have access to abrasive wheel safety training? People working with sources of radiation or lasers? Staff working with waste products (including biological waste and hazardous chemical waste) – is there job specific training available for them? With job specific training you are relying, to a large extent, on the processes of risk assessment and analysis at a local level by managers and recruitment staff, and also on communication between departments. These are challenges that all institutions face, so how do you strike the right balance?

We go back again to what we talked about in earlier sections – the relationship your health and safety team have with the different departments across your university.

It's important here that we acknowledge the importance of these relationships and the insight they offer. Perhaps you should consider having regular meetings with the different departments throughout the year – not your standard audits and inspections, but informal chats and coffee mornings that allow people to talk about their safety and training needs more informally. The authors' personal experience of this set up has shown them to be useful and informative for all involved, improving communication and the relationship with health and safety overall.

Now that you've made a plan for your compulsory and job specific health and safety training (you may find Table 5.2 useful to assist you with organising this plan) you can move on to considering what else is possible. In addition to your compulsory and job specific elements you can design and deliver a multitude of different training programmes from multiple platforms, increasing accessibility and uptake.

Table 5.2 Safety training needs assessment sheet (example).

	Target group	Refresher frequency	Number due to attend training 20XX
Compulsory training			
Fire safety	All staff, all postgraduates	Every 3 years	
Chemical safety	All staff	Every 3 years	
Manual handling	All staff	Every 3 years	
Display screen equipment training	All staff	Every 3 years	
Risk management	All staff	Every 3 years	
Job specific training			
Autoclave safety	Staff and students using autoclaves	2 years	
Abrasive wheel safety	Staff using abrasive wheels	3 years	
Asbestos awareness safety course	Staff whose work might entail contact with historic asbestos products *e.g.* buildings maintenance	3 years	
Biological agents safety	Staff and students working with biological agents	2 years	5
GM safety	Staff and students working with genetically modified organisms	2 years	5
Centrifuge safety	Staff and students using centrifuges	2 years	
Confined space	Staff who work in locations where the atmosphere can become compromised and requires particular controls and emergency procedures	3 years	
Compressed gases	Staff and students using compressed gases	2 years	9
Gases detection systems	Staff and students using compressed gases	3 years	
Electrical awareness safety	Senior technicians, office managers, *etc.*	3 years	
Fume cupboard safety	Staff and students using fume cupboards	3 years	
PAT testing course	Staff designated to carryout Portable Appliance Testing in their unit	3 years	0
Radiation course	Staff and students using any form of ionising radiation	3 years	
Liquid nitrogen/cryogenics workshop	Staff and students handling cryogenic materials	2 years	7
List others if appropriate	HF training	3 years	2
Safety responsibilities in research – Principal Investigators (PIs)	All PIs	3 years	3

5.5.5 Additional Training and Guidance

The job specific training can sometimes overlap with the additional training, as was mentioned above. This isn't a bad thing and allows people the opportunity to learn something new and perhaps outside of their normal role. As an institution you may be looking to expand your training portfolio and to offer additional courses to help your employees to progress. There are a number of options here depending on what it is that you do. There is the option to deliver courses that have been developed in-house – courses designed and delivered by the health and safety team; to "buy-in" external trainers for specific topics (radiation safety; first aid training; emergency spillage responses) – sometimes we need to be able to look beyond our own expertise and its potential limitations and utilise external sources; or to expand beyond that and deliver externally accredited training (IOSH, NEBOSH *etc.*). At some organisations, the approach to safety training incorporates a mixture of these.

With these key aspects in mind you should be able to begin to plan and deliver a strong repertoire of safety training at your organisation. The benefits of these detailed and structured provisions are that they help you to improve safety overall across your organisation, encouraging people to discuss safety issues and in turn raise the profile of your safety team. It is well established that a strong positive safety culture leads to greater productivity, as fewer hours tend to be lost due to sickness and accidents when staff are appropriately trained.

And with that, you're done!

Only joking – if only it was that simple!

In a more straight forward organisational setting, or somewhere with a more defined employer/employee structure you could stop there, you would be providing a strong, balanced approach to health and safety training – but this is just the beginning in higher education. This chapter started out discussing the educate *vs.* indoctrinate of it all, and where, you may ask, does that fit in to everything we have just discussed?

The very nature of our organisations means that there are layers of complexity interwoven into the fabric of our structures – both in terms of organisational arrangements, and physical structures and locations. Innovative design and research, by its very definition, presents unique and often very niched challenges to safety management. Who among us hasn't had a conversation that included the phrase "it's never been done before, so how can I risk assess it?" or "so-and-so used to deal with that part of the risk assessment/safety guidance and they're gone now".

The "educate" bit that we mentioned at the beginning comes in here. We need to be able to ensure that, when presented with these challenges our staff, and particularly our researchers, understand the process, and its necessity and that they are equipped to deal with it (with our support of course). In order for this to happen it's not enough to have the compulsory elements, and the interesting add-ons – you need to ensure that our academics, postgraduates and undergraduates are taught to adapt, innovate and apply safety measures as an inherent part of their planning.

In a perfect world we would again be able to say to you: "and here's how you do that", but we're all very much grounded in reality – rose coloured glasses do not come with our standard safety kit!

Rather than telling you how you should do this, we'll give you a combined overview of how other people are approaching this challenge and let you work out for yourselves what might work best for you. (Feel free to pick and choose from the advice below and work out what might meet your needs).

5.5.6 Some Approaches you May Find Useful

5.5.6.1 General Delivery to Senior Academics and Technical Staff

One key element of the process is to try to bring the staff or student members with you on the "safety journey", you do this during the training process to get buy-in from them and to help them to see themselves as part of the system rather than outside of it – to foster a sense of belonging and ownership. It is often the case that new members of any institution (staff or students) may find it difficult to adapt to new systems that are either different from their previous employment (in the case of staff) or a totally new concept altogether (in the case of students). The new safety system can seem cumbersome and appear to involve never ending forms and hurdles to their research or learning. It is important therefore that the process is kept as simple as possible, that forms and documents require only what's necessary and where possible are succinct and comprehensive. Spending time on reviewing and improving your documentation to keep it relevant can be invaluable. Conveying this to the researcher or student helps them to understand what you are trying to achieve and why. It is important that, as part of the training process you engage in discussion about the process involved in developing the procedures as much as you do about its application to their work. This helps them to understand the work that has gone into refining the process to be as streamlined as possible to accommodate their needs and time constraints – this can be useful for pre-empting questions about necessity and relevance of the paperwork.

Discussion is a key part of the education process in health and safety training. We all appreciate that any training has to pass on a certain amount of information of course but it should evolve into discussion (and sometimes a debate). This debate/discussion helps people to realise the role that they play in the process – it can often help them to realise that they have specific responsibilities – it is sometimes the case that people think that the "university has that covered" or there is someone else who has the responsibilities that are actually attributable to the individual themselves. Reinforcing the message that we all have a part to play and we have specific health and safety responsibilities is key.

Often, in our experience, people don't necessarily realise the direct responsibility that they have in terms of safety. In some cases, particularly with academic staff supervising students or postgraduate researchers, they realise that they are there to guide these students and budding academics on their research journey, but not perhaps the responsibility that they as individuals have in terms of safety, risk assessment and training. A useful document to help you to explain these responsibilities is the IOSH publication "Responsible Research – Managing health and safety in research: guidance for the not-for-profit sector".[6]

For new academic staff it is always beneficial if you can try to help them early on to adapt to your processes and procedures, helping them to understand the documentation and safety responsibilities when their research group is small, thus ensuring that the systems in place are appropriate and scalable for when groups grow.

You can use the training plan you have been developing with the guidance above to help with the process of engagement, and teaching people how to apply safety knowledge to their work. Take for example a basic risk assessment course – normally this will include the five steps to assessing risk (1. Identify the hazard; 2. Who can be harmed and how?; 3. Identify the risk and put mitigations in place; 4. Record your findings and 5. Review), and maybe a sample generic risk assessment. Delivering this basic content will mean that yes, you have taught attendees about the process – but have you really educated them? Have you given them the tools and information to do this for their own work? An approach taken by one of the authors here was to develop two different risk assessment courses at his institution – the first a basic overview of the risk assessment process for staff who need to understand what it is and how to write and review a standard risk assessment, and the other – a risk assessment course aimed specifically at academic staff, technical staff, researchers ad postgraduates. This second course covered the basics of risk assessment in the first half, and then delved into the different things that staff might need to think about when applying this process to their own work. The content expanded from the five steps to include considerations surrounding blood borne pathogens; fit testing for RPE; fieldwork and the challenges therein; and different licencing arrangements (with a bit added about the length of time it can take to get an external licence). The feedback from the attendees was extremely positive as people appreciated the examples given and found themselves able to apply the learning to their personal situations.

We have also found it beneficial to include links to cases where something has gone wrong to illustrate the importance of planning and safety considerations, *e.g.* students being detained, assaulted or killed while studying abroad; laboratory accidents resulting in death or injury – somewhat macabre, yes, but effective.

That's a key element to the educate part of this chapter – people tend to respond better to examples rather than just facts. Quoting accident statistics is important but may not mean anything to your attendees without context. Accident or injury may lead to staff absences – most people get this, but for some reason they don't get the impact that it will have directly on them. Sometimes it is helpful to talk about what this might mean in terms of the organisation and their own research – unfortunately we sometimes have to bring it back to how it might affect them as individuals to hit the point home. Asking questions like: "How would you manage if you couldn't go out and get your shopping due to an injury at work?" or, "if you couldn't pick up your children due to an injury at work, how would that impact your life?" forces people to think about the real implications of inadequate safety measures. It is also helpful to mention that this may mean reduced support within the laboratory or workshop, or delays in meeting deadlines and milestones. Sometimes you need to join the dots for people.

This approach helps people to contextualise the safety aspects – and enables them to start thinking about the real life impacts of inadequate safety, and hopefully helps them to plan more efficiently to avoid these situations.

5.5.6.2 Delivering to Postgraduates

The principle here is similar to that adopted for the above section, the key difference being the level of explanation required and the timing of delivery. If you're lucky enough to have taught postgraduate students then the timings can easily be managed to coincide with the beginning of the academic year; for other postgraduate students it can be useful to have at least two or three sessions available throughout the year to capture the late joiners.

It is very easy to forget that our budding researchers lack the experience and expertise of their more senior counterparts – quite often the confidence exuded can belie their current abilities. It is with this cohort that we may need to delve deeper into our processes and procedures to ensure that they appreciate and understand what it is that is required of them.

You may wish to start the training process by covering the basics with them:

- organisational safety policies and associated guidance documents
- safety infrastructure within the organisation
- the adapted risk assessment course
- their responsibilities to other researchers and undergraduates – some of these postgraduate students could act as demonstrators or teaching assistants

Giving them a firm base and a clear structure is the first half of the process – Figure 5.4 is an example of the type of structure you might consider presenting as it will introduce them to the concept of projects and subprojects and the information and procedural documents that go hand in hand with planning and safety.

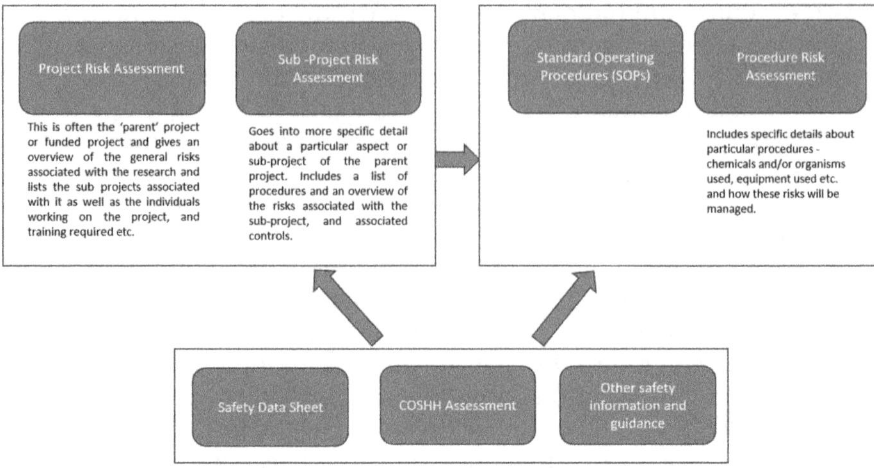

Figure 5.4 Reference framework for safety management within a project. The overarching Project Risk Assessment outlines the project (and subprojects) as a whole and addresses the associated risks from a strategic level. The Subproject Risk Assessment refers to individual projects supporting the central project and focuses on specific risk associated with different aspects of the work. The Standard Operating Procedures provide instruction on undertaking the different elements of the research (cell culture; wood cutting *etc.*) and the procedure risk assessment provides information on the risk associated with these procedures and how they are mitigated. The Safety Data Sheet, COSHH assessment and other safety guidance documents inform the entire process.

There are a number of ways this delivery can be structured; one option is to embed this training into their required core skills modules (or compulsory modules) – taken by all new postgraduate students. This gives you the opportunity to set assignments to assess their engagement with the material presented to them and allows you to measure their ability to apply the safety knowledge imparted to various situations. Another option is to set up postgraduate specific workshops or training days (it helps to have support here from your academic colleagues to ensure attendance!) where you can work through various scenarios and tasks that they might need to complete.

While these are all necessary aspects of the training programme, we need to make sure that we don't drown them in paperwork and policies – the training needs to address these things of course, but remember that your aim is to educate them about safety as a whole. We need to remember that learning and engagement have evolved and quite often for postgraduates and students we need to ensure that we deliver the training that they need and respond to – instead of what we're comfortable with and what has been in place since the beginning of time (so to speak). This may mean using videos, animations *etc.* to engage a tech-savvy generation. The basic delivery concept remains the same – teach them how to apply aspects of safety and risk management to their work – help them to understand the process and the rest will hopefully follow.

5.5.6.3 *Delivering to Undergraduates*

Quite often, we have found, the direct safety training of undergraduates within the laboratory or workshop is devolved to the academic, technical staff, or teaching assistants. Here you are relying on the people you have trained directly to communicate the safety message efficiently and accurately within their particular environments. The actual delivery is often out of your control – this is where those strong relationships we have encouraged you to build earlier in the chapter come into play. If you have a good working relationship with the people delivering the training, then they quite often will appreciate your input and guidance as to content.

On most degree courses undergraduate students undertake very defined tasks in their laboratory work (years one to three) and as such the work is well planned and the hazards easily identifiable and largely controllable. Risk assessment is devolved to the academic co-ordinator for the course in question and the technical staff who will also be present during the lab sessions. Historically safety briefing has been undertaken during a lab briefing directly before the lab and during the lab *via* announcements reminding students of key safety points, with safety information also included in the lab documentation that the students receive before the course commences. Nowadays it's not uncommon for students to be able to undertake initial safety briefings online – using various online platforms that can be used to check that the student has watched the tutorial and where applicable, gained an appropriate pass mark in the quiz following the slides. Where students haven't completed this training their laboratory or workshop privileges can be removed, ensuring that they only begin working in research spaces when suitably informed.

How you chose to design and deliver your undergraduate safety courses is entirely dependent on what it is you are delivering, and the other chapters of this book will touch on some more examples of this for you.

Earlier in the chapter we mentioned training records and the challenges posed here. This seems to be a cross sector challenge. If you are lucky enough, as some of us have been (a big shout out to the Brunel University London Staff Development Team!), to have a fully supportive staff development team who coordinates all of the safety courses you deliver and updates individuals records, then you are very lucky!

But even with this there is the challenge associated with local and job specific training that isn't captured *via* this route. There are numerous challenges associated with this:

- Out of date staff lists – when they aren't updated by the responsible individuals or departments, when someone with a key skill (fire warden/first aider *etc.*) has left.
- Attendance at external training – often we are relying on the provider to send us a sign in sheet and any certificates of attendance. Sometimes these can be missed.

I'm sure you can add numerous other issues in here.

We appreciate that not all organisations have the capability to deliver centralised training and the following simple suggestions might help.

There are various options of training record keeping – electronic records and/or online spreadsheets. The one that we like best (and is often the simplest) is the implementation of training records. This is a basic folder which contains a list of all of the non-centralised courses that they have attended, along with a course completion certificate or a certificate of attendance that the individuals update on a regular basis themselves. It is a very simple and effective solution and has been in operation within the civil service for a number of years. The folder is then auditable and can provide evidence of competency.

5.6 Step 5 – Are We There Yet?

The simple answer is no.

This is a chapter about developing and delivering training in a high turnover workplace *e.g.* a higher education establishment, and as such you are never going to be "there". We don't mean this to be defeatist, but the reality is that this process is never ending, it must continuously evolve and adapt to meet the needs of those we are delivering to and the ever changing health and safety landscape.

So, what do we do?

We grow and we innovate.

In order to be able to deliver effective training we must learn to be adaptable and to utilise different platforms to engage. I'm not suggesting that we ditch all of the established training delivery methods – workshops, classroom based training, demonstrations – these serve a purpose and we need to remember that in the environments in which we work the diversity inherent therein, also means a diverse group of people with different learning needs and expectations.

A paper by Michael Urick[7] discusses the importance of providing a varied approach to training based on the fact that people like to learn or be trained in different ways and not based on the generalisation that these differences are solely generational. He proposes that the best approach, to accommodate the largest number of people is to give people options.

It's important, as trainers and educators in safety that we stay informed around the different methods we can employ to deliver training and their relative effectiveness.

I recommend reading a paper by Martin *et al.* 2014 [8] that looked at thirteen different training methods: "case study; games-based training; internship; job rotation; job shadowing; lecture; mentoring and apprenticeship; programmed instruction; role-modelling; role play; simulation; stimulus-based training, and team-training". These were then categorised by the authors into different modes of learning: "learning by doing, learning by seeing, and learning by hearing". The paper is an extremely useful tool that offers advice and recommendations on the methods of delivery most suited to particular situations. While this is not a hard and fast rule, it is one of the most comprehensive reviews that we have come across and we think you will find it useful when planning your training programme.

So, what should you take away from this chapter?

Our overarching message is to plan and design your training. Think about what it is you are trying to achieve. If you spend the time and energy doing this then the training programme that you offer will allow your staff and students to better engage and rather than just ticking boxes, you will help them to develop their ability to apply the knowledge that you impart through your various methods and approaches to their day-to-day work.

The aim of any safety training is to ensure that staff/students understand the hazards and risks associated with their work, and are equipped, with your help, to do that work safely, and for them to take this knowledge and improve the safety culture of their future workplaces.

Abbreviations

HSE	Health and Safety Executive
IOSH	Institution of Occupational Health and Safety
NEBOSH	National Examination Board in Occupational Safety and Health
SHP	Safety and Health Practitioner

Acknowledgements

The authors would like to thank the Health, Safety and Environment team, the Staff Development team, and the wonderful technical staff at Brunel University London for their continuing support. We would also like to acknowledge the staff and students of the School of Chemistry in NUI Galway for their ongoing work and support and also Prof Michael Hynes (emeritus) for the organised system he developed in his time in the role and his continued advice and support.

References

1. I. Ioana and C. E. Cracsner, Education *vs.* Indoctrination, *Sci. Res. Educ. Air Force*, 2016, **18**, 561–574.
2. D. Machles, *Why Safety Training May Not Be the Answer*, Occupational Health and Safety. Available from https://ohsonline.com/Articles/2007/03/Why-Safety-Training-May-Not-Be-the-Answer.aspx?m=1&Page=1, last accessed 26 June 2020.
3. Safety and Health Practitioner, *The Modern Health and Safety Professional: Interactive Infographic*. Available from https://www.shponline.co.uk/infographic/interactive-the-modern-health-and-safety-professional/, last accessed 26 June 2020.
4. A. A. Milne, The Old Sailor, in *Now We Are Six*, Methuen & Co. Ltd, London, 1927.
5. Health and Safety Executive, INDG345(rev1) 'Contains Public Sector Information Published by the Health and Safety Executive and Licensed Under the Open Government Licence', last accessed 26 June 2020.
6. Institution of Occupational Safety and Health, Responsible Research – Managing Health and Safety in Research: Guidance for the Not-for-profit Sector. Available at https://iosh.com/media/1541/responsible-research-managing-health-and-safety-in-research-advice-guide.pdf, last accessed 26 June 2020.
7. M. J. Urick, Adapting Training to Meet the Preferred Learning Styles of Different Generations, *Int. J. Train. Dev.*, 2017, **21**, 53–59.
8. B. Martin, K. Klodiana and T. C. Lam, Training Methods A Review and Analysis, *Hum. Resour. Dev. Rev.*, 2014, **13**, 11–35.

CHAPTER 6

Laboratory Safety for Undergraduate Chemistry Students

L. M. HANCOCK[a], M. J. HOLLAMBY[a] AND G. R. JONES*[a]

[a]School of Chemical and Physical Sciences, Lennard-Jones Laboratories, Keele University, Staffordshire, ST5 5BG, UK
*E-mail: g.r.jones@keele.ac.uk

6.1 Introduction

Health and safety in chemistry laboratories is coming under increasing scrutiny, as recently highlighted in a seminal article by Ménard and Trant.[1] Clearly, the approach towards health and safety that a training chemist encounters at an undergraduate level will directly influence their own future approach and therefore it is crucial to get this right. The Royal Society of Chemistry[2] places an emphasis on safety as part of its degree course accreditation system and the American Chemical Society[3] has clear guidelines on health and safety in academic institutions. Creating the right health and safety environment requires a culture of positive reinforcement for safe practice, clear lines of responsibility, and an open culture around reporting accidents and incidents to be fostered, that when combined places the safety of all laboratory workers at the highest priority.

In approaching this chapter it was surprising to find, considering the importance of the topic, how relatively little advice is available in the literature concerning safety in undergraduate laboratories. Consequently, herein

Challenges for Health and Safety in Higher Education and Research Organisations
Edited by Olga Kuzmina and Stefan Hoyle

Published by the Royal Society of Chemistry, www.rsc.org

we have attempted to enhance this chapter with references judged as useful, including those not only to journal articles but also to some web resources. The citations arise from a range of cultures: predominantly the US, but also including the UK and other nations. We also attempted to gather our own data by undertaking a survey of UK institutions with significant chemistry programmes. The survey background is presented here in the introduction, and the results will be referred to throughout the chapter.

We also thought it is important to present ourselves, the environment in which we teach and the Keele approach to teaching health and safety in chemistry undergraduate degrees here, to provide a clear perspective of where we come from. We are not claiming that ours is an exemplar course and indeed in the writing of this chapter we have noted many flaws and improvements that can be made to our own approach. The principle aim of this chapter is to collate the numerous facets of laboratory safety, and thereby allow the reader to formulate an approach applicable to their own laboratory.

6.1.1 The Authors

Over recent years the chemistry curriculum at Keele has undergone radical transformation and this has involved a complete restructuring and refreshing of our laboratory courses. Laura Hancock and Martin Hollamby[4] have been directly involved in driving these changes. Graeme Jones, also involved in the development of the course, is currently the Chemical Safety Advisor within the School. This is a role fulfilled by an academic and is conducted in collaboration with the University Chemical Safety Advisor and the University Safety Officer who are there to advise the Head of School on safety matters.

6.1.2 Undergraduate Chemistry Courses Safety Survey

During 2019/2020 we have undertaken a survey of Undergraduate Chemistry Courses Safety (UCCS) within the UK by posting a Google form on ChemSafNet, the email list of the University Chemical Safety Forum.[5] This requested participation from Chemistry Safety Officers from university departments that run straight BSc and MChem chemistry degrees. It was our hope that we would be able to use this to inform the chapter and have a greater diversity of views and practices represented than just our own. We received nine responses (including our own) out of a potential of approximately 65, which limits the conclusions that can be drawn, but the results do give an idea of the level of consistency in practice across the sector. It is difficult to judge why uptake was not greater, but it may have been due to the timing (December/January) or perhaps the survey length. With hindsight, it may have been better to have a number of more focused surveys. What was pleasing was that all types of institutions were represented: three Russell group, two pre-1992 and three post 1992, and there were a range of respondents from academics (4) to technicians and managers (5), all bar one of whom had a specific safety role within their department.

6.1.3 Keele Teaching Laboratories

Chemistry at Keele sits within the School of Chemical and Physical Sciences. As of September 2019, all of our laboratory teaching facilities are part of the Central Sciences Laboratories (CSL). The Chemistry labs are primarily shared with Pharmacy and to a lesser extent Forensic Science and Environmental Sciences. The CSL labs presently cater for all Chemistry teaching from Foundation Year (FY) to Year 3, with the MChem students in Year 4 working in research laboratories sited within our own Lennard-Jones building.

There are three main Chemistry teaching laboratories with a spectroscopy suite (IR, UV-vis and fluorescence spectrometers), an NMR suite (2 × 400 MHz instruments) and an analytical suite (GC, GC-MS, HPLC, MS, ICP, IR, UV-vis) all on one floor. The name and capacity of each laboratory is shown (see Table 6.1). We classify different activities into different hazard levels, high, medium and low. Low hazard activities are those that do not involve experimentation, such as data analysis or poster sessions. Medium hazard activities include experimentation that may be carried out on the open bench and/or may require the use of a fume cupboard, but that are scripted in a lab manual. For some medium hazard experiments students work in pairs and in others they work individually. However, capacity is restricted when students require the use of fume cupboards. High hazard activities are either scripted experiments using particularly hazardous chemicals or experimentation associated with an unscripted research project. For high hazard experiments we require students to work at a maximum of two to a double fume hood.

6.1.4 The Keele COSHH System

Compared to most COSHH systems the one we have at Keele would best be termed unsophisticated but effective. It was only introduced in September 2019 and is based on a paper-based system but now works with electronic forms that are filled in by experimenters, checked and signed by academic

Table 6.1 Lab capacity and working practices.

				Scripted experiment		Unscripted
Laboratory name	Hazard level	Capacity[a]	Number of double FC[b] equivalents	Working in pairs (4 per FC)	Working individually (3 per FC)	Working individually (2 per FC)
Multi Lab	High	70		—	—	32
	Medium	90	16	64	48	—
	Low	120		Not used	Not used	Not used
Heavy lab	High	60	26	—	—	52
Light lab	High	80		—	—	16
	Medium	80	8	28	22	—
	Low	80		Not used	Not used	Not used

[a]On the grounds of fire safety
[b]FC = Fume cupboard.

staff and then submitted to an email repository for storage. Using the email system's search facility the data can be easily searched for specific chemicals, experimenters, laboratories *etc.* The system includes a list of restricted chemicals, the use of which has to be cleared by the chemical safety advisor.

The H and P phrases are gathered manually from chemical sales websites and PubChem. From this data experimenters determine the hazard of a chemical which, together with the circumstance of its use, add up to an associated "COSHH number" that determines the appropriate precautionary measures to be taken (see Table 6.2). A hardcopy of the final COSHH form is displayed adjacent to the experiment, so that this data is available to others working in the same environment.

6.1.5 Research Competency

All researchers entering the research laboratories at Keele must undergo a competency assessment (see Table 6.3). This system was introduced with the arrival of MChem students in the research laboratories four years ago and was designed to prevent academics from just telling students to "go into the lab and get on with some practical work" without ensuring that they really know what they are doing. Researchers progress through the levels as they gain experience. Most MChem students move on to level 2 after two to three weeks, having proven that they can work safely in the laboratory and know their limits.

One advantage of this system is that there is now a demotion sanction for students who start to behave inappropriately in the lab, carrying out unsafe procedures or leaving the laboratory potentially unsafe for other users. We have threatened, and even actioned this, but in the main it has been an effective deterrent.

6.1.6 The Keele Chemistry Laboratory Pathway

At Keele we have developed a clear pathway (Figure 6.1) to training practical chemists and have categorised the progression of a student at every stage of their journey from novice practitioner to competent researcher. This reflects not only the sophistication of experimentation being undertaken, but also various aspects of safety in the laboratory including levels of supervision, detail provided in the laboratory manuals, COSHH form preparation and safety instruction. The following section chronicles the progression of students through this pathway and describes how this specifically maps with matters of health and safety training.

6.1.6.1 Foundation Year (FHEQ Level 3) (Novice Practitioner)

Foundation year students at Keele currently study chemistry as part of a science or health foundation year. Those progressing from the former will go on to study chemistry or related degree subjects (*e.g.* forensic science) while the latter will study medicine or pharmacy degrees. Students may have studied

Table 6.2 COSHH numbers table.

Substance properties	Process carried out in a fume cupboard		Process carried out in an open laboratory environment	
	Non-volatile solid	Volatile liquid or gas	Non-volatile solid	Volatile liquid or gas
Non-toxic Not harmful Low hazard	**1** Safety glasses Lab coat Fume cupboard	**2** Safety glasses Lab coat Fume cupboard	**3** Safety glasses Lab coat	**4** Safety glasses Lab coat
Irritant Toxic Flammable Corrosive	**5** Safety glasses Lab coat Fume cupboard Gloves	**6** Safety glasses Lab coat Fume cupboard Gloves	**7** Safety glasses Lab coat Gloves Consider move to fume cupboard	**8** Safety glasses Lab coat Gloves Consider move to fume cupboard
Very toxic Highly flammable Health hazard Oxidant Highly corrosive Mutagen Carcinogen Explosive Pyrophoric	**9** Safety glasses Lab coat Fume cupboard or glove box Protective gloves Possible additional measures See Restricted Chemical List	**10** Safety glasses Lab coat Fume cupboard or glove box Protective gloves Possible additional measures See Restricted Chemical List	**11** Safety glasses Lab coat Protective gloves Possible additional measures See Restricted Chemicals List Urgent move to fume cupboard unless Safety Advisor consulted	**12** Safety glasses Lab coat Protective gloves Possible additional measures See Restricted Chemicals List Urgent move to fume cupboard unless Safety Advisor consulted

Table 6.3 Research laboratory competency levels.

Competency level	Expected research experience	Key aspects
1	MChem OR non-UK masters OR without chemistry degree	High level of supervision, supervisor or proxy-supervisor (with competency 3/4) present at all times.
2	1st Year PhD, MChem with experience	Intermediate level of supervision. Requires supervision by supervisor for medium risk experiments. Competent to undertake straightforward tasks with experience.
3	Experienced PhD, Post-doc w/o specific experience	Low level of supervision. Requires direct supervision by supervisor for high risk experiments. Most laboratory procedures can be undertaken without supervision.
4	Post-doc	Researcher requires oversight supervision with agreed experimental plan.

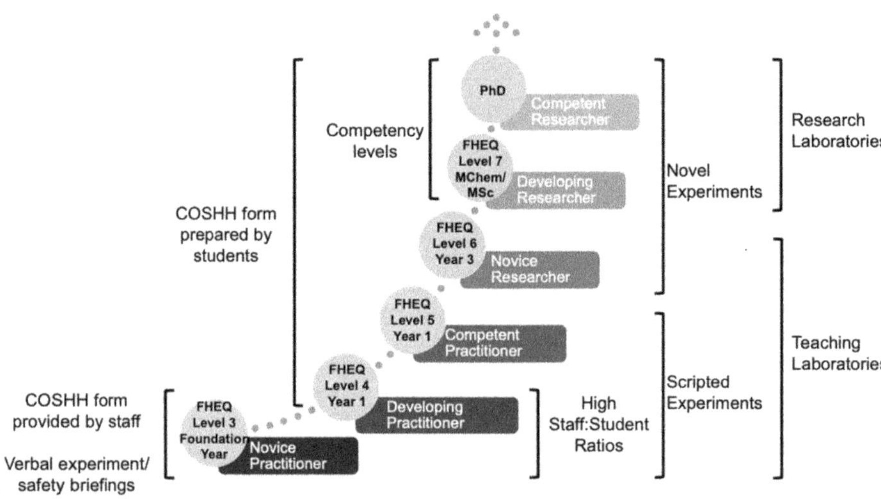

Figure 6.1 The Keele Chemistry Laboratory Pathway.

chemistry at Level 3 (*e.g.* A-level, BTEC) but it is not a requirement. It is likely that students will not have experienced a university laboratory setting before.

Before laboratory work commences, students are given a detailed safety induction. They are required to sign to state that they have understood the laboratory safety rules with regard to appropriate behaviour, dress *etc.* Each laboratory has an associated multiple choice test that students are required to complete in advance of the laboratory session, with questions relating to procedures and safety. This ensures that students are familiar with what they will be doing before entering the laboratory. At the start of each laboratory session there is a detailed briefing led by the academic in charge of the lab that includes all aspects of health and safety associated with the experiment.

At this level, all COSHH risk assessments are completed by the staff supervising the laboratories. The experiments that students complete are deemed to have a low level of risk, such as simple acid base titrations, following a sodium borohydride reduction of a ketone using TLC and measuring the first order kinetics of the HI/H_2O_2 reaction.

6.1.6.2 First Year (FHEQ Level 4) (Developing Practitioner)

The aim of our first year practical module is to develop professional chemistry skills. This is assessed through submission of pre-lab work, laboratory reports, laboratory diaries and a practical exam. At the beginning of the first year of a chemistry degree course at Keele, students are treated similarly to the foundation year students described above, since they possess a similar level of prior experience.

Students are given a detailed safety induction at the start of the semester, which includes some active participation by students answering multiple choice questions using a voting system. The induction includes explicit discussion of lab safety rules, for example those pertaining to clothing and behaviour, and informs students of the safety features of the lab, such as the drench shower and eyewash stations. During this session our Safety Teams initiative (Section 6.4.4) is also introduced and guidance given on correctly filling in our bespoke COSHH form (Section 6.4.5). In addition to this induction a safety briefing is given at the start of each practical session throughout the academic year.

The first semester of the practical course is focused on skills development (with safe working practices considered an important skill). To this end, we assign relatively high staff : student ratios (1 : 10/12) to address the limited experience students will have in a chemistry laboratory (and hence ensure safe working practice) and to allow adequate supervision for skills development. All experiments contain clear instructions, with no room for deviation from the script. At the end of the first semester, a practical exam[4] assesses core competency in practical skills.

In the second semester, given students' developing competency in laboratory skills and safety, staff:student ratios are decreased (1 : 15). Typically, ratios remain at 1 : 15 until students enter research laboratories in their third year. Students carry out scripted practical sessions in pairs, largely independently. In this semester, they complete three COSHH forms independently, which are marked and contribute to their module mark, with the remaining COSHH forms supplied to them.

In the first year, the experiments that students complete are considered to have a medium level of risk. We do not avoid using high risk chemicals, but instead find ways to limit exposure (for example, oxalyl chloride is dispensed by demonstrators rather than students) or generate these reagents *in situ* (bromine is generated from sodium bromide and sodium perborate).

6.1.6.3 Second Year (FHEQ Level 5) (Competent Practitioner)

In the second year, students are introduced to air-sensitive reagents, including those that potentially react violently with water in the atmosphere *e.g.* $TiCl_4$. Laboratory scripts, in general, still provide a detailed list of instructions, however towards the end of the year students follow scripts in the form of experimental sections from research articles, in preparation for their independent research project in the third year.

In advance of commencing experiments, students are given a 'lab skills refresher', which includes practise of skills already developed (*e.g.* in the first year) alongside the introduction of new techniques, such as the handling of air-sensitive reagents – including the use of needles and syringes and flushing flasks with an inert gas. This session includes a safety induction that recaps key points from the first year and also, due to the enhanced risks of the reagents that are to be used, emphasises the use of the eye wash and safety shower: a staff member feigns spillage and demonstrates their use, admittedly not for the full fifteen minutes. We also make it clear to all students that we have a duty of care to use the safety equipment if they have had an accident. However, no student is expected to tackle a fire and so they do not undertake fire extinguisher training.

During the second year, as part of the pre-laboratory exercises, students are required to complete COSHH forms in advance of the laboratory session, which must be signed-off by staff at the beginning of each laboratory session, prior to the start of any experiment. If the COSHH forms are deemed to have not assessed risk adequately, students are asked to make appropriate amendments before being allowed to commence their laboratory work. When working with particularly hazardous chemicals or novel technical procedures a safety briefing is additionally given.

6.1.6.4 Third Year (FHEQ Level 6) (Novice Researcher)

All students studying for a single honours chemistry degree at Keele complete an independent research project in their third year. Supervisors design projects to be appropriate to a novice researcher, (*i.e.* they typically follow established work) and hence monitor the level of risk.

Students are given a research level safety induction and are introduced to the system of uploading COSHH forms to the electronic departmental safety drive (Section 6.1.4). In general, supervisors expect students (with guidance) to find and follow procedures from research articles. Supervisors will discuss safety aspects of procedures with students.

Students work in the teaching laboratory (Heavy lab, Table 6.1) supervised by the project module leader and a demonstrator. Supervisors will also often be present in the laboratory, particularly when difficult or hazardous procedures are undertaken.

6.1.6.5 *Fourth Year (MChem, FHEQ Level 7) (Developing Researcher)*

All students in their MChem year complete a substantial research project, often carrying out novel chemistry. By the completion of this year they are expected to reach the level of competent researchers. Consequently, they are treated as researchers and carry out their work in research laboratories, ideally alongside PhD students and post-doctoral research scientists.

MChem students receive a minimum of three safety inductions which are recorded on their competency forms. The first is given by the Chemical Safety Advisor, who explains the competency system, COSHH form procedure and standard operating practices (SOPs). The second is given by the workplace safety advisor, who undertakes a full induction to safety in the building. The third is typically given by the academic supervisor and covers safety in the research laboratory in which they are working. In addition to these, other safety briefings may be required for the use of specific equipment (*e.g.* NMR, X-ray instruments).

Aside from the location of students' work, a key difference is the use of the research laboratory competency system (Table 6.3). As noted above, this system assigns a competency level to all research staff, and researchers at different levels require a different level of supervision. At the beginning of their MChem year, students are a competency level 1 and require direct supervision by someone with competency level 3 or 4. This level of supervision is required until they have shown that they can work safely in carrying out the standard laboratory procedures they are expected to use during the project, at which point they can be promoted to competency level 2. At level 2, students can work independently alongside other students with this competency level, but lone working is not permitted.

6.2 PPE and Dress Codes

6.2.1 Eye Safety

In the UCCS survey two thirds of institutions (including Keele) provide students with their own safety glasses. Others provide communal safety glasses in the lab and a minority require students to purchase their own safety glasses. At Keele we run a hybrid system, we provide standard safety glasses free of charge which are functional, but some say uncomfortable, so students can choose to purchase a more expensive brand if they wish. At Keele, we provide students wearing prescription glasses with standard safety glasses that fit over them; we do not provide prescription safety spectacles. We strongly discourage students from wearing contact lenses on safety grounds. While this is difficult to police, in general we have found a description of the potential risk of a substance getting trapped between the eye and the lens to be a reasonable deterrent.

Safety culture now receives a much higher profile in schools and the wearing of safety glasses is standard practice there so we have little problem enforcing the rule of 'safety specs on at all times in the lab'. Occasionally, students enter the lab without them on or go for the 'beach look' with them perched on the top of their head but it is our experience that they tend to be quick to rectify this after it is pointed out by laboratory staff or, as is often the case, fellow students.

6.2.2 Lab Coats

Like safety glasses, a majority of institutions provide lab coats free of charge to students, though in two of the UCCS institutions surveyed students are provided with communal lab coats that are stored in the lab. In one institution, space was provided in the laboratory for students to individually store their PPE, however in a majority of institutions students were expected to store the lab coat themselves in between lab sessions. Clearly this is far from ideal, given that the purpose of the lab coat is to stop chemicals getting onto clothes and hence stop them being spread around areas outside of the lab. If students are placing lab coats in a bag then there is the potential for getting chemicals on its contents, including computers, papers, food, coats *etc.* not forgetting the ubiquitous water bottle. All of these could then go on to contaminate either the student directly, or other students who they share these items with. In addition, there is also the danger that students on joint and combined honours degrees, for example in chemistry and biology, could be wearing their chemistry lab coat in the biology laboratory and *vice versa*. One of the problems to overcome here is different approaches in different disciplines across an institution, which can be difficult to resolve, coupled with a tendency to 'make as much as possible out of the space' and not have it 'wasted' on cloak/locker rooms. It is noticeable, however, that where legislation drives practices, for example in some biological hazard laboratories, cloakroom space for lab coats, changing (and laundering) is routinely catered for.

At Keele, we have recently grappled with the various issues involving lab coats. Like many institutions we only have limited space and individual storage in the lab was not a viable option. Similarly, communal lab coats were also not possible because of the storage problem. It was also felt that students might be less prepared to wear communal lab coats and that this might not only lead to significant complaints but also a reduction in pride in and ownership over lab coat appearance that naturally comes from possession. Potential other problems were identified associated with poor practice, for example glass, needles, tissues or other materials being left in pockets. While this could have been solved with pocket-free lab coats, the reality of the sector is that lab coats have got pockets for a reason. Our solution was to provide all students with a stout plastic zip lock bag along with their standard issued polycotton lab coat and to highlight the problem of spreading chemicals around with anecdotes about students spreading chemicals onto the sensitive parts of their nearest and dearest. This approach is also used to encourage lab coat removal and hand washing before going to the toilet.

On the topic of cleanliness five out of the nine institutions offered a free cleaning service for lab coats, which is to be applauded. We stress to our students that they must not wash their lab coats in their domestic washing machine and certainly not with other clothing. However, we admit we were not one of the five who offered free cleaning, which has prompted the following reflection: Is it reasonable to tell students not to do something when they have no alternative? What message does this give out about our real attitude to safety culture? Thankfully, the days when a dirty lab coat was worn as a badge of honour are fast receding and they are starting to be treated as a practical piece of safety equipment. With the introduction of poppers, and verbal reminders to ensure their closure prior to entering the lab in the early degree stages, open lab coats also seem to be less of a problem.

Coloured lab coats are widely used by laboratory staff (eight out of nine institutions) so that they can be easily identified.

6.2.3 Gloves

In the USSC all institutions reported that they provide disposable nitrile gloves. In one third of institutions, students are required to wear them for all procedures, whereas in the other two thirds gloves were worn for some procedures. In our experience, glove wearing policy can be a bit of a minefield. From a health and safety perspective, not all chemicals require you to wear gloves when handling them and those that do require gloves may require the gloves to be changed at regular intervals. Reflecting on this, we conclude that as part of their chemistry education, students should be taught to wear the right kind of glove for the right kind of chemical and receive training on glove removal, breakthrough times and how long gloves should be worn for. However as multiple groups use teaching laboratories using a wide range of chemicals it is often uncertain what chemicals may be present on surfaces, indeed we had one incident in the recent past where a student had a skin reaction to a chemical which we suspect was on the bench. Therefore, we ask our students to wear gloves when handling all chemicals in the laboratory.

Students also need to receive training on when to replace gloves. We have the simple message if you get chemicals on the gloves then they should be changed. Students with visibly dirty gloves are also asked to change them. This stops students contaminating themselves and also the equipment and surfaces around them with chemicals they may have on their gloves. One problem with gloves arises from handling materials that pass between lab and non-lab spaces, such as laboratory diaries, laptops and perhaps phones. Laboratories are often easier to control if they are clearly marked as a place where you either wear specific PPE such as gloves or you don't. The wearing of gloves while using spectrometers is more ambiguous.

The problem of wearing gloves around buildings that contain laboratories, especially where chemists are moving between laboratories (for example to retrieve a chemical or to deliver a sample) is one that we are encountering. We are introducing a PPE "one glove" policy for moving around the building carrying chemicals, chemicals in the gloved hand, bare hand for opening doors. Secondary containment for samples, as employed for larger volumes, was considered but just seemed impractical. We currently operate a 'wear gloves when handling chemicals' policy in synthetic labs, and in the NMR suite we have a no lab coat policy but in here students arrive with a sealed NMR tube with a gloved hand, place it in a spinner, throw away the glove, pick up the tube with a tissue and clean it, adjust the depth and then put it on the autosampler. However, we also have spectroscopy and analytical labs which contain spectrometers and chromatographic equipment where wet chemicals in the form of aqueous and organic solutions are handled but also keyboards are extensively used. We are still debating should this be gloveless or gloved? Some have a strict "no gloves on keyboards" rule but does this make sense in this sort of multi-purpose lab?

Aligned with the wearing of gloves is the provision of hand washing facilities. With our new laboratories that have extensive hand basins, we are bringing in a new regime reminding first year students to wash their hands before they leave the laboratory, to hopefully instil good future practice that has somehow been lost with the proliferation of glove wearing. We do foresee some practical difficulties with this strategy applied throughout the building though, as despite a capacity of over 100, our Multi Lab has only three hand basins, which can lead to bottlenecks if all students complete experiments at roughly the same time. Unfortunately, that issue has recently been further compounded by a change in location of the main entrance to the lab and now there is only one sink by the door, which we hope in the medium term will be rectified by the creation of a new locker room area which will contain handwashing facilities. This is not ideal as the handwashing will be outside of the lab but it is the best that can be done under the circumstances.

6.2.4 Dress Code in Addition to PPE

All UCCS institutions surveyed did have additional dress code regulations. No sandals, ballet shoes or flip-flops was the most common (8/9), followed by no high heels (5/9) and no bare skin below the hem of the lab coat (5/9). One interesting comment was 'the footwear is a rule, the bare skin is advisory'. There is a debate to be had here. If having no bare skin below the hem of the lab coat is a requirement for safety purposes then there has to be appropriate PPE provision by the institution. On the other hand, it is not unreasonable to request people to wear certain, easily catered for, clothes. At Keele, we are clearly quite strict here as we enforce a "no bare skin" rule along with no tights or leggings. Demonstrators, academic staff and technicians all enforce the dress code local rules and one institution volunteered that this is done as the students enter the lab which is probably best practice. We have operated a similar policy at Keele, particularly early on in the first year, to encourage compliance. Six out of nine institutions turn away students who are inappropriately dressed, three institutions just warned their students, and some institutions, like Keele, make provision for students to purchase items. In 2011 Crockett[6] reported their dress code sign off sheet which had punitive sanctions for not following the rules. On the first violation, violators are sent away to dress appropriately, on a second violation they receive zero for the lab.

6.2.5 Headphones

The use of headphones in undergraduate teaching laboratories at Keele is forbidden. This is primarily because the laboratories constitute teaching sessions, but also for safety reasons, as they may stop the wearer hearing important safety information, suspicious noises indicating a fault or misuse, and/or alarms. The arrival of the Bluetooth 'ear bud' has brought problems in our laboratories; in the last two years we have noted students wearing them in just one ear. Depending upon hair style, this can be difficult to spot.

6.2.6 Electronic Devices

A question on the use of electronic devices in laboratories was unfortunately omitted on the UCCS survey. While all would agree that engaging with social media would be an inappropriate use of a phone, taking a photo of a TLC plate is a useful way of recording information. Similarly, using a device to look up safety information on a chemical or procedure might be invaluable. The real worry comes from students becoming distracted (for example, by incoming messages), and for this reason our policy in general is that phones should be left outside of the lab, although we note that it is difficult to police this. Relevant to this is a concern over mobile phones becoming contaminated by chemicals which then come in contact with the users face and then potentially could cause skin irritation or even be ingested.

Maybe we are going to need to be pragmatic about devices and accept that they are here to stay. It is notable that most researchers carry their phone on them, indeed this is encouraged at Keele where people working out of hours are required to use the 'SafeZone' App. However, to allow this we would need strict rules about usage such as notifications should be switched off, never handle a device when wearing gloves and only use a device for a scientific reason – taking a picture, looking up information *etc.* It is usually better to educate than to ban.

6.2.7 Rules Enforcement

A no tolerance approach to breaking the rules is the best approach and all staff academic, technical and demonstrators in laboratories need to police and enforce them (see Section 6.5.5 for comments on demonstrators). This was consistent with the findings of the UCCS survey which showed that all institutions had the same response to rules enforcement. In terms of dress code, having systems whereby students can borrow lab coats, laboratory glasses or buy socks, makes life easier. In our system a £1 fine is administered for each infringement of our PPE rules and all profits go to charity.

Problems can arise where there are multiple users from different departments or schools, with different approaches or different demonstrators interpret the rules differently or are sometimes lenient. These are particularly pertinent when both sets of users occupy the laboratory or adjacent laboratories at the same time. Another issue arises when rules are repeatedly broken by postgraduates (or even academic staff) in research labs, who then set a poor example to undergraduates. This raises a broader issue of safety culture within a department and a clear line being set by the head of department or school, or the head of laboratories is invaluable.

6.3 Student Health

Increased access to HE has widened the range of people entering universities.[7] At Keele, an annual average of around 20% of students report as having physical or mental disabilities compared to 13.2% nationally as reported by the office for students.[7] Many of these students' needs can be more straightforwardly addressed by improved accessibility for all (in all senses), but annually our intake comprises individuals with specific needs to facilitate their safe working that can be more challenging to address in a manner satisfactory to all stakeholder parties.

6.3.1 Physical Health

At Keele, we ask that students undertake a GDPR-conforming paper-based health survey at the start of their degree. This allows us to quickly identify any risks, in particular concerning allergies to certain substances, and to collate key information (*e.g.* doctor's name and address) in case of emergency. In contrast with this in the UCCS survey, it was found that other institutions either didn't survey health, noted that students were not under an obligation to tell them anything about their health, or in one case, during an induction session, encouraged students to inform the Health and Safety officer of any relevant health issue. Most institutions reported dealing with problems such as asthma, dermatitis and skin sensitivity on a case by case basis and one institution took the sensible action of suggesting that asthmatics have their inhalers on them while in the laboratory.

Over the years we have had a number of women declare during the course of the year that they are pregnant, and in these circumstances they have been excused from the laboratory classes and given other work *in lieu*. Breastfeeding women are allowed access, but the chemicals they are exposed to are carefully monitored.

Being in the laboratory can lead to physical injury and in 1986 Hellman *et al.* undertook an epidemiology of laboratory accidents in the US which is a sobering read.[8] The final summary of attitudes from academics is revealing of the time including the contribution 'We have a good relationship with our students. They would never start a legal action against the professors'. It is hard for us to believe that this was an attitude at that time in the US and it is hard to imagine anyone in the UK believes this now either.

6.3.2 Physical Disabilities

Good advice is given in two literature case studies by Pence *et al.*,[9] one of a student with reduced mobility and the other who was blind. These situations are also helped by the installation of fume hoods and benches with adjustable working heights. At Keele, we have had situations of students with various physical disabilities working in our laboratories, including restricted mobility, hearing impairment and chronic fatigue. We have a range of methods by which we support these students to work safely and effectively, including:

- Management of seating to reduce walking distance to equipment and/or to position students near the central lectern.
- Use of an adjustable height fume hood.
- Provision of specific equipment such as lab stools with back supports and/or rubber aprons to protect against spills while allowing procedures to be carried out while seated.

In more severe cases, our students may be provided with a support worker by the University. It is our enforced expectation that this is somebody with laboratory experience, so that they themselves do not pose a risk, and that they follow the instructions of the student. It can be difficult to find appropriately qualified support, but we work closely with the University Disability and Dyslexia Support service to facilitate this process.

In another communication we have been told of a student who was unable to work in the laboratory so instead they directed a demonstrator using a video and audio link on how to complete the experiments and explained what they needed to do and why. The system worked well and the learning outcomes were achieved and the student felt that they were able to participate in the practical class as well as their peers.

6.3.3 Learning Disabilities

As a consequence of having a learning disability such as social communications impairment or autism spectrum disorder, several of our students require support workers in the laboratory. This may require the allocation of a support worker to act as a scribe and/or to aid with organisation. As with a physical disability, we require that this support worker is able to operate safely in the lab and not pose a risk to themselves or others.

6.3.4 Mental Health

Of growing concern are the number of students with various mental health issues. Obvious risks might, at an extreme level, extend to a potential for self-harm and/or harm to others, but more commonly students having panic attacks during laboratory classes may, through accident, put themselves and others at risk. A major worry here is that staff are often unaware of the seriousness of an individual's condition and its potential implications without discussing it with them directly, which is not always easily done. Such situations are hard to monitor. While to-date there have been no serious incidents or accidents at Keele involving chemicals and lab equipment that might be attributable to a mental health condition, there have been instances where we have taken steps to avoid such incidents (for example, by allocating students taking our level 4 practical exam to a smaller group). At present, we do not record instances of students having panic attacks as part of our procedures, but in writing this we are reflecting on the need to do so in the future. We do now have specialist mental health first aiders in addition to our physical health first aiders and have policies in place relating to the potential need to exclude students unfit for lab work, which require that they are judged safe by an external professional prior to recommencement of lab work.

6.3.5 Students under the Influence of Drugs

This was not a question specifically posed in the UCCS survey, but it is our understanding that most institutions have a similar no tolerance approach to not allowing students in the laboratory believed to be under the influence of drugs (legal or illegal). At Keele, if a student is suspected of being under the influence of drugs or alcohol, they are asked to leave the laboratory with immediate effect. We have had two such incidents in the last three years and both students were removed from the laboratory before experimentation began.

6.3.6 Exposure of Students to Chemicals

This is clearly a pertinent question to ask but very few have attempted to answer it. Clearly there are specific incidents which are reported but what is the safety of students who have an "incident free" experience: How much exposure do they receive? We shied away from asking such questions in the UCCS thinking that the answers would be too subjective to be of use.

However, we do have one study of our own. As part of measuring work place exposure limits (WEL) to experimenters in teaching laboratories here at Keele a contractor was brought in to monitor exposure during an experiment where a liquid–liquid extraction was being undertaken using dichloromethane. The solvent was chosen because of its high volatility and ease of detection, and the experiment was fully undertaken in the fume cupboard – apart from solvent removal, which was undertaken using a rotary evaporator on the open bench but vented into the air extraction. Exposure (to dichloromethane) was measured for the undergraduate students, post graduate demonstrators, technicians and academics. This concluded that the individual with the highest exposure (the academic) still experienced a level that was 100 times lower than the workplace exposure limit. All others were nearly 1000 times lower. From this we concluded that under normal operating procedures we are well within WEP exposure. In the UCCS, one other institution mentioned that they had also undertaken such a study but did not report the result.

As part of an experiment presented by Gallion *et al.*, in which students made their own fluorescent silica TLC plates, the cleanliness of how the students worked was assessed.[10] Spills and chemical contamination were revealed by turning off the lights at the end of the experiment and then revealing all the fluorescent dye using a UV lamp. Dye was found throughout the fume cupboard areas where the students had been working and on students' hands, even though they had been advised to wear gloves (clearly this had not been enforced). Interestingly, this did have the effect of students notably wearing gloves for the remainder of the experiment and in them conducting a thorough clean down after each subsequent experiment.

In a similar paper, Stark *et al.*[11] recently undertook an experiment to find out how much first year undergraduate students spill. Once armed with this data they then undertook a randomised controlled trial to see if reporting the spillage rates back to students made them more careful when carrying out further procedures. Based on their spillage rate, set phrases were reported back to the students all the way from 'If you spilled this volume of chemicals routinely then... you would be able to handle high hazard chemicals safely such as 1 M potassium cyanide' right the way through to '...you would not be able to handle even low hazard chemicals safely such as ethanol'. They found that there was a 50% reduction in the volume spilled when the students were provided with such feedback. As the paper concludes, providing feedback to students has an effect on their behaviour in the laboratory. Interestingly, this study was carried out by students undertaking a third year research project.

The real take-home message from these two studies is that if you can demonstrate to the students their poor handling of the chemicals, they tend to take more care in the immediate future. It is of course however difficult to know the permanence of this extra vigilance.

6.3.7 Safety Incidents

In 2013 Sieloff *et al.*[12] published a survey of injuries and respiratory irritation in US undergraduate chemistry students. Using an anonymous online survey, they found 40% of injuries were chemical burns, 26% inhaled irritating gas, 20% cuts or lacerations, and 6% heat burns with the remainder not reported. Interestingly the response rate of women was nearly double that of men.

From a review of our own incident reports (2017/18/19, forty incidents in total) the majority recorded for undergraduate students were cuts (33%) and needle sticks (15%). Eleven incidents (27.5%) involved splashes/spills of chemicals, all of which were treated immediately by washing with copious amounts of water. Two resulted in minor burns/skin irritation that did not require medical attention beyond first aid, and thankfully the two which were to the face resulted in no injury, although one did get past eye protection.

At Keele, we have had issues with students either not disclosing chemical spills or not disclosing the full extent of spills. We have redacted the full details, to avoid any individuals or individual circumstances being revealed, but these issues required medical intervention, and resulted in scarring. As a result, we now have a clear procedure for first aiders treating people after spills: the report must be agreed and signed by the student. It must state the areas that were affected, all must be checked and, once the incident is over, the student is given the MSDS sheet for the chemicals involved and told if there are any further side effects to go to A&E and present it there.

In our review period, we only had three reported adverse reactions to chemicals, two were to the skin, with untraced sources (chemicals on benches, lab coats?) and one was because of the inhalation of fumes from a rotary evaporator. All of our rotary evaporators are exhausted to a fume extraction, however concentrating a chemical and then exposing it to the open laboratory environment when taking the flask off a rotary evaporator, however briefly, is a risk, particularly if it is a lachrymator; sometimes this is recognised too late. This relatively low incidence of inhalation of fumes, compared for example to the Sieloff report,[12] is probably because of the movement of all synthetic chemistry off the bench and into the fume cupboard.

The final category of incidents was not associated with the use of chemicals or laboratory equipment and accounted for 17.5% of all incidents (for example, nose bleeds, feeling faint due to a lack of food, late manifesting sporting injuries).

6.3.8 In Emergency

In the UCCS survey the response to the question ‘what are your instructions to students in the event of an accident such as a major spill or fire’ was a unanimous – seek help from the demonstrators and evacuate the area. We think that this is generally well observed but we do have two stories which illustrate what happens when things go well and when things go wrong.

The first involved a sure-seal bottle of titanium chloride ($TiCl_4$) in dichloromethane which a Year 2 student was dispensing in the presence of a postgraduate demonstrator who delivered it from the storage fume hood. For such reagents, we have a policy that the students must dispense the chemical in the presence of the demonstrator. When the needle was introduced into the bottle, a lot of smoke was generated around the perforation. The demonstrator immediately recognised there was a problem as this was not what they had experienced in the past. They immediately alerted other laboratory staff and evacuated the immediate area. More staff were brought in so that the lab class could continue and the bottle of $TiCl_4$ was destroyed by syringing into a large bath of ice. The following investigation concluded that the student must have accidently put the syringe and needle down in some water (the reaction is conducted at 0 °C; an ice bath was in use), inadvertently introducing water into the bottle. What was crucial was having the demonstrator there who knew what things should look like and was able to recognise that something was not right. Handling air and moisture sensitive chemicals is part of the Indicative Learning Outcomes (ILO's) of this module and therefore we continue to use this reagent and see this as a fundamental part of an undergraduate training.

It is essential that students react in the right way when an emergency is called. We had a situation where MChem students failed to evacuate a laboratory when asked to do so by a postgraduate researcher. Although there was no injury, we took this matter very seriously and as part of the recommendations of the incident investigation it was deemed that re-education on these matters of safety was required. It is in situations like this that having the ability to demote students from competency level 2 down to level 1, thereby requiring additional supervision and restricting what they can do and when they can do it, is invaluable.

6.4 Safety Training

There are clearly many ways in which safety training can be taught and our aim here is to show a wide breadth of approaches. Much of this is focused on the when and how to test for understanding but there is less about the content.

Clearly there are a number of approaches. Some take a 'shock-tactic' approach showing series of pictures of injuries from lab accidents such as the effectiveness of eye protection in flash fires, burns from liquid nitrogen and the effect of needle sticks, with the hope of making safety messages hit home. While we do occasionally employ this tactic, the Keele approach is generally more muted and educational, and presented in a 'if you want to be a professional chemist then this is what you need to do' way. There is a focus on careers and the future and building up professional skills of which safety is one. Our shock tactics are much more focused around what will happen in emergency situations and what the sanctions will be for not following health and safety rules. It is not very clear if these different approaches lead to the same desired outcome – safer chemists.

6.4.1 Laboratory Safety Induction

Bradley described a whole curriculum approach to integrating health and safety, comprising lectures, quizzes, implementation of green chemistry principles of minimising waste and choosing less hazardous materials or experiments, and that includes research supervised by a professor.[13] This approach has been mapped against the ACS Committee on Professional Training guidelines, which highlights specific areas for health and safety training.[3] It is interesting to note that the Royal Society of Chemistry in its degree accreditation scheme[4] writes about health and safety as part of degree programmes but there is not a suggested mapping document against which programmes can be evaluated.

In an interesting study Withers *et al.*[14] found that there was no difference in the effectiveness of classroom and computer based teaching of health and safety information. Both were equally effective in terms of initial engagement and most importantly retention of the information. However, there should be caution about a dash to online and consideration of the nature of the safety information that is needed to be conveyed, the learning objectives and if this is better face-to-face. In the UCCS survey the majority of institutions (78%) provide all undergraduates with a safety handbook/guide/manual either as an electronic copy (56%), hard copy (11%), or both (33%). In other institutions the health and safety information is incorporated into the lab manuals. All those that do issue manuals require the students to sign a form agreeing to abide by health and safety rules, presumably for similar reasons to Keele, so that the institution is covered and has something to fall back on in the case of a legal action. One of those who do not issue manuals requires students to sign a form during a safety induction lecture.

All institutions run laboratory safety induction sessions for undergraduates. Some run this once in Year 1, some every year, and some hold specific additional ones when students move into undertaking independent research in Year 4. These cover a wide range of things such as how to use a fume cupboard, how to use a fire extinguisher, an eyewash and even a safety shower. Some commented that they point out safety features in the laboratory but they are not demonstrated. In addition to this, additional specific safety briefings are given at the start of each module, and for some at the start of each practical, which are delivered by a range of academic staff, technicians and demonstrators.

As can be seen in Section 6.1.6, The Keele Chemistry Laboratory Pathway, we have a very clear and structured approach to laboratory inductions depending on the level of the student; at novice and developing practitioner level there are additional safety briefings. What is also clear from writing this section is that terms used mean different things to different people. For Keele, Safety Inductions are activities that happen at the start of the year or a semester which are usually to the whole class and are either carried out in the laboratory or a lecture theatre. We contrast this with Safety Briefings, which happen in the laboratory on the day of the experiment.

6.4.2 Assessment

Multiple Choice Questions (MCQs) with a high pass mark is a common approach to assessing if students have a good understanding of the health and safety information presented in inductions. Kennedy and Palmer[15] reported that they provide students with a series of introductions and lectures on health and safety, after which students are asked to pass an MCQ exam, with a pass score of 85%, part of which is based on the comprehension of an MSDS sheet. The pass rate is usually around 90% and if failed, students are given the opportunity to take another test (different questions) which similarly has a pass rate of 90%. In their experience it is those with English as a second language who tend to fail, as well as those who just do not take the activity seriously.

Crockett[6] reports a similar system which consists of two 75 minutes safety sessions which are followed by a fifteen question MCQ test when students first enter chemistry laboratories. Every subsequent time they start a chemistry module the students then sit a second thirty question MCQ test to ensure that they haven't forgotten the safety information. If this second test is failed then the student has to attend the two 75 minutes safety sessions again and then retake the tests.

This focus on MCQ and high pass scores was interrogated by Withers and Freeman[16] who concluded that institutions are lulling themselves into a false sense of security thinking that a high pass score means that they students have good health and safety knowledge. They used a series of sixteen question MCQ tests which were developed by a panel of experts and which were given to students before training, after training and one year after training. The tests were structured so that they contained three questions that tested particularly specific knowledge which was essential to being safe in that laboratory. If students were able to answer these questions, they should then be able to answer all of the questions. However, it was found that students struggled to answer these difficult questions but still came out with high pass rates. In its conclusions the paper talks about health and safety knowledge in terms of learning outcomes – a rare nod to an educational approach to health and safety training which often focuses more on process.

An interesting study undertaken by Reniers, Ponnet and Kempeneers[17] attempted to discriminate between the effect of safety interventions on students' knowledge perceptions and attitudes towards safety and the students' behaviour and thus resulting consequences. They found that there was a positive effect on the students' knowledge of giving safety briefings on topics such as PPE, clean working environments, knowledge of the lab, chemical hazard signs *etc.* at least in the short term, but that these interventions did little to change the students' actual behaviour in the lab. They concluded that safety behaviour is difficult to change and that this can only be actioned by a system of punishment or reward. This could neatly fit in with the recent vogue for badging promoted by Seery *et al.*[18] where students are rewarded with badges for acting safely in laboratories.

It was difficult to get a clear picture from the USSC about what assessment students undertake in health and safety. The general comment is that it is assessed as part of the laboratory write ups and two responders made the comment that write ups are not marked if the COSHH form is not present. A number of responders also commented that at a Year 3/4 level COSHH forms were assessed as part of the project and that these were checked as part of the marking scheme.

All of this said, in our experience it is difficult to get students to engage in an activity unless they receive credit for it. Interestingly though, health and safety is the only area in which we administer sanctions on the students from charging notional fines (£1 per infringement) for a lack of PPE, through to stopping them doing labs (and hence losing credit), to demoting them on competency levels.

6.4.3 Pre-laboratory Work

Pre-laboratory work[19] is designed to ensure that the undergraduate arrives at the chemistry laboratory prepared for the planned experiment. As part of the UCCS we did not specifically ask about pre-laboratory work associated with health and safety however we are aware that many institutions use online video teaching resources, such as those collated by the Royal Society of Chemistry, to allow students to see how specific procedures, manipulations and laboratory techniques are undertaken.[20] At Keele, we combine these with online MCQ's at foundation year level to check students' engagement with the resource, and as part of paper-based pre-lab work in the first and second year. A number of virtual lab simulations are also accessible, which may be of use,[21–23] but we are unaware of any institutions subscribing to these.

It is difficult to describe here, but Lin *et al.* reported their experience of creating a laboratory safety video that uses clips from popular TV and films, interspersed with safety information to create a faster paced and more engaging experience.[24] They tested this against a 'traditional' information video and it was found to be more appealing to their student audience, however they did not report if it had been effective in conveying the health and safety measures. In a recent project with the ACS, Bristol University put together a cartoon video about safety which is appealing.[25]

One novel approach reported in the literature came from Kumasaki *et al.*[26] who created graphic novels (manga) to teach students in Japan, Taiwan and Thailand about the risk of explosions and fires in the laboratory. They found that students were better able to identify potential hazards in a photograph after reading the manga than before. The manga were well received by the participants, who found this an engaging way in which to undertake active learning in health and safety.

6.4.4 Safety Teams

Active Safety Teams in laboratories is an approach that has been pioneered by Alaimo *et al.* to encourage students to engage in health and safety matters.[27] For each laboratory session, an allocated team of students prepare a pre-lab safety briefing in which the hazards associated with a specific procedure are highlighted. This must be submitted for vetting (and grading) by the instructor a minimum of two days before the session, and is then delivered by the team to others at the start. During the session, the team members monitor their cohort on their adherence to the safety instructions and are again graded on their participation. Finally, there is a post-lab inspection and clean-up followed by a debrief. This placing of responsibility on the students increased their awareness of health and safety issues compared to those who did not have the safety team experience.

Here at Keele, first year students are allocated to a safety team for one practical session in semester 1. Those on the safety team must prepare a COSHH form (see below, Section 6.4.5) for the relevant session and additionally be involved in a lab patrol during the session that uses a checklist of common lab safety issues to both facilitate discussion around safety issues within the team and to allow for interventions by the team aimed at improving other students' practices.

6.4.5 COSHH

The UCCS had a variety of comments about COSHH and it was very much dependent upon the circumstances – undergraduates undertaking scripted experiments or research in the latter years of a degree – and also on how the practical work is supervised. Some institutions take the approach of having COSHH in the background and provide all safety information in their lab manuals alongside the instructions for the experiment. In this case, the academic responsible for the lab is also responsible for completing the COSHH form for that experiment. In others, COSHH forms are given to the students who are asked to read them (sometimes beforehand online) and when the academic introduces the experiment, they highlight specific health and safety issues. In institutions where demonstrators take responsibility for a group of students then it is up to the demonstrator to communicate the COSHH information to the students. The majority of institutions use paper forms whereas three reported using electronic forms. The majority of institutions said that students do not start to complete their own COSHH forms until third year at the earliest and in one case not until fourth year.

As noted above, as part of our safety induction to first year undergraduates at Keele, we introduce the students to our bespoke COSHH forms and give them guidance on filling them in. In semester 1, we focus on training the students in the use of the various techniques they will require competency in later in their degree, including the completion of one COSHH form as part of a safety team system (Section 6.4.4). This is marked, feedback is provided, and the students are also able to compare their submissions with the printed "model" COSHH forms provided to the full cohort before commencement of all semester 1 practical sessions. In semester 2, students are required to complete three COSHH forms, which they again receive feedback on. In all cases, the marks contribute to their final module tally, but are not high stakes, each comprising less than 1% of the total module mark. In addition to this, as part of planning for their group project assessment item (which is an open activity, based around one of about ten different topics), the group must submit a COSHH to cover their planned experiments. Feedback on this is both given in a thirty minutes face-to-face group meeting with their supervisor to discuss the plan prior to starting investigations, and in written marks and comments. In the past, COSHH form filling was additionally explicitly assessed as part of the practical exam.[4] Although this was dropped in 2016 when cohort size dictated a change in exam structure, we kept COSHH form filling as something that the students need to prepare for and therefore they have an expectation that this could be on the exam.

In second year, students are expected to complete COSHH forms for each experiment, which are checked by demonstrators at the start of the lab. This gives the demonstrator the opportunity to question the student and reassure themselves that the student is prepared for the experiment. For particularly hazardous experiments, COSHH forms are also provided to ensure all the information is detailed: Demonstrators can use the discussion to highlight aspects of health and safety. In third and fourth year, students engage in research and therefore they use the full COSHH system with the forms checked by academic staff and submitted to our repository by email.

A new alternative to completing COSHH forms has been recently implemented by Gaynor in Liverpool.[28] In the new system, COSHH forms are provided to students once they have received a 100% pass mark for an online MCQ safety test specific to that experiment. The rationale was that this ensured that students received accurate and consistent information in the form of the COSHH form but that they had to actively engage in the health and safety information and understand it before receiving it. This new system was received positively by the students and by the University (who awarded it with a safety prize). However, when it was objectively compared with the original system that involved completing a COSHH form, the jury appeared to still be out on whether the new system improved on the old.

6.4.6 Health and Safety Teaching Outside the Context of the Laboratory

Many practical chemistry books have a chapter on health and safety, and it is standard that such aspects of a specific experiment are discussed as part of the introduction to it. There is however only one book which can be considered a textbook in laboratory safety: Laboratory Safety for Chemistry Students by Hill and Finster.[29] It is centred around the acronym RAMP; Recognise hazards; Assess the risks of hazards; Minimise the risks of hazards; Prepare for emergencies. Uniquely, it is not a laboratory manual of experiments but chemistry taught from the perspective of health and safety. The example stories behind accidents in the laboratory are particularly illuminating. All too often safety in laboratories can take on a negative tone, a list of “Don't do this” and “Don't do that”, with students being reprimanded for poor practices, but this book brings a refreshing perspective.

A recent report by Huston *et al.* takes an interesting approach to developing a health and safety course, which is a full syllabus.[30] Not only is health and safety discussed from the view point of being safe in the practical lab, it brings a focus on health and safety in industry. Topics such as chemical waste, storage of chemicals and emergency response are delivered by a range of practitioners including medics, toxicologists, and health and safety professionals. Doing so places laboratory health and safety in a wider context, giving the students additional perspective. Assessment of the course is based on attendance and participation in the activities (45%), evaluation on written assignments (45%) and on a 'capstone' project in which small groups of students present to the whole class on a health and safety topic of their choice (10%). The course was devised with the help of students and was received very positively by the participants.

In the UCCS survey there was no institution that reported actively teaching health and safety outside of the various aspects already covered (inductions, COSHH systems, undertaking MCQs) above. Perhaps there is an institution out there who delivers a third year topic on chemical health and safety, but we are as yet not aware of this.

6.5 The Undergraduate Laboratory

It can be taken for granted, but a well organised lab, clear scripted instructions and staff who have been fully briefed on the experiments will minimise the hazards students are exposed to. There is a lot that can be done before the students even enter the laboratory.

6.5.1 Lab Preparation

It is essential to have clear signage in the laboratory so that students know where to find equipment. The importance of the ergonomics of a lab was recently brought home to us when we started to use our new Heavy Lab (Table 6.1). Minimising the number of movements around a laboratory to collect equipment and chemicals should be considered, as each movement increases the risk of an accident.

6.5.2 Chemicals

Preparation of the chemicals before the laboratory session is essential, with the clear labelling of all chemicals and areas. It is not uncommon for students to use the same chemical – for example we have an experiment in which students use ethyl acetoacetate which they can confuse with ethyl acetate that is set out for another concurrent experiment. Fortunately, in this situation the mistake is not dangerous, it just means the reaction doesn't work but to stop the confusion we moved over to use the IUPAC name ethyl 3-oxobutanoate for ethyl acetoacetate. For us, this issue has recently been reduced as the new

Heavy Lab comprises four bays, each of which can have its own set of chemicals. That noted, students could still potentially mistakenly help themselves from the communal solvent cupboards.

Dispensing of chemicals into smaller manageable volumes is good practice to prevent contamination and to minimise volumes if there is a spill. In some circumstances, particularly hazardous chemicals are provided in the quantity required for the experiment. For example, the first time our students handle sodium metal it is pre-cut into the quantity they require and the first time they handle concentrated sulfuric acid it is similarly distributed in the correct volume to the fume hood in which the students are using it. It should be noted that consideration needs to be given to the health and safety of technical staff dispensing larger amounts of chemical.

We have a policy now that all liquids should be dispensed by volume and not by weight. This is because we had several incidences of students knocking over bottles of liquid reagents on the balance tables. Although the balances are on bunded disposable sheets they are on the open bench and the increased surface area following a spill leads to a rapid build-up of fumes, whereas a volume can be measured using a syringe or autopipette in a fume cupboard.

When using chemicals under inert atmospheres and in sure-seal type bottles that will be dispensed by syringe we have the vessel/bottle clamped onto a short lab stand so that it cannot move. For particularly hazardous chemicals that can only be dispensed in the presence of a demonstrator this is clearly written on the base of the stand. We also use colour-themed inert gas balloons so that all the same reagent has the same coloured balloon.

It is essential that chemical containers are appropriately labelled with the correct chemical safety information. At first glance it would appear that this is a futile exercise, because in two investigations in Greece and Thailand[31,32] students' understanding of chemical labelling was tested and was found to be poor. However, in these studies students were asked to place a hazard symbol against the name of a chemical which is rather different to identifying a hazard label and understanding what it means.

Chemical disposal must also be considered here. We didn't tackle this as part of the UCCS but from our own experience we know this is an area which students struggle with. For example, how many students want to put HCl in the 'halogenated waste', we now label our waste containers organic halogenated and organic non-halogenated waste. Careful labelling, clear waste streams in lab manuals and identified on COSHH forms, prompting questions as part of pre-labs are all approaches that try to make it clear but this is unlikely to be an issue that will go away or for which there is a perfect solution – perhaps if there was it would have been published by now. As part of any disposal station it would seem prudent to have a sign saying "if unsure, ask".

6.5.3 Laboratory Script

Clear instructions are essential to running a safe lab and constant updating and correction is a system that we employ, using shared documents on a drive so all laboratory staff can make suggested changes. The amount of specific safety information depends on the approach of the institution, with some putting all the information in, whereas others use separate COSHH forms and just highlight specific safety and hazard issues.

6.5.4 Safe Operating Procedures

In the UCCS, six institutions do provide students with safe operating procedures whereas three do not. All institutions have specific training/close supervision for students using equipment for the first time. At Keele, we have started placing safe operating procedures adjacent to the equipment in our undergraduate laboratories in A4 clear acrylic 'menu holders'. This means that it is easily visible to the student when they are using the equipment. Preferably, these comprise just one side of A4 but in some cases it needs to go over both sides.

6.5.5 Demonstrators

Keele uses a mixture of personnel as demonstrators, from postgraduates, technical staff and academics. This was mirrored in four other universities in the UCCS survey, however, in some universities academic staff do not demonstrate and in others technicians do not demonstrate and in one, only postgraduates demonstrate. Unfortunately, the staff to student ratio was poorly worded in the UCCS as it was unclear if 'staff' meant academics, technicians or postgraduates or all demonstrators, which is what we meant. However, it appears from this question and the next on varying ratios, that average staff to student ratios range from 1:10 to 1:20. These ratios also vary depending on the experience of the class, with lower ratios for less experienced experimenters and riskier procedures, however a number stated that these ratios were generally not fixed but considered as part of the risk assessment for the experiment.

The laboratory staff are the first line of defence in health and safety and making sure they are well-trained and familiar with the chemistry is essential. For particularly hazardous experiments we ask new demonstrators to carry out the experiment in advance of the lab so they know what to expect. With students working at the practitioner level, demonstrators are expected to be active within the laboratory, engaging with the students. Not only do we encourage them to be hands-on demonstrating how to undertake procedures but also encouraging the students to keep their environment clean and tidy.

6.5.6 Sharing Laboratories

Extra hazards arise when two groups of students with different levels of experience share a laboratory. From the UCCS survey, this rarely arises, is clearly avoided and the risks are real. At Keele we had our own near miss when a Foundation Year (FY) course over-recruited, leading to a spill-over group of students in another laboratory. There was plenty of space and adequate supervision, but there was a class of master's level students using the end bench in the laboratory to prepare some solutions for analysis. The FY students' experiment was to make copper sulfide from heating copper and sulfur and, by mass differences, to work out its empirical formula. Fortunately, before the FY students heat the crucible a demonstrator checks their apparatus and chemicals. When they did this, it was noted that the yellow substance in one crucible was more crystalline than the powdery sulfur in all the other crucibles. One pair of FY students had picked up a bottle next to a balance, seen a yellow substance and assumed it was sulfur, whereas it was in fact dinitrobenzene that had been left out by the other class. All the COSHH forms and risk assessments were in place for both experiments, however no one had done the risk assessment if both groups of students were working in the same laboratory at such different levels – developing researchers and novice practitioners. We now have a specific laboratory risk assessment that is completed when two groups are sharing the same laboratory space.

6.6 Undergraduates in Other Laboratories

6.6.1 Undergraduates in Research Laboratories

In the UCCS survey the movement of undergraduates into research laboratories marks a step change in health and safety training. Many report additional health and safety inductions at this juncture, fire extinguisher training and the movement towards a research culture of health and safety where academic supervisors are responsible for COSHH and SOP systems. This is dealt with extensively in the other chapters in this book but it is just worth remembering that the culture and philosophy the students have been exposed to prior to this will influence their engagement. McGarry *et al.*[33] have recently reported a scheme to improve safety in research intensive laboratories using graduate students pairing up with more experienced researchers influenced by researchers from industry. If undergraduates enter a research laboratory that is already rather lapse on health and safety, standards tend to drop to the lowest common denominator.

6.6.2 Undergraduates on Placements

It is not uncommon for undergraduates to engage in industrial and study abroad placements. Seven of the institutions in the UCCS survey ran courses up to MChem level and most had industrial and study abroad placements as options. There were two approaches to the safety of students on placement. Four institutions took the view that while on placement students came under the health and safety jurisdiction of the host organisation. However, five felt that the institution has a duty of care to the student while they were away from the university and at a minimum reported carrying out checks for appropriate health and safety measures, which are undertaken as part of a risk assessment associated with the placement.

6.7 Conclusions

The chapter starts with the quoting of an excellent paper by Ménard and Trant[1] published in Nature Chemistry, which is normally noted for publishing cutting-edge chemistry research. However, it should be noted that this is not the first time a call to arms[34–37a,b] to tackle the issue of health and safety in academic laboratories has been made and one fears that it will not be the last.

Having completed this chapter it is now an opportunity to reflect on what we have been able to achieve. Overwhelmingly for us this has been a very positive experience because this has allowed us to review how we undertake health and safety training across the whole syllabus. It has allowed us to identify what we do well and what we should improve and how to use some of the ideas published in the literature. We believe a clear strength is our training pathway and the clear categorisation of students into different levels allows us to tailor our safety inductions and safety briefings. We hope that by reading this chapter you will be prompted to reflect on and review your own approach to health and safety of your undergraduates.

Some points that were really striking when writing the chapter were:

- Health and safety is generally referred to as 'training' as if it is something different and distinct to the 'education' that the students are receiving as part of their degree course.
- Health and safety training should be more explicit in the ILOs of modules and courses.
- The vast majority of assessment is referred to as "testing", and it is telling that no institution referenced an exam in health and safety.
- We didn't uncover any feedback from employers on their opinion of the attitude and approach of newly graduated students to health and safety.
- What place is there for a health and safety option topic for third year undergraduate chemistry?

Maybe these factors are part of the problem of health and safety not receiving such a high a profile as it should with students. It also makes you think how many employers specifically ask about health and safety in their request for references and how many students have an explicit statement about it on their CV?

Finally, we would like to thank those members of the ChemSafNet community who took part in the UCCS. It wasn't a huge number but it was really useful to get other perspectives and we hope to attempt more targeted surveys in the future so watch your inbox.

References

1. A. D. Ménard and J. F. Trant, *Nat. Chem.*, 2020, **12**, 17.
2. Royal Society of Chemistry, *RSC Accreditation of Degree Programmes*. Available at http://www.rsc.org/Education/courses-and-careers/accredited-courses/bsc-accreditation.asp, last accessed 03 February 2020.
3. ACS, *ACS Guidelines and Evaluation Procedures for Bachelor's Degree Programs*. Available at https://www.acs.org/content/dam/acsorg/about/governance/committees/training/2015-acs-guidelines-for-bachelors-degree-programs.pdf, last accessed 05 February 2020.
4. L. M. Hancock and M. J. Hollamby, *J. Chem. Educ.*, 2020, **97**, 972–979.
5. University Chemical Safety Forum. Available from http://www.chemsaf.org/, last accessed 05 February 2020.
6. J. M. Crockett, *J. Chem. Health Saf.*, 2011, **18**, 16.
7. Office for Students, *Equality and Diversity*. Available at https://www.officeforstudents.org.uk/data-and-analysis/equality-and-diversity/equality-and-diversity-data/, last accessed 05 March 2020.
8. M. A. Hellman, E. P. Savage and T. J. Keefe, *J. Chem. Educ.*, 1986, **63**, A267.
9. L. E. Pence, H. J. Workman and P. Riecke, *J. Chem. Educ.*, 2003, **80**, 295.
10. L. A. Gallion, M. J. Samide and A. M. Wilson, *J. Chem. Health Saf.*, 2015, **22**, 28–31.
11. A. M. Tsokou, A. Howells and M. S. Stark, *J. Chem. Educ.*, 2019, **96**, 2180.
12. A. C. Sieloff, D. G. Shendell, E. G. Marshall and P. Ohman-Strickland, *J. Chem. Health Saf.*, 2013, **20**, 17.
13. S. Bradley, *J. Chem. Health Saf.*, 2011, **18**, 4.
14. J. H. Withers, S. A. Freeman and E. J. Kim, *J. Chem. Health Saf.*, 2012, **19**, 47.
15. S. Kennedy and J. Palmer, *J. Chem. Health Saf.*, 2011, **18**, 26.
16. J. H. Withers and S. A. Freeman, *J. Saf. Health Environ. Res.*, 2013, **9**, 89.
17. G. L. L. Reniers, K. Ponnet and A. Kempeneers, *J. Chem. Health Saf.*, 2014, **21**, 4.
18. M. K. Seery, H. Y. Agustian, E. D. Doidge, M. M. Kucharski, H. M. O'Connor and A. Price, *Chem. Educ. Res. Pract.*, 2017, **18**, 403.
19. H. Y. Agustian and M. K. Seery, *Chem. Educ. Res. Pract.*, 2017, **18**, 518.
20. Royal Society of Chemistry, *Learn Chemistry - Health and Safety Tutorials*. Available at http://www.rsc.org/learn-chemistry/collections/health-and-safety/, last accessed 05 February 2020.
21. R. A. R. Blackburn, B. Villa-Marcos and D. P. Williams, *J. Chem. Educ.*, 2019, **96**, 153.
22. T. D. Clemons, L. Fouché, C. Rummey, R. E. Lopez and D. Spagnoli, *J. Chem. Educ.*, 2019, **96**, 1491–1496.
23. N. Veiga, F. Luzardo, K. Irving, M. N. Rodríguez-Ayán and J. Torres, *Chem. Educ. Res. Pract.*, 2019, **20**, 229.
24. M. L. Matson, J. P. Fitzgerald and S. Lin, *J. Chem. Educ.*, 2007, **84**, 1727.

25. American Chemical Society, *Why Should We Do Laboratory Chemical Risk Assessments*. Available from https://www.youtube.com/watch?v=mYTa-ITYUEKE, last accessed 05 February 2020.
26. M. Kumasaki, T. Shoji, T.-C. Wu, K. Soontarapa, M. Arai, T. Mizutani, K. Okada, Y. Shimizu and Y. Sugano, *J. Chem. Educ.*, 2018, **95**, 584.
27. P. J. Alaimo, J. M. Langenhan, M. J. Tanner and S. M. Ferrenberg, *J. Chem. Educ.*, 2010, **87**, 856.
28. J. W. Gaynor, *New Dir. Teach. Phys. Sci.*, 2019, **14**(1), https://doi.org/10.29311/ndtps.v0i14.3221.
29. R. H. Hill and D. C. Finster, *Laboratory Safety for Chemistry Students*, Wiley, Hoboken, 2nd edn, 2016.
30. E. M. Huston, J. A. Milligan, J. R. Powell, A. M. Smith, D. Neal, K. M. Duval, M. A. DiNardo, C. Stoddard, P. A. Bell, A. W. Berning, P. Wipf and G. C. Bandik, *J. Chem. Educ.*, 2018, **95**, 577.
31. T. D. Karapantsios, E. I. Boutskou, E. Touliopoulou and P. Mavros, *Educ. Chem. Eng.*, 2008, **3**, e66.
32. R. Artdej, *Procedia Soc. Behav. Sci.*, 2012, **46**, 5058.
33. K. A. McGarry, K. R. Hurley, K. A. Volp, I. M. Hill, B. A. Merritt, K. L. Peterson, P. A. Rudd, N. C. Erickson, L. A. Seiler, P. Gupta, F. S. Bates and W. B. Tolman, *J. Chem. Educ.*, 2013, **90**, 1414.
34. D. A. Nelson, *J. Chem. Health Saf.*, 1999, **6**, 43.
35. N. Langerman, *J. Chem. Health Saf.*, 2009, **16**, 49.
36. K. P. Fivizzani, *J. Chem. Health Saf.*, 2016, **23**, 18.
37. (a) R. H. Hill, *J. Chem. Educ.*, 2016, **93**, 1495; (b) R. H. Hill, *J. Chem. Health Saf.*, 2018, **25**, 12.

CHAPTER 7

The Medical School: Managing Safety in Medical Education

DIANE THOMASON*[a], TRISHA BROWN[a] AND HEATHER COMBE[a]

[a]Faculty of Medicine, Imperial College, Level 2, Faculty Building South Kensington Campus, London, SW7 2AZ, UK
*E-mail: diane.thomason@imperial.ac.uk

7.1 The MBBS Programme

The Bachelor of Medicine and Bachelor of Surgery (MBBS) programme is the primary medical qualification of the United Kingdom and one of the requirements to practise as a doctor.

The structure and length of the MBBS qualification is a significant variation from other typical undergraduate programmes. The MBBS curriculum can run from four to seven years depending upon the institution. Graduate or fast track programmes are typically four years for those with a prior science degree. Most traditional programmes are five years but can be extended by a further year to enable the student to intercalate and take an additional BSc qualification. This is typically in a science or technology based subject although business and humanities may be common at some institutes (Oxbridge). Most students intercalate at their home institute, but a smaller number may apply to another for their intercalating year. Those intercalating remain registered at their home university for purposes of their MBBS qualification but will enrol at the second for their intercalated degree. Some programmes can include a foundation year which provides a route into medicine for students from non-traditional backgrounds.

Challenges for Health and Safety in Higher Education and Research Organisations
Edited by Olga Kuzmina and Stefan Hoyle

Published by the Royal Society of Chemistry, www.rsc.org

Another factor which differentiates medicine from other programmes is that it is regulated by the General Medical Council (GMC) who sets the standards and outcomes required for UK undergraduate and postgraduate training. The GMC quality assure the MBBS programme to ensure their standards and outcomes are being delivered and it involves all aspects of the programme including:[1]

- curriculum content
- structure and delivery
- assessment
- facilities and resources
- student support

It is therefore essential that H&S staff working within a medical school be familiar with these standards and requirements to support compliance. Some of the facilities and resources are provided by partner NHS Trusts or other non-governmental organisations (NGOs) and it is incumbent upon the university to ensure that these partners are meeting these requirements.

The cost of providing undergraduate medical education to students within the clinical domain is covered through a complicated grant funding mechanism from Health Education England (HEE) directly to the Trust or placement provider (*e.g.* GP surgery). HEE requires the medical school to quality assure the delivery of this clinical education on its behalf which would ensure that the placement provider is working to the standards set out by the GMC. This is often monitored through a quality assurance cycle and it is best practice to monitor Trust compliance with medical school H&S policies and procedures during this process. This might include confirmation that inductions are completed, and students are aware of reporting pathways for sharps injuries. This can be challenging, and schools must often compete with overwhelming financial pressures within the NHS Trusts when issues are identified which require resolution.

Medical schools must ensure the education their medical students receive meets the standards set out by the GMC.[2] A rigorous quality assurance process is in place to ensure these standards are met and maintained. The GMC can intervene and implement enhanced monitoring if they do not feel an issue is being dealt with appropriately by the school. Some examples of enhanced monitoring include yearly monitoring for new medical schools. In established schools it might include how raising concerns is dealt with.

It is also important to note that the requirements set out by the Office for Students oversees quality for all students and medical schools must continue to adhere to their guidance; the GMC is specific to medicine. This may also be like other healthcare programmes such as nursing, midwifery and the Physicians Associates.

7.1.1 The Early Years Curriculum at Imperial College (Phase 1)[3]

Years 1 and 2 of Phase 1 are primarily delivered through a model known as the 'flipped classroom' which uses a range of large and small group teaching which is supplemented by pre-sessional materials. The large group teaching includes interactive lectures, team-based learning and skill-based learning, with clinical skills and scientific practicals. The small group teaching will involve tutorials, clinical skills and scientific practicals and situational learning such as clinical placements. A proportion of the teaching will be delivered by case studies that enables students to integrate their knowledge of the scientific and clinical elements from different modules and domains.

Phase 1 patient contact is delivered through the module 'Patients, Communities and Healthcare'. This enables students to see patients in their homes, in general practices and in hospital settings, and to interact with community health providers.

Year 3 comprises extended clinical placements within medicine and surgery which are hospital-based, and Medicine in the Community which is based in primary care. Each module finishes with two weeks of campus-based teaching to enable students to consolidate their learning from their clinical placements.

7.1.2 Phase 3 (Years 5 and 6)[3]

This phase is characterised by extended clinical placements which are based in hospitals and in community settings. These cover the main clinical specialties within medicine and surgery and are designed to integrate students' clinical knowledge and prepare them for practice as a Foundation Doctor. An additional range of structured teaching and simulation sessions enable students to experience unfamiliar situations, manage uncertainty and to enable them to develop their decision-making skills. After Year 6 examinations and the Medical Licensing Assessment students will undertake a longitudinal clinical apprenticeship which is based upon a single site and will enable students to spend time shadowing Foundation Doctors to prepare for practice after graduation. The programme will end with an extended elective period where you design a programme tailored to your own interests. Most students choose to spend this period in a placement abroad and robust management of this process is essential for managing student safety in what can sometimes be challenging environments.

7.1.3 Electives

An elective is a popular experience for medical students. It is an extended period (six to twelve weeks) which is designed and arranged by the student in a field of their choosing. The timings of the electives vary depending upon the institutions. Some programmes contain a number of mini-electives or student selected components, which are interspersed throughout the MBBS programmes. In others, the elective is a substantial placement which takes place at the end of their studies. Some medical schools may use both approaches.

Students may elect to stay in the UK and work in a different healthcare setting that they may not have had much experience in (*e.g.* pre-hospital care or advanced procedures and/or equipment such as surgical robotics). Others chose to work with NGOs in developing countries or wish to experience a different healthcare model (*e.g.* the USA or Australia). Students commonly underestimate the time required to arrange their elective and overseas projects as various checks need to be undertaken before their travel. Therefore, it is essential that students be very clear on what they must do to satisfy the School's H&S requirements for electives. Key issues to consider include:

- Risk assessments: students should complete a placement risk assessment which outlines their intended activities during their placement and the type of facilities that students will be working in. It is essential that this be assessed by a competent person within the School who has a designated H&S responsibility. The RA must be discussed with the student to ensure that students are aware of measures to reduce the risk of injury/illness. This includes consideration of their personal safety and travel to/from their accommodation to their placement facility.
- Location: some students may wish to gain experience in environments which present medically challenging opportunities. However, those within conflict zones or where the UK Foreign Office advises against travel are unlikely to be approved or gain insurance to travel. Visas to these countries may also be restricted.
- Occupational health clearances for working in an overseas medical facility. This is country dependant and the timelines to achieve this can be significant; some require that tests are only completed within a certain time before departure and others may have a more onerous process to complete. Service pressures arising from peak business as usual activity may also impact on timelines for clearances.
- Insurance and indemnification for electives: the university and the organisation hosting the student must be clear on the requirements for insurance and professional indemnity during an elective. Most medical defence unions will provide student cover, but some may not cover certain countries and it is essential that these are checked carefully before the student leaves the UK.
- Visa application and approval must be completed before travel.

There has recently been a rise in medical elective tourism and companies often arrange many aspects of a student elective to countries and regions such as the Caribbean, Malawi, Tanzania, and Sri Lanka. Students and universities should ensure their due diligence is conducted before engaging with these organisations as the range of administration and support varies considerably.

7.1.4 The Medical School and the Hospital

The traditional medical curriculum typically splits the MBBS programme into two or three phases (depending upon the culture of intercalation). The first two years of a typical five year programme were often spent in 'pre-clinical' study with the latter three years in 'clinical' training. This may then be interrupted by a one year BSc at some point between the 'pre-clinical' and 'clinical'. This approach is still prevalent in some UK institutions but there is increasing recognition of the benefits of the spiral curriculum. This involves an iterative revision of topics, subjects and themes through the course[4]. This enables students to build upon knowledge and skills across vertical and horizontal domains which become increasingly complex as they progress through the course. As a result, there is greater early years clinical exposure that enables their development of professional values and behaviours and integrates a student's understanding of how basic and clinical sciences relate to their clinical practice. However, the majority of clinical placements still occur in Years 3–6 of the programme.

The introduction to the clinical environment is considered through a lens of patient safety and student H&S considerations. This places responsibility upon the medical school to ensure that risks of early years clinical exposure are managed and that students are made aware of the boundaries to their skills and knowledge. It is best practice to ensure that students are well versed in a range of topics that will enable them to operate safely in a clinical environment. This should be completed before the first clinical placement takes place and although a crowded timetable can limit face to face opportunities, E-Learning modules can be used to good effect. The suite of e-Modules *via* the E-Learning for Healthcare initiative[5] is an excellent resource for this.

- Basic Life Support (BLS)
- e-Modules *via* ELFH
  - Data Security Awareness
  - Equality and Diversity and Human Rights
  - Moving and Handling
  - Safeguarding (adults and children)

In addition to the above training, students have taught sessions which cover the following:

- Introduction to clinical placements which covers training on hand washing, confidentiality, communication, vulnerable patients, and sessions with nutritionists, physiotherapists and other healthcare professions.
- As part of their teaching they will also have had preparatory sessions on communication skills (including simulated patient interviews) and on professional values and behaviours.

They will have been engaging with patients at GP surgeries before they go into hospital and will have had work-based assessment signoffs around their interactions, examples of which include Direct Observation of Procedural (or Clinical) Skills (DOPS/DOCS), Community Survey Project and End of Placement assessments.

7.1.5 The Clinical Placement

The GMC guidance on clinical placements[6] was originally written to supplement the GMC standards in place in 2009. However, it remains relevant today and is a comprehensive document which defines what is meant by clinical placements, student assistantships, shadowing and induction. In addition, it discusses medical schools' responsibilities in relation to both clinical placements and student assistantships. The guidance also draws on examples of local practice in order to illustrate how medical schools are delivering clinical placements and student assistantships at present.

The guidance[6] states that a clinical placement "can be defined as any arrangement in which a medical student is present in an environment that provides healthcare or related services to patients or the public. Placements can take place in primary, secondary or community healthcare or social care settings. Students can be actively involved in patient care or they can be observing health or social care processes."

Clinical placements take place in a variety of different settings which likely include GP practices, hospitals and community outreach centres, NHS walk-in centres, Non-Governmental Organisations (such as charities for the homeless, hospices and care homes) and ambulance stations. The UK has robust H&S legislation to protect student, staff and patient wellbeing and services providing medical and social care are subject to scrutiny by the Care Quality Commission as an independent regulator. When utilising the services of a health/social care facility, it is incumbent upon the School to ensure that they are able to provide appropriate facilities and services which meet the needs of the students. Therefore due diligence should always be completed ahead of establishing new placements. Good practice would include a physical visit to the facility, a risk assessment for the placement and activities and checks with relevant regulators to ensure that the facility and staff are appropriately licensed and or registered to conduct its stated activities. Insurance and indemnity checks should also be completed and a formal agreement be entered into.

Our experience of monitoring clinical placements within trusts is that the student experience is always safer, the placement more effective and the student experience more enjoyable when the following are in place:

- The students have clear guidance on when to arrive, what they should wear and to whom they should report. They understand the standards required by the Trust/Placement Provider.
- The students are given an induction programme which introduces the Trust staff who will be supporting them during their placements (*e.g.* Teaching Coordinators, Clinical Teaching Fellows). This should include information on how to raise concerns or deal with data governance issues and incidents such as needlestick injuries.
- They are assigned to a junior (Foundation) doctor who can mentor and provide informal learning and shadowing opportunities and will also support students' induction to the Trust and clinical environment. A Foundation Year 1 doctor (F1) is the first training grade of a postgraduate medical student and is able to practice medicine under the supervision of a Consultant.
- The students have a timetable which has been agreed with Firm Leads and which provides tutorials, Consultant led structured teaching, clinical skills sessions and identifies clinics and ward rounds which they must attend as part of their placement.
- There is clear guidance on attendance and on-call requirements and the procedures for signoffs for their placements.

Students will typically undertake a range of activities on their placements although it will depend on their environment, phase and purpose of their placements. Some examples include:[7]

- Consultant ward rounds. Students will accompany a Consultant around a ward and observe how Consultants deal with patients, their examination techniques and their clinical reasoning skills.
- Attending ward rounds with doctors in training or nurses and other healthcare professionals helps students understand more about the day-to-day tasks of working in a hospital.
- Tutorials on topics associated with their placement (*e.g.* common presentations of a range of conditions) or clinical skills practice sessions such as learning how to take blood (phlebotomy).
- Students may attend operating theatres to observe surgical and anaesthetic procedures.

- Students may attend outpatient clinics assisting Consultants or junior doctors.
- GP surgeries and home visits – students will learn how GPs manage their time and make diagnoses, decide when to refer patients to specialists *etc.*
- Observing Multidisciplinary Team (MDT) meetings to observe how health care professionals (clinicians, nurses, social workers, physiotherapists) work together to support a patient's care and treatment pathway.
- Developing clinical skills such as examinations and learning to communicate effectively with patients and the MDT.
- Clerking patients, which entails taking a detailed history of a patient and examining them. Students may then present the patient in exam conditions which should include a summary of the case and a plan of investigation and treatment.

7.1.6 Developing Skills

Clinical work exposes students to the risk of infection and injury. Infection with HIV or other blood-borne viruses (BBVs) can restrict them from working in some areas of surgery. Blood-borne viruses are carried in human blood, tissue and secretions. Accidental exposure can result in infection and in some individuals cause severe disease. Although occupational transmission of these viruses is rare, all unscreened blood should be treated as potentially infectious. Infection is not only damaging for a student's health – it could affect their future career.

Training and competence are key aspects of accident prevention. Learning how to perform clinical skills correctly and safely is crucial in reducing the risk of occupational injury and infection *via* accidents such as needlestick injuries. Needlesticks are fairly common in a clinical setting; Public Health England received 1216 reports of SOEs (Significant Occupational Exposures) from England, Wales and Northern Ireland in the period January 2015 to December 2017.[8] A SOE is an incident in which a healthcare worker receives a sharps injury contaminated with a patient's blood or body fluid, and that patient is positive for a blood-borne virus.

7.1.7 Clinical Skills

Year 1 and 2 students have patient contact but do not carry out any invasive procedures and are mainly observing.

The first clinical attachments are in Year 3, from September to May, and during this time they develop a set of specific clinical skills, required by the GMC. Students are attached to both a medical and surgical firm in hospital and also a GP practice (MICA = medicine in the community). There is a defined Clinical Methods and core competency programme. Clinical skills teaching takes place in small groups during each attachment.

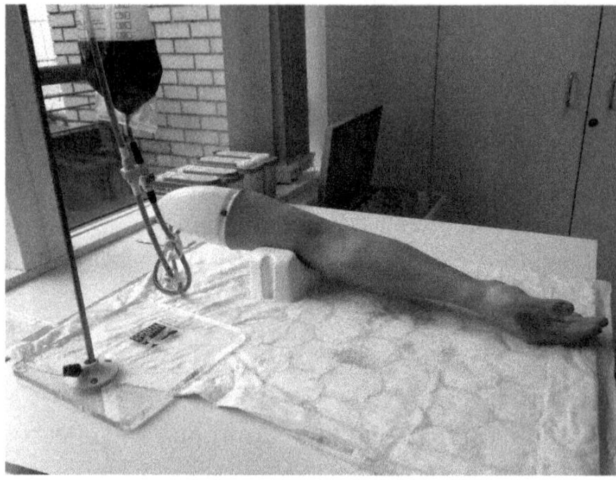

Figure 7.1 Arm simulator used for skills teaching.

The students' first experiences of skills such as cannulation, suturing, and rectal examinations, are practised in a skills lab using simulators – anatomical models such as the artificial arm shown in Figure 7.1, used for teaching injection, venepuncture, cannulation and infusion.

Each week there is a different skill taught. The first attachment (clinical placement) is when all the skills will be new to the students, although they have a venepuncture session in Year 1 when they take blood from each other in practical classes, so have had an introduction to taking blood. Session times are 1.5 hours and the clinical skills tutor sees groups of students in their firm groups (clinical teams) – usually around six to eight in each session.

For their first time on the ward, students are expected to know the skills but not to be competent. Clinical skills that they undertake in the clinical area must be closely supervised, and there is a DOPS (Direct Observation of Procedural Skills) sign-off for most of them.

There is a clinical skills manual, on a virtual learning environment (Blackboard), which accompanies the practical sessions as well as clinical skills videos on MedLearn. Students therefore have online resources that they can refer to at any time.

Taking suturing as an example, the session for the first attachment covers:

- Sterile gloving techniques
- Correct selection and use of instruments and suture material
- Interrupted suturing and knot tying

This consists of a 1.5 hours teaching session plus a video and use of the manual. This is repeated in attachment 2, and in attachment 3 there is a revision session. This allows the students to build their competence. At the same time they see the learning put into the context of patient care in their clinical environments. Any students who struggle are given extra support, and they can request additional sessions.

During the training sessions the correct safety procedures are used as they would be in the clinical setting, *e.g.* safe handling and disposal of sharps. Hence when they are in the ward or clinic they already know and have practised these procedures.

When students are on attachment and have their first hands-on experience with patients, they are no longer under the control of College teaching staff but instead are instructed and supervised by the NHS staff treating the patients. Primary responsibility for the management of health and safety of students on placement lies with the placement provider, *i.e.* the NHS (under Section 2.1 of the Health and Safety at Work *etc.* Act 1974). The student is treated as equivalent to an employee with regard to health and safety matters.[9] However, College retains a duty of care and statutory duties for students' health and safety while they are on attachment, since it is part of their programme of study. This is largely discharged by the checks carried out by the programme management team. Communication between course staff and the clinical staff who will be supervising students is essential, for assurance that the students are being supported as we expect. Attachments are monitored by the programme management team.

Student wellbeing is considered as well as physical health. In the early clinical attachments students experience distressing events, which most of us would not encounter in any other life experience. They witness ethical conflicts, human suffering and death. Pastoral care is an essential part of medical education.

During Year 5 and 6 the students continue to practise clinical skills with assistance and supervision. In Year 6 they reach the stage where they are competent in the required skills. At all stages the students are given boundaries within which they must work, for which they have personal responsibility.

7.2 The Dissection Room

In the vast majority of medical schools, students will be taught anatomy using cadaveric material. This can take the form of student-led dissection, study of prosected material, or a mixture of both techniques. This experience is unique to medical and dental education and can begin as soon as the first term of the first year.

7.2.1 Prosection *Vs.* Dissection

The choice of teaching methods for anatomy is based on academic-led opinion, and may not take into consideration the levels of safety of the different approaches. There is currently no consensus as to which is the most effective method of teaching anatomy,[10,11] but there are differences in the hazards and risks involved.

Dissection is student-led, with small groups working on parts of an embalmed whole cadaver. Demonstrators and academics are on hand to guide students through the procedure, but the actual manipulation of material is carried out by students.

Prosection, on the other hand, involves students examining body parts which have been prepared in advance by trained staff, showing the optimum amount of information possible presented in a clear manner.

When considering the different methods of teaching anatomy, it is not just the risks to students which must be considered. The following tables illustrate the main hazards and risks involved in dissection (Table 7.1) and prosection (Table 7.2), with reference to both the students and the anatomy unit staff.

Table 7.1 Primary hazards and risks from dissection.

Hazard	Risk(s)	Control measures	People
Dissection equipment *e.g.* scalpels	Cuts to fingers, hands.	Always pass instruments in dishes, not by hand. Observation of correct techniques by demonstrators.	Students
Embalming chemicals (*e.g.* Formalin, industrial denatured alcohol, polyethylene glycol, phenol)	Skin/eye irritation, sensitisation, exposure to carcinogens, chemical spills.	Extract ventilation. Correct PPE – thumb loop gown, nitrile gloves, eye protection or face shield, suitable shoes. Appropriate chemical storage and spill kits.	Students, staff
Lifting and moving cadavers	Musculoskeletal injuries from poor manual handling.	Paired lifting, hydraulic lifts, tailored training in manual handling techniques.	Staff

Table 7.2 Primary hazards and risks from prosection.

Hazard	Risk(s)	Control measures	People
Dissection equipment *e.g.* scalpels	Cuts to fingers, hands.	Always pass instruments in dishes, not by hand. Work carried out by trained prosectionists.	Staff
Embalming chemicals (*e.g.* Formalin, industrial denatured alcohol, polyethylene glycol, phenol)	Skin/eye irritation, sensitisation, exposure to carcinogens, chemical spills.	Extract ventilation. Correct PPE - thumb loop gown, nitrile gloves, eye protection or face shield, suitable shoes. Appropriate chemical storage and spill kits.	Students, staff
Lifting and moving prosections	Musculoskeletal injuries from poor manual handling (lower risk than whole cadavers).	Tailored training in manual handling techniques.	Staff
Ergonomics	Musculoskeletal injuries from poor workstation setup or long periods of uninterrupted work.	Work rotation. Regular breaks. Adjustable height seating and footrests.	Staff

The other safety consideration, which cannot be ignored, is the balance between crafting the safest environment possible for an individual student and creating safe doctors. Students should be given the space to learn and make mistakes in a controlled environment where patient safety cannot be compromised, even if this introduces slightly more risk to the individual.

7.2.2 Virtual Teaching

It is important to note that alongside the debates regarding prosection and dissection, a broader discussion about the nature of medical teaching is taking place.[12] An increasing number of academics are of the opinion that cadaveric material is no longer necessary for teaching anatomy at all, and could be replaced with surface and living anatomy, and the use of medical imaging technology.[13] Other methods include the use of virtual reality and other computer assisted learning techniques.[14] While these methods would undoubtedly be safer as they remove the need for embalming and dissection, it remains to be seen whether these approaches provide the same benefits to students, and for now the focus at Imperial remains on cadaveric material.

7.2.3 Student Wellbeing

In Western society, it is increasingly likely that undergraduate medical students will start their course without ever having seen a dead body.[15] A decrease in open casket funerals and traditional mourning rituals means that for many students their initial encounter with a dead body will be in the context of the dissection room. If this interaction is an unpleasant or distressing experience, then the impact can have a lasting effect on the mental wellbeing of the student. In addition, students come from diverse backgrounds and cultures, many of which will have their own traditions and beliefs surrounding death. These can cause tension with the learning that they are expected to carry out with cadaveric material, as well as with the beliefs of their fellow students. For example, Jewish traditions relating to the handling of dead bodies, and the importance of immediate burial, can conflict with parts of the process of using donor bodies for dissection.[16]

An induction and awareness session for new students must take into account these factors, as well as the Human Tissue Act 2004 and all relevant local safety rules, so all students attend an introduction day prior to attending any classes. This includes a lecture from teaching staff covering the correct way to put on and remove PPE, rules for working with donor bodies, and the ethics of working with cadaveric material, as well as introducing some photographs of relevant material. The introduction day also includes an hour long session in the dissecting room, where students are shown a full body cadaver, some prosections, and some individual organs and bones, and given time to familiarise themselves with the setting. Studies at a variety of universities have shown that anxiety decreases after the first anatomy session takes place,[17–19] and that students tend to view the experience as positive, so the familiarisation session is designed to demystify the process as much as possible and reduce apprehension.

However, even with a comprehensive induction and ongoing support, incidents will still occur during anatomy classes. The heightened emotions experienced by students can bring to the surface other problems such as exam stress, relationship issues, or underlying mental health conditions, as well as recent or more distant bereavements. Specific trigger points identified by anatomy staff included work with hands, faces and feet, but individuals can be affected at any time. Students will often present to the unit staff with problems such as dizziness, light-headedness, or feelings of nausea, but when they are encouraged to talk, an alternative underlying cause of their physical symptoms may become apparent. Staff make sure that they are a visible presence during student practicals and encourage a "sit down and chat" culture in order to encourage openness and communication. Several unit staff have also been trained in Mental Health First Aid,[20] in order to help them notice the signs of poor mental health and equip them with skills such as non-judgemental listening. Individuals can be signposted to resources such as personal tutors, the university medical centre, or the student counselling service.

7.3 Accidents and Incidents

The Faculty of Medicine at Imperial College London has eight academic departments as well as the School of Medicine (which provides the undergraduate medicine programme). It is located across many buildings on seven London hospital sites as well as the non-medical Imperial College campuses. As in any university faculty, the activities are centred on research and teaching, with much of the research being laboratory-based. The Faculty Centre's team of safety advisors monitors and investigates reports of accidents, near misses and other incidents such as spills of hazardous chemicals or biological materials, throughout the whole Faculty. Reports are made online to a central College system, which serves as a repository of reports that can be downloaded. The numbers are too low to allow meaningful statistical analysis, but we look out for trends of similar incidents, and hotspots where a lot seems to be going on in a particular area. We can then have discussions with people in the relevant departments and make targeted recommendations for improvement.

As you would see anywhere, there is likely to be under-reporting, and reporting rates vary between different parts of the organisation. There are some inaccuracies and ambiguities as well, with some reports being made under the wrong "responsible part of the organisation" (of which there are many in the Faculty of Medicine's departments, divisions, sections or other units) or category. If spotted these are corrected but some errors inevitably remain. The analysis here is based on our best efforts.

Figure 7.2 is a summary of the incidents reported by one academic department over a period of three years; the pattern is typical and is likely to be very similar in other departments.

The majority of injuries and ill-health reports in the Faculty as a whole are of accidents or work-related ill-health suffered by staff members or research postgraduates. Of the total injuries, the ones we see most often are:

- Needlestick injuries
- Sharps injuries other than needlesticks (*e.g.* cuts from glass, scalpels, and microtome or cryostat blades)
- Injured while handling, lifting or carrying
- Exposure to, or contact with, a harmful substance (*e.g.* cryogenic burns from liquid nitrogen, splashes with corrosive or irritant chemicals)

The highest risk from needlestick and other sharps incidents is from a sharp contaminated with human blood or other human tissue, or cultured human cells. This means that there is a risk of infection with a blood-borne virus (BBV), the highest risks being HIV and Hepatitis B and C. Before starting a research project involving any human tissue, the investigator has to carry out a biohazard risk assessment, and the risk controls and emergency actions are identified. For body fluid exposures the affected person would

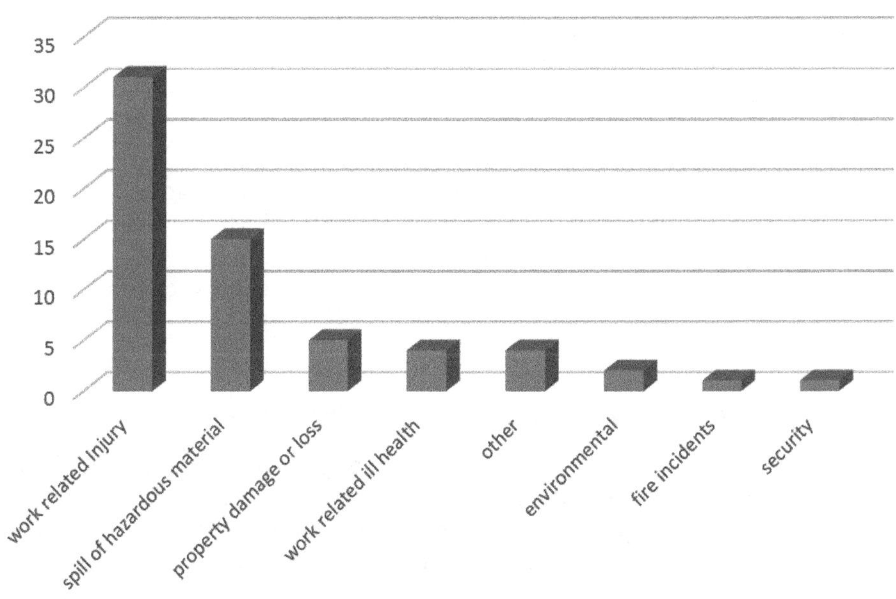

Figure 7.2 Incident reports from an academic department of the Faculty of Medicine over three years.

need to get prompt advice from the College Occupational Health service which is the primary source of support for staff and students across all campuses. Occupational Health have arrangements with hospital Occupational Health Services on other campuses to provide emergency assistance for exposures to human biological materials. This is important for Faculty of Medicine departments, which have groups on all the hospital sites, often with people working on a number of different campuses. A protocol is followed and depending on the type of material involved, and the injury, the injured person may undergo blood tests and receive post-exposure Hepatitis B immunisation or HIV post-exposure prophylaxis (PEP). Staff and students need to know the emergency arrangements on all the campuses where they may work with biological material.

Many of the ill-health reports from staff or students are in fact not work-related, or at least not directly. A lot of these are fainting or "feeling unwell" incidents attributed to the effects of underlying illness, prescription medication, dehydration, fatigue or not eating, or last night's party. Others are due to simply being unlucky and falling ill at work. All of these have been excluded from the figures. The instances that are work-related are much fewer and include allergies and chronic ill-health due to ergonomic factors.

We see very little injury and ill-health in non-medical undergraduates, but what is reported tends to fall into the same pattern as above.

7.3.1 School of Medicine Incidents

In comparison with a typical scientific academic department within the Faculty of Medicine there is a higher number of incident reports involving undergraduates from the School of Medicine. These incidents fall mainly into three categories:

1. Incidents during general teaching activities, such as practical classes
2. Incidents during human anatomy teaching, which are mostly faints and cuts in the dissecting room
3. Incidents during clinical attachments, mainly needlestick injuries and contact with body fluids.

7.3.1.1 Incidents During General Teaching Activities

These are uncommon but tend to be of the type you would see in any undergraduate programme that includes lab work. They are investigated by the Faculty Safety Advisor (FSA) for the campus. Staff running lab practicals have to carry out risk assessments, as for any other activity, and these should be reviewed by the Faculty Safety Advisor for the campus. Here is an example of an accident report from such an incident and the outcome:

"During a routine laboratory practical, a student broke a small glass capillary tube. Part of the broken glass pierced his glove and the skin on his thumb. The blood in the capillary tube had been taken from a year six student demonstrator who had donated a sample following failed venepuncture.

Careful emphasis was given during the safety briefing about risk of needlestick injury and disposal of sharps, although nothing specifically about the risk from fragile capillary tubes.

To mitigate future risk, the safety briefing and pre-session demonstrations were adjusted for subsequent sessions, with emphasis on the fragility of capillary tubes."

Anyone who has handled small capillary tubes will know how easily they are broken – the staff did well to avoid this happening in any of the sixty plus previous times this practical was run. The risk of injury from a blood-filled glass tube was not considered, possibly because of the prominence given to the risks from needlestick injury. The teaching staff put this right for future sessions.

7.3.1.2 Incidents During Human Anatomy Teaching

These are the student incidents reported from the Faculty's Human Anatomy Unit (HAU), where students observe and take part in dissection of a human cadaver. These too are investigated by the Faculty Safety Advisor. The overwhelming majority of incidents are fainting and/or vomiting, with a smaller number of cuts. Unlike the fainting/not feeling well general reports which are excluded from the statistics, these are considered here as being work-related, if they directly result from the student's response to the sights or smells in the dissecting room. Reports of just feeling unwell but not actually fainting have not been included. These responses can be emotional, or provoked by the smell of embalming fluid, although the dissecting room is well-ventilated. The impacts of human dissection classes on students were discussed in Section 7.2.

The HAU staff are well-prepared for and attuned to people fainting or feeling ill and have become expert at "catching" them before they fall. A sofa in the adjacent office serves as a safe spot to recover with help from the staff. Often the student recovers quickly and can rejoin the class. Occasionally they are injured or are affected too badly to continue. Here is an example of such an incident:

"A student was taking part in the Year 1 "Introduction to the human body" teaching session in the Dissecting Room, where they fainted and sustained a head injury (no broken skin). The student was attended to by the HAU staff and demonstrators (junior doctors) and given first aid.

The student was brought to the office after recovering and reported head and jaw pain, a headache and blurred vision. Within 20 minutes the student vomited and was taken to A&E as a precaution. The student was assessed and treated at A&E and released to the care of their family later in the day."

Cuts are not common but are one of the primary risks of dissection. Year 1 students will have little or no experience of deftly handling sharp instruments, and do not yet have a feel for the textures of the tissues – though notably, one of our students was unlucky enough to get a cut from a cadaver's rib. In the past, safety personnel in universities have had discussions about the rate of cuts in medical school dissection classes, which seemed to vary wildly between universities. It was not possible to identify any particular risk factors that no one had thought of before, or any novel measures to reduce the risk. It was notable that students at one medical school no longer carried out dissection of cadavers, but instead studied specimens prepared by technicians. At another, dissection was removed from the syllabus except for students specialising in certain areas.

The number of incident reports from the HAU is an accurate reflection of the number of incidents, because there is a high awareness of the need to report, and a fairly high rate of incidents (mostly very minor) so that staff in the unit are familiar with the reporting system. This is likely to give an impression that there is a much higher rate of incidents here than elsewhere in the Faculty, which we believe is almost certainly false because of underreporting elsewhere.

7.3.1.3 Incidents During Clinical Attachments

During their clinical attachments or electives, which may be overseas, the students are no longer under the control or direct supervision of the College teaching staff. They have direct contact with patients and perform procedures on the wards and in community healthcare settings, supervised by clinical staff. They are only allowed and expected to carry out a defined set of procedures for which they have been trained. In their Year 3 attachments they are expected to know this set of skills but are not expected to be competent and must be closely supervised.

Lack of practical experience is always going to increase the risk of an accident. Students are informed that they must report any accidents or near misses to College and are told how. If in doubt they can report by email and a staff member will make the report on their behalf. In the past this has led to reports appearing under different categories or being entered a long time after the event but there is an ongoing effort to improve reporting rate, timeliness of reporting and accuracy. Clinical staff are very aware of the need to report needlesticks on the NHS system and students usually report that this was done. Most also report to the College as they have been instructed, but not always, and this has sometimes led to delays in investigating. Since the students are away from the College, and sometimes overseas in the case of electives, the onus is on the student to report; their supervisors and colleagues are unlikely to do so.

The NHS Trusts at the various medical campuses where the Faculty has a presence are represented by their safety managers on the Faculty Health,

Safety and Environment Committee. This provides an opportunity for formal discussion of safety issues involving our students in their hospitals if necessary.

A difference between "lab" needlesticks and those that happen on clinical attachments is that most of the lab accidents involve needles that are clean, or have been used with low risk biological material such as uninfected animal tissue, low risk cell cultures, or low risk chemicals. Some of the needlesticks reported from clinical areas involve only clean needles but some are "dirty" needlesticks involving a patient's blood. As mentioned, there is a defined process for follow-up of these; the student must consult Occupational Health or A&E, and the needlestick protocol is activated. A risk assessment will be undertaken, the student's blood taken, and depending on the circumstances the student may need a Hepatitis B booster or HIV PEP. This can cause great anxiety to the student.

Few incidents other than needlesticks/sharps injuries and body fluid exposures are reported by students on clinical placements, apart from unusual events that they realise need to be reported, such as violence or the threat of violence (there were two such incidents in the three years we looked at). We believe that there is probably under-reporting of other incidents and near misses.

Investigations of these clinical incidents are normally carried out by the Faculty Education Office (FEO) safety officer and involve a dialogue with the clinical staff supervising the student. When an accident happens on College premises, or hospital premises occupied by the Faculty's departments, the Faculty Safety Advisor can visit the area if necessary, review records, and speak to the affected person and their manager or supervisor. This is almost impossible in the case of clinical incidents and is not appropriate. The FEO staff are familiar with the clinical settings, know who the relevant clinical personnel are, and are generally better placed to understand what happened and make recommendations. The FSAs, who are health and safety professionals, can take part and provide support as necessary. Here is an example of a clinical incident:

"I had scrubbed in for a surgery alongside another medical student. We have both received formal instruction and/teaching and been signed off for scrubbing. I looked briefly in the scrub room for the box of masks with visors attached. I found the box empty and instead opted for a regular mask as all the other surgical staff were wearing. The patient was in for excision of a large testicular mass. When this mass was being handled within the body cavity, an incidental hydrocele [fluid-filled sac in the scrotum] suddenly burst. This resulted in a splash to both of my eyes."

Whether or not there was much infection risk from this exposure, this must have been unpleasant and quite a shock. To a non-clinician this sounds like the kind of surgery that does pose a risk of splash or spray, but the view of the surgical team was clearly that there wasn't too much of a risk. The student expected to find surgical masks with visors (these are intended for use when there is a significant risk of being splashed or sprayed with body fluids). When these were not available he opted for a mask without visor, only because the other theatre staff were wearing them. He did not question the decision to wear normal masks. Why? There are common factors in incidents like this:

- Reluctance to question the actions of senior people
- Feeling nervous, too embarrassed or too junior to speak up and risk disapproval from others
- Unsure whether you are being too cautious in feeling you should have a visor – questioning your own judgement
- Reluctance to cause any delay
- The need to conform and be "part of the team"

The students are expected and taught to take responsibility for their actions and decisions while they are on clinical attachment, but these human factors will always be there. Here is another example of an incident:

"I was on my first week on-call on my first night shift, and I was asked by a nurse to cannulate an elderly lady. I was trained previously at another hospital, but not at this one, so I began to insert the needle. As soon as the needle entered the patient's arm, she pulled her arm away from me in pain. I had not managed to insert the cannula, so I was holding the dirty needle (with the patient's blood on the end) while trying to stop the patient bleeding by trying to undo the tourniquet. It was not a tourniquet I had used before, so I was struggling to undo it, and the patient was not cooperating, as she kept on writhing around. In this process, I sustained a needlestick injury to the base of my left thumb. I did not ask the nurse to supervise me as I had done cannulations previously at another hospital.

I went to A&E (as it was out of hours), had my blood taken, then returned to the ward to fill out a DATIX form [NHS incident report]. The patient was consented and her blood was taken, which came back negative for Hep B Antigen, Hep C antigen and HIV. I went to see Occupational Health on my next day back to work and they reported that my blood results were also negative for all the above BBV screens."

In this case the student felt that his training in the previous attachment was enough to allow him to work unsupervised. Unfortunately inexperience caught him out. Other factors of note:

- The patient was very elderly and the cannulation was hurting her, so she pulled back her arm. This presents an additional challenge to an inexperienced person – such a patient is going to be more difficult to work with than a helpful fellow student in a training session, or even a calm patient.

- This was a Year 3 student who should have been closely supervised when performing any clinical procedure. The nurse should have known that as well.
- It happened on a night shift, when there are fewer experienced staff members around to supervise and advise.
- During clinical skills training students have a variety of tourniquets demonstrated to them. If a student is unsure of their ability to use the equipment provided, or run into difficulty, they should ask for assistance beforehand or ask another practitioner to carry out the procedure.
- The wording of the report was that she was "not cooperating", which gives the impression of partially blaming the patient and not having much empathy.
- While not a serious incident, this is a bad experience for both parties. The patient also had to have a further blood sample taken for screening.

7.3.1.4 *Incident Types Overall in the School of Medicine*

Figure 7.3 shows the reported injury and ill-health incidents over a three year period, 2016–2018. It does not include near misses or non-work-related illness and injury. The majority of these incidents are faints/vomiting during dissection classes, and sharps/needlesticks and body fluid contact, mostly during clinical attachments or electives. The two violence reports were from

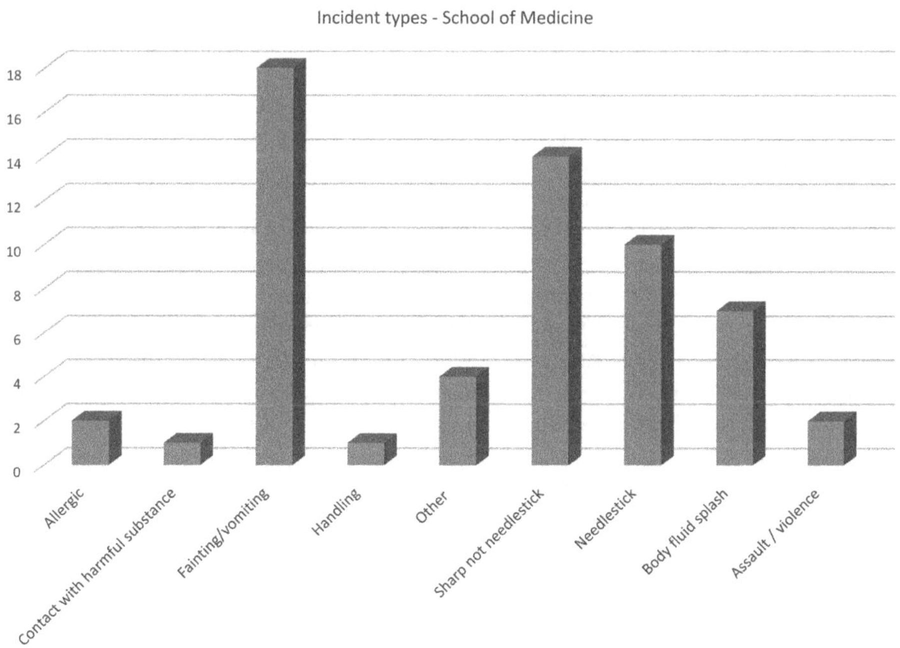

Figure 7.3 Incident reports from the School of Medicine over three years.

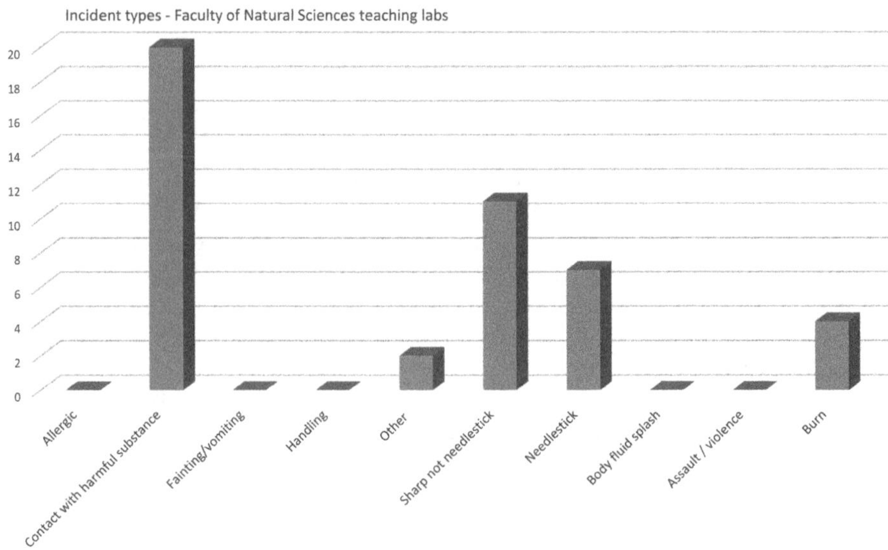

Figure 7.4 Incident reports from the Faculty of Natural Sciences teaching labs over three years.

clinical settings; one involved students being threatened by a patient who attacked one of their family members and was wielding a makeshift weapon, and the other was a student attacked by a patient.

The breakdown of reports from the teaching labs in the Faculty of Natural Sciences gives a very rough comparison with non-medical undergraduates. The numbers were filtered in a similar way, to exclude non-work-related illness and near misses (Figure 7.4).

This is typical of what is commonly seen in a life science teaching environment. Contact with a harmful substance is an obvious hazard in a faculty that has chemistry and biochemistry practicals. Needlestick injuries happen but there is rarely a risk of blood-borne virus infection.

Despite the differences in incident types and reporting rates within and between faculties, the principles of accident prevention are the same, and include training and competence, assessment and appropriate control of risks, encouraging informed decision making during a student's education, and good leadership. Good awareness of this benefits our future doctors (and their patients) just as it benefits our future scientists.

Acknowledgements

The authors would like to thank Helga Koch, Eddie Hartrick and Olga Kuzmina for providing accident statistics, Rachael Waddington for background information on the Anatomy Unit, and Joanne Jones and Steve Marchington for information on clinical skills teaching.

References

1. Standards of UK Medical Education, https://www.gmc-uk.org/education/becoming-a-doctor-in-the-uk/standards-of-uk-medical-education, last accessed 26 February 2020.
2. How we quality assure, https://www.gmc-uk.org/education/how-we-quality-assure/sharing-good-practice, last accessed 06 May 2020.
3. Imperial College London MBBS Programme Specification (2019–2020) https://www.imperial.ac.uk/media/imperial-college/study/programme-specifications/school-of-medicine/1920/MBBS_BSc-Medicine-Programme-Specification-2019-20.pdf, last accessed 26 February 2020.
4. R. M. Harden, What is a spiral curriculum? *Med. Teach.*, 1999, **21**(2), 141–143.
5. E-Learning for Healthcare Programmes, https://www.e-lfh.org.uk, last accessed 26 February 2020.
6. Clinical Placements for Medical Students, https://www.gmc-uk.org/-/media/documents/clinical-placements-for-medical-students---guidance-0815_pdf-56437824.pdf, last accessed 26 February 2020.
7. https://www.healthcareers.nhs.uk/explore-roles/doctors/medical-school/clinical-placements-medical-students, last accessed 26 February 2020.
8. Public Health England, *Eye of the Needle Report*, 2020.
9. USHA, *Guidance on Health and Safety of Placements*, Universities Safety and Health Association, 2018.
10. C. E. Dinsmore, S. Daugherty and H. J. Zeitz, *Clin. Anat.*, 1999, **12**, 110.
11. K. S. Topp, *Anat. Rec.*, 2004, **281B**, 12.
12. P. G. McMenamin, J. McLachlan, A. Wilson, J. M. McBride, J. Pickering, D. J. R. Evans and A. Winkelmann, *Med. Teach.*, 2018, **40**(10), 1020.
13. J. C. McLachlan, *Anat. Rec.*, 2004, **281B**, 4.
14. C. Triepels, C. Smeets, K. Notten, R. Kruitwagen, J. Futterer, T. Vergeldt and A. Van Kuijk, *Clin. Anat.*, 2020, **33**, 25.
15. C. M. Parkes, P. Laungani and B. Young, *Death and Bereavement Across Cultures*, Routledge, London, 1997.
16. N. Notzer, D. Zisenwine, L. OZ and Y. Rak, *Clin. Anat.*, 2006, **19**, 442.
17. H. K. Lempp, *Med. Educ.*, 2005, **39**, 318.
18. V. Bernhardt, H. J. Rothkotter and E. Kasten, *GMS J. Med. Educ.*, 2012, **29**, 10.
19. C. O. Grochowski, M. Cartmill, J. Reiter, J. Spaulding, J. Haviland, F. Valea, P. L. Thibodeau, S. McCorison and E. C. Halperin, *Clin. Anat.*, 2014, **27**, 835.
20. Adult Mental Health First Aid, https://mhfaengland.org/individuals/adult/, last accessed February 2020.

CHAPTER 8

Safety Outside of the Classroom

P. COULAHAN*

University College Dublin, Dublin 4, Republic of Ireland
*E-mail: peter.coulahan@ucd.ie

8.1 Managing Safety in Sporting Activities

8.1.1 Introduction

A cursory online search of the websites of Irish and UK Higher Education Institutes (HEIs) reveals a large number of sports clubs and sporting activities associated with such organisations. The benefits of sport for both physical and mental health are well documented and sporting activity amongst the student body is to be welcomed. Within a HEI sports clubs and sporting activities can be competitive or participatory in nature; an individual or team activity; novice or advanced; such that there exists a sport that almost every student can engage in. However, when considering the benefits that students obtain by engaging in sporting activities, the risks associated with sport must also be considered, in particular those associated with higher risk sports.

If a HEI supports and encourages its students to engage in sporting activities, then it must put in place an appropriate safety management system and oversight regime that allows students to partake in sport without being exposed to undue risk. The elimination of all risk from sport is neither possible nor desired, indeed in some sports it is the exposure to the risks that the activity generates that are a key attraction of the sport. However, that does not mean that a HEI should prohibit students

Challenges for Health and Safety in Higher Education and Research Organisations
Edited by Olga Kuzmina and Stefan Hoyle

Published by the Royal Society of Chemistry, www.rsc.org

from engaging in higher risk sports or decline to support such activities. With the implementation of an appropriate safety management system students can be supported and facilitated in undertaking a wide range of sports without undue risk to themselves or their contemporaries. Notwithstanding this a HEI must be cognisant of the very real and ever present risks that are inherent in some sports, some of which require careful management, and decide if as an organisation it has the capacity, the competency and the support structures to manage these risks. In some cases, the institution may choose to discharge this duty of care through a third party, such as *via* a Students Union.

If a HEI considers that the risks posed by a sport are too difficult to manage or cannot be reduced to an acceptable level, then they may decide to place restrictions on that sport or not to facilitate or support that sport. In the past I have placed restrictions on students in my university from engaging in activities such as cave diving and ice climbing due to the risks associated with these sports. If a HEI is to accept and support potentially high risk sporting activities then it is incumbent upon them to ensure that the risks associated with these sports when undertaken by their students are assessed, remediated where possible, and appropriately managed.

Whilst this section on *Managing Safety In Sporting Activities* is written with HEIs and students to the fore, the guidance and advice it contains also applies to Research Institutes (RIs) that support sporting activities and to the safety of their employees and researchers.

8.1.2 Classifying Sports and Sports Clubs

From a participant safety perspective I suggest that sports can be divided into two very broad categories. These categories are based on the severity and extent of the risks that these sports pose to participants, and whilst qualitative in nature, can be used to guide a HEI in assessing the level of oversight and support an individual sports club will require.

8.1.2.1 Mainstream Sports

These are sports that exhibit a low level of risk to participants, or where contact and exertion risks (*e.g.* person to person, person to ball/implement/environment) are inherent and accepted by participants. Examples include cricket, snooker, football/soccer, athletics, badminton, basketball, tennis, volleyball, netball, *etc.* Managing safety in mainstream sports is by definition relatively straightforward. In such sports participant safety can be achieved in the main by adhering to the rules and regulations laid down by the sport's governing body. For most mainstream sports this should be sufficient to address most if not all risks associated with playing the sport.

Some mainstream sports may also exhibit what can be described as 'special risks'. In this context 'special risks' refers to risks which are intrinsic in the sport and which have the potential to cause very serious harm over and above the typical sporting injuries associated with simple exertion and competitor to competitor/environment contact. These would include sports such as rugby union, boxing, American football, *etc.* As well as exhibiting the typical risks associated with mainstream sports, sports such as rugby union and boxing also expose competitors to the risk of serious head and brain injury and whilst this risk can be mitigated to some degree and steps can be taken to manage such injuries when they arise or are suspected, these risks cannot be completely eliminated as they pertain to the core activity of these sports. Hence it is vital for these sports that the rules and regulations laid down by the relevant sport's governing body are adhered to, including in the case of boxing for example, the requirement for the presence of physicians at bouts, competitor medicals and mandatory rest periods between fights.

8.1.2.2 High Risk Sports

High risk sports are defined as those sports where there exists substantial risk or the potential for the existence of substantial risk to the safety of the participants, either from the activities being undertaken, the equipment being used, or the environment in which the sport is being carried out. In some high risk sports there exists a critical dependency on safety equipment, the failure of which could lead to serious injury or death. Some sports take place in locations where environmental changes can occur suddenly, generating a risk to participants. Examples of high risk sports include kayaking, caving, sub aqua, sailing, mountaineering, windsurfing, *etc.* Ensuring participant safety in high risk sports is often complex and based around very technical safety rules and regulations issued by that sport's governing body. For some minority sports there may not be an agreed set of rules and regulations and in these cases the HEI itself may have to design, introduce and validate safety standards. For example, in the past I have designed and implemented minimum training and competency standards for members of a mountaineering club engaged in belaying activities in the absence of the availability of appropriate standards. Indeed, even in cases where safety rules and regulations do exist a HEI may decide to add to these in order to improve participant safety.

There exist many examples of persons losing their lives whilst engaged in sporting activities, including in sports that would not usually be considered to be high risk in nature. For example, participant deaths have been recorded in hockey,[1,2] rugby union,[3] boxing,[4] cricket[5] and soccer.[6] Consequently, whether a sport is classified as being mainstream or high risk, the safety of participants requires some form of oversight and support from the HEI.

8.1.3 Sports Club Oversight and Support

I suggest that in order to ensure the safety of participants there are two levels of oversight and support necessary with respect to a HEI sports club:

i. At an institutional level the HEI must provide some level of central oversight and support for approved sports clubs, or arrange for same to be provided by a third party, such as *via* a Students Union.
ii. At a local club level, a club committee or similar is required to manage the day to day activities of the club and to ensure that the club functions in a safe manner.

8.1.3.1 Central HEI Oversight and Support

Once again, a cursory online search of Irish and UK universities shows the use of Sports Offices and Student Unions by HEI's to oversee their sporting activities. The level and nature of the support provided will of course depend on the resources made available by the HEI, but it is suggested that at a minimum the following be centrally organised and provided:

- A centralised club member registration system which allows the HEI to track member numbers and hold next of kin contact information.
- A high level club operations handbook outlining both the organisation's expectations and available supports with respect to safety.
- Advice on safety and risk assessment, to include relevant templates and guidance documentation.
- Provision of a minimum level of insurance for club activities and club members. In my own university sports clubs are provided with third party liability cover, personal accident insurance for club members and travel insurance when engaged in activities abroad. Clubs that own high value equipment, such as the sailing club, are also provided with property insurance for this equipment.
- A central location for the recording of accident details and subsequent investigation and follow up.
- Training for club committee members. In particular, training for the key positions outlined in Section 8.1.3.2, along with training in how to produce a club safety statement and associated risk assessments (Section 8.1.4).
- The continuous proactive oversight of club activities.
- Final approval of all club activities and club safety documentation.

8.1.3.2 Local Club Management

At the club level, each club should have in place an oversight or club committee. The membership of this committee should be drawn from experienced members of the club who have the necessary experience and skills to manage the day to day activities of the club. From a safety perspective there are three key roles on such a committee whose importance increases as the risks associated with the sport increases. These are the Club Safety Officer; the Club Equipment Officer and the Club Training Officer (or their equivalents).

- Club Safety Officer: The role of this committee member is to coordinate the management of safety within the club. They will be responsible for developing the club safety statement and associated risk assessments, maintaining records, recording accidents, auditing club activities, *etc.*
- Club Equipment Officer: In clubs where there is a critical dependency on equipment for safety or where the failure of a piece of equipment could lead to a serious injury then the role of the equipment officer is key. They will maintain an inventory of equipment along with a register of testing and maintenance of same. They will ensure that club owned equipment is fit for use before being placed into service and examine equipment for damage when it is returned to the club. In respect of a participant's own sporting equipment the club equipment officer will lay down minimum standards that have to be met, which can be based on requirements as set by the sport's governing body, applicable industry standards or the club itself. Of particular importance is the role that this person plays in ensuring that a record is kept of what equipment is loaned to members and for what purpose. The club equipment officer must ensure that only persons trained in the use of equipment may borrow same from the club and that a record of any such borrowing is maintained. Once returned to the club by the borrower the club equipment officer will also inspect the equipment to ensure that it has been returned in an undamaged state.
- Club Training Officer: Again, as the complexity and risks associated with a sport increase it becomes essential that all persons participating in that sport receive appropriate training. Training may cover issues such as how to use equipment, what to do in an emergency, how to recognise environmental hazards, *etc.* The training officer will be responsible for ensuring that all club members receive appropriate training, and will also collate and manage the club's training records, including obtaining and validating training certification from outside of the club.

These three roles may be carried out by the same person in smaller or low risk mainstream sporting clubs, whilst in more complex, higher risk sports clubs they may be split between persons. What is critical is that these roles coordinate their activities so that for the higher risk sports in particular, club members can engage in their sport in a planned way with the risks having been identified and assessed, whilst they utilise appropriate and maintained equipment that they have been trained to use.

By working closely together in an integrated and collegiate fashion the club committee and the central resources provided by the HEI can implement a club safety management system that enables sports clubs to undertake activity in way that does not dilute the attraction of the sport but which is done in a manner without undue risk to participants. By doing so the club safety management system can form part of a wider integrated sports risk management programme that addresses parallel issues such as insurance, the protection of vulnerable participants, large event management, *etc.*

8.1.4 Club Safety Statements and Risk Assessments

8.1.4.1 The Club Safety Statement

In order to clearly outline how safety within a club is managed each club should produce a club safety statement. In this context a club safety statement is a written document, ideally based on a template supplied by the HEI, which clearly details how each sports club manages safety and addresses the key issues as outlined in Section 8.1.5 where they are relevant. By providing clubs with a template the HEI can ensure that all of the issues that it considers pertinent with respect to safety are considered and documented.

8.1.4.2 Risk Assessments

Each club safety statement should include a set of robust risk assessments. In the case of sporting activities, a risk assessment is a careful examination of what could cause harm to club members, third parties or property during club activities (*i.e.* a hazard), and the identification of control measures that the club can take to minimise the risk from these hazards. The purpose of these risk assessments is to reduce the risk to participants to as low a level as possible by implementing risk control measures that are reasonable and which reduce the risk to an acceptable level, whilst also recognising that some sports are always going to carry an inherent risk of injury to participants, which in some cases is the appeal of the sport.

Whilst risk assessments lay down steps to minimise risk to participants, they should not restrict the initiative of individual club members 'on the ground' in responding to the prevailing conditions or circumstances. Especially for high risk sports, senior or more experienced participants should have the necessary experience and training such that they can respond to changing conditions or circumstances by engaging in dynamic risk assessment. In this context dynamic risk assessment is defined as the continuous process of monitoring and identifying emerging hazards and taking action to eliminate or reduce risk in a changing environment. This is key in sports such as sailing, sub aqua, caving, *etc.*

The central oversight function provided by the HEI has a key role to play in ensuring that each club produces a suitable club safety statement and set of risk assessments. It is suggested that both the club safety statement and risk assessments be reviewed and updated on at least an annual basis.

8.1.5 Higher Education Institute and Research Institute Sports Clubs – Key Safety Considerations

The following outlines some key safety considerations that should be taken into account when managing safety within a HEI/RI sports club and constructing a club safety statement and risk assessments.

8.1.5.1 Governing Body Rules

Many sports have governing bodies that lay down rules and regulations that must be adhered to when persons are engaged in the sport. Adherence to and the implementation of such rules and regulations forms a key part of ensuring participant safety in many sports. For many mainstream sports with no special risks this is often sufficient to minimise risk.

8.1.5.2 Approved Activities

Sports clubs should be clear as to what sports they promote and partake in, and this must be approved by the HEI. Whilst for many sports this is very straightforward, *e.g.* soccer clubs train for and play soccer matches, some sports have nuances that can increase the risk. For example if a HEI has a sub aqua club, does it restrict that club to diving on air only or does it allow it to use mixed gases and accept the increased complexity and potential risks that this practice can pose? Similarly, in a sport like kayaking, does the HEI place restrictions on the grade or difficulty of the river that club members may tackle?

8.1.5.3 Definition of a Club Activity

The HEI and the sports club needs to clearly define what constitutes an approved club activity, and what constitutes a club member's private sporting activity. For example, if a small group of caving club members borrow some club equipment and meet up with some non-club members to undertake caving activities, does that constitute an approved club activity, and hence is it subject to the HEI's rules on such trips? From a safety perspective it must always be clear what activities are being undertaken under HEI oversight.

8.1.5.4 First Aid

The availability of access to trained first aid personnel can be a key safety requirement in many sports. In particular, as the risk of injury from the sport increases and the remoteness of the activity increases, immediate access to such assistance can become more important. In some cases, participants who are engaged in activities remotely may need to depend on themselves or from within their group for first aid assistance. Consequently, minimum training requirements for first aid may be required for some clubs.

8.1.5.5 Transport Issues

How sports clubs and their members move between locations is an important consideration. Does the HEI provide access to a vehicle, are commercial services used or do participants use their own vehicles? Is club equipment transported between locations and is it done so in a safe manner? If trailers are in use for transporting equipment then there may be load stability issues, maintenance requirements, driver licencing and other road traffic issues to be considered. Minimum standards in this area must be set down as part of any safety management system.

Clear rules with respect to the use of private vehicles, institutional vehicles or other third party vehicles must be in place for sports clubs as part of any safety management system. It may be appropriate for the institution to apply any wider policy on vehicle safety in the workplace that it may have to a sports clubs use of vehicles given the hazards associated with driving.

8.1.5.6 Use of Third Party Facilities

Some sports require the access to, and the use of, facilities provided by third parties, for example access to dry ski slopes. When selecting a third party facility to utilise, a club must satisfy itself that the facility is safe to use, and in some circumstances may seek to use the services of a competent person to make an assessment of its safety and suitability *e.g.* when assessing and reviewing range safety provisions at a gun range.

8.1.5.7 Coaches and Trainers

How a club selects and recruits coaches and trainers is important. What is the minimal standard that coaches trainers in a sport must have achieved? Is a coach's or trainer's competency based on their qualifications, experience, or a combination of both? Must a coach/trainer in a particular sport be registered with an oversight body? In some sports, such as martial arts, the use of qualified and/or competent coaches is a key risk reduction measure.

8.1.5.8 Leader-participant Ratios

For some sporting activities more senior 'leaders' take control of the activity and direct and oversee the activities of others. For example, in hillwalking a group may be led by a more experienced member(s). In such scenarios, leader-participant ratios can be important. How many persons can one group leader manage? Is a deputy required? Is the ratio affected at all by the nature of the activity and the experience of the followers? Do novice participants require a higher level of supervision and a higher leader-participant ratio? Selecting the right leader to lead the right size group of followers can become very important if an emergency situation were to arise and the group were to reply upon their leader's competency and experience to protect their safety.

8.1.5.9 Equipment Issues

For some clubs the use of certain equipment can be vital to ensuring participant safety; for example in mountaineering karabiners are a key piece of safety equipment, the failure of which could lead to catastrophic injuries. Club equipment needs to be maintained in a safe and useable condition by a competent person and stored in a suitable manner. The use of equipment must be restricted to trained and authorised persons. Unsafe or poorly maintained equipment should not be used. Equipment registers and records of pre use visual checks and similar may be deemed necessary. A club must also have in place clear rules around the loaning of equipment to members, how that is authorised, managed and recorded.

8.1.5.10 Emergency Planning

A critical part of sports safety management is having a protocol in place to be followed in the event that there in an incident, injury *etc.* For many mainstream sports the injuries likely to be incurred by participants are relatively minor and appropriate first aid and/or contacting the emergency services may be sufficient. However some sports have the potential to generate injuries that require immediate medical attention, hence the need for example for a doctor to be present at boxing bouts; whilst for some complex risk sports specialist equipment and training may be necessary to render emergency assistance, *e.g.* the delivery of oxygen to treat suspected 'bends' in sub aqua. Emergency planning links closely in with the identification of a clubs first aid requirements.

In addition to medical type emergencies some clubs may need to develop plans that address the response to a damaged or malfunctioning piece of safety critical equipment, *e.g.* if a boat engine fails whilst at sea; whilst other clubs may need to consider issues such as their response to a 'lost' participant, *e.g.* caving and hillwalking clubs.

Ensuring that a club has in place an emergency response plan that is easily enacted, understood by all participants and commensurate with the risks of the sport is an important part of any safety management system.

8.1.5.11 Club Trips and Expeditions

A system of recording and overseeing sports clubs' trips and expeditions may also be required. It is important that a HEI knows where its club members may be at any one time in the event that it needs to contact them, whilst the organisation may wish to lay down norms of behaviour for club members who are engaged in such trips around issues such as alcohol consumption, member to member interactions, *etc.*

8.1.5.12 Event Planning

On some occasions the activities undertaken by a club will fall outside the scope of the club safety statement and risk assessments. For example, when planning and managing larger sporting events there may be a need to consider wider campus impact, crowd control, installation of temporary structures, *etc.* The HEI should be clear in the guidance that it gives to its sports clubs as to when specific event management plans are required to complement a club's existing safety documentation.

8.1.5.13 Member Training and Competency

As has already been alluded to, as the complexity and risks associated with a sport increase it becomes increasingly important that all persons participating in that sport receive appropriate training. Where appropriate a club must lay down the minimum training that a member must have achieved before they can engage in specific activities, *e.g.* in sub aqua there is a system of graduated training as divers move from standard type shallow diving to more technically complex deeper diving.

8.1.5.14 Insurance

Whilst not strictly part of the safety management system a HEI may wish to ensure that appropriate insurance cover is in place for club operations. Depending on the jurisdiction the HEI may wish to ensure that public or general liability cover, along with personal accident cover for club members is in place. Some clubs also utilise relatively high value equipment, *e.g.* sailing clubs; and again a HEI may wish to ensure that such equipment is adequately insured against loss or damage.

8.1.6 Sources of Information

In addition to the information that is available from a sports governing body with respect to managing safety within a sport, there are also a number of other resources that may be of use to clubs, administrators and safety practitioners. These include:

- UK Health and Safety Executive Advice for amateur sports clubs[7]
- Royal Society for the Prevention of Accidents (RoSPA) Guidance on sport, leisure and water safety[8]

8.1.7 University College Dublin Sports Club Management – A Case Study

In 2007 a student from University College Dublin (UCD) was killed in a kayaking accident whilst on a trip with the University's Kayaking Club. Full details of this incident can be obtained from the statutory authority's investigation into same.[9]

Following the incident the University undertook a review of the management of safety within its clubs and developed a structured safety management system for its over 60 Sports Clubs. The system has evolved over time to become integrated with other aspects of risk management in sport but in respect of safety the system works in the following way.

Sports clubs are overseen by a central Sports Office which is resourced by the University. This office and its personnel provide support and advice to clubs. At a high level this takes the form of:

- The provision of a club operations manual, which contains advice on safety, sets minimum standards for coaches and leaders, lays out norms of behaviours for club members, and outlines disciplinary processes for club members who may breach safety rules.
- Training for all club committee members in respect of safety.
- The provision of a club safety statement and risk assessment templates.
- The provision of large event management templates for larger sporting events that fall outside the scope of a standard club safety statement and risk assessments.
- Detailed training for club safety officers in how to complete the aforementioned documents and in particular in how to identify sporting hazards and assess risk.
- Ongoing face to face advice for clubs regarding safety and associated issues.
- A central repository for accident and incident reporting and investigation.
- Oversight of sports event planning.
- An online system for the recording of overnight trips by sports clubs and all activities of designated high risk clubs.

In conjunction with the support provided by the University each individual club is required to ensure the following:

- A club committee is in place to oversee the operations of the club.
- That critical committee roles are assigned and fulfilled.
- That the requirements of the University regarding sports safety are implemented at all times.
- That on an annual basis the club submits an updated club safety statement and risk assessments for review.
- That at all times the club operates in accordance with the provisions of its club safety statement and risk assessments.

The risk assessment process undertaken by UCD's sports clubs is outlined in Table 8.1.

All club safety statements and risk assessments are reviewed by a member of the University's central health and safety team on an annual basis. The reviewer advises both the sports office and the club as to the efficacy of their documentation and in particular their risk control measures.

Table 8.1 University College Dublin – sports club risk assessment process.

Step 1: Identify potential hazards	A hazard is anything that has the potential to cause harm. Consider: • The environment *e.g.* the weather, the location of the activity, *etc.* • The hazards within the sport *e.g.* body contact, dependency on critical safety equipment. • Hazards associated with the equipment used *e.g.* weapons in martial arts, boats and trailers, *etc.* • Use of vehicles *e.g.* private bus operators towing of trailers, *etc.* • Travel and equipment transport to and from venues. • Use of third party facilities. • *Etc.*
Step 2: Identify what control measures are already in place	• Specialist training for participants • Special safety equipment for the sport • Participant safety briefings • National governing body guidelines *e.g.* rules, use of referees, good practice guidelines *etc.* • Use of qualified and insured coaches only • *Etc.*
Step 3: Identify if the hazard is likely to cause harm	• **High** – could occur quite easily • **Medium** – could occur sometimes • **Low** – unlikely, although conceivable
Step 4: Identify if additional risk reduction measures are practical/necessary	What more can reasonably be done to reduce the likelihood of an accident happening? For example, can you: • Try a less risky option or another way of doing things • Provide or undertake additional training • Purchase specialist equipment • Carry out 'safety briefings' with all members • Provide clear procedures for club members to follow • *Etc.*
Identifying Risk Control Measures **Mainstream Sports** – Risks are managed primarily by adherence to rules laid down by a sports governing body, *e.g.* use of referees, implementation of rules, wearing of protective clothing, availability of first aid/medical assistance, *etc.* **High Risk Sports** – Risks are managed by adherence to rules laid down by a sports governing body, minimum competency standards for participants, use and maintenance of and training in the use of life protecting equipment, robust emergency planning, development of own safety standards where none exist or those that do exist are considered inadequate, limitations on some very high risk activities, *etc.*	
Step 5: Record risk assessment findings in the club safety statement/review annually	

A core component of UCD's sports risk management programme is that unless all club safety documentation is approved by both the sports office and the University's central health and safety function a club will receive no funding from the University and, depending on the nature of the club, may be prohibited form recruiting members and engaging in any activities.

Sample extracts from UCD sports club risk assessments are provided in Table 8.2.

Table 8.2 Sample extracts from University College Dublin sports club risk assessments.

Hazard/activity with potential to cause harm	Hazard/risk control measures	Risk (high, medium, low)
Extract from UCD Archery Club Risk Assessment		
Getting shot by/shooting someone with an arrow.	Prior to shooting an advanced archer/s is responsible for setting up a safety zone for shooting to take place. This includes the placement of signs at all external entry points to the zone and the establishment of a 'shooting line' and 'equipment/waiting line' to alert people that a shoot is taking place. No bow is to be nocked with an arrow until the designated safety zone is clear of people. An arrow is said to be 'nocked' when it is clipped onto the bowstring. A nocked arrow may only be drawn when on the shooting line facing the targets. The draw must commence and complete with the arrow pointing at the target. All members should be alert to the risk of someone entering the safety zone and are responsible for raising the alarm if they observe someone entering the safety zone. In the event of someone entering the safety zone after shooting has commenced or of any other risk arising from shooting, the words 'stop shooting' are to be shouted by any and all people who have noticed this risk. In this event everyone on the line must immediately lower their bow, remove any nocked arrows from their bow and not recommence shooting until the advancer archer is made aware of the risk and is satisfied that it is no longer present. If the advanced archer is either not satisfied or unsure as to whether it is safe or not to begin shooting again, shooting may not resume until the club's safety officer is consulted and only when he/she is satisfied that shooting recommence.	Low
Extract from UCD Kayak Club Risk Assessment		
Drowning – As with every water based activity drowning is an underlying hazard that must always be considered. Drowning may occur due to being pinned under an obstacle, a kayaker becoming incapacitated, or for some reason that leads to a person being held under water for a prolonged period of time.	Every club member must wear a buoyancy aid and helmet. If either of these items is missing they are excluded from the activity.	Low – although constant vigilance is required.
	All club safety equipment is inspected on a regular basis to ensure it is of an adequate standard.	Refer to other relevant club risk assessments.
	Designated safety personnel on the water monitor other kayakers (especially novice paddlers) and carry multiple rescue tools such as throw ropes and slings to aid in rescue should an emergency occur.	
	It is a club rule that a member trained in first aid is present on every club trip.	

University College Dublin considers that the safety management system as described is functioning well. Clubs have become accustomed to the requirement to risk assess their activities and have developed a culture of putting safety to the fore. Whilst the system does require a not insignificant investment in supporting resources from the University, it is deemed to represent a worthwhile investment in the safety of students.

8.2 Managing Safety on Student and Researcher Work Experience Placements

8.2.1 Introduction

Many Higher Education Institute (HEI) and Research Institute (RI) programmes now incorporate a professional work experience (PWE) placement or an internship of some kind. Under such programmes students and researchers can be placed with external hosts where they gain experience of a professional working environment. It is suggested that when such PWE placements are a mandatory or credit bearing component of a course of study or research, that the organisation takes steps to ensure, in so far as is reasonably practicable, the safety of participants whilst on such placements.

8.2.2 Selecting Work Placement Hosts

In my own institution PWE hosts may be sourced by the university or by the student/researcher themselves. The sourcing model used will differ between HEIs/RIs, and may differ between programmes within the same organisation (as it does in my own University), and will likely be linked to the number of such placements made by the organisation and the availability of resources to manage same. Notwithstanding this it is suggested that when such PWE placements are a mandatory or credit bearing component of a course of study or research that the organisation make reasonable enquiries to determine that the host meets a minimum safety standard.

When vetting a potential host the HEI/RI may consider:

- Internal Stakeholder Feedback – what have previous persons hosted at the same location reported, do other units within the organisation utilise or have relationships with the host (*e.g.* research links), *etc.*
- Reputation – does the host organisation have a reputation as a good employer, do other HEIs and RIs utilise this host, has it hosted persons from other organisations in the past, does it have a strong commitment to corporate social responsibility, is it a registered member of a trade or professional association, does it hold recognised quality or safety management standards, *etc.*

- Suitability – will the host provide the necessary learning or research environment required by the individual on placement in order to allow them to achieve the learning or research objectives of their placement?
- Safety record – has the host a poor or a good safety record?
- Geographical location – is the host located abroad in a politically or environmentally unstable region, are the statutory health and safety safeguards for employees in that jurisdiction comparable to those in the HEIs and RIs home country, *etc.*

A HEI or RI must always consider its duty of care to its researchers and students and put in place an appropriate vetting and approval system for PWE hosts.

8.2.3 The Safe Student or Researcher

At all times the safety of the individual on the PWE placement should override all other considerations for the HEI. The safety of the student or researcher should not be compromised in the pursuit of research goals or academic objectives. A key component in ensuring the safety of the person on placement is the person themselves. Knowledgeable, safety aware students and researchers are the ideal candidates for PWE programmes, and when such programmes are mandatory it is incumbent on the HEI to ensure that students and researchers are safety aware before allowing them to enter a professional working environment.

In addition, students and researchers whilst on placement are ambassadors and representatives of their home institutions. As well as behaving in a safe manner such persons will be expected not to engage in any behaviour that could bring their home institution into disrepute. In my own university students on PWE placement are subject to the University's Student Code of Conduct at all times and must abide by the provisions of this document.

The concept of the 'safe student or researcher' is introduced here in the context of professional work placements. I would suggest that at a 'safe student or researcher' is:

- One whose stage of learning and knowledge is commensurate with the nature of the work activities that they are likely to undertake as part of their placement. This becomes increasingly important when one considers work placements in technical subjects such as engineering, agricultural sciences, medicine, *etc.* and high risk placements.
- One who has been given the necessary training and/or information by their HEI/RI so as to be able to identify workplace hazards associated with the types of environments that they are likely to be exposed to when on placement. Again, as the complexity of the hazards that

participants may be exposed to increase, the level of safety training and/or information that the HEI or RI should provide the student/researcher increases.

In respect of any safety training or information that an organisation may provide to its students and researchers in advance of a PWE placement the following is suggested as a minimum:

- Clarification on the students' and researchers' legal status in the workplace as a PWE participant with respect to health and safety legislation and the duties that the host organisation have to the participant under same; but also what statutory obligations may be placed on the student or researcher by the same legislation. This may differ as persons move between different legal jurisdictions (and indeed in some jurisdictions may not be relevant). As an example, under health and safety legislation in the Republic of Ireland students on work placements are considered to be employees of their host organisation for the purposes of health and safety management. This places statutory obligations on both the hosts and the students, with the potential for legal repercussions for either party if it fails to adhere to the requirements of the legislation.
- An overview of the key hazards that exist in the industry or working environment which the student or researcher is entering and an introduction to the concept of hazard identification, risk assessment and risk control measures. Again, as the complexity of the hazards that participants may be exposed to on work placement increases the extent of the knowledge required by them will increase. Whilst on placement the HEI or RI will expect that the host provides the student or researcher with a safe working environment, it is still important that the student or researcher comes to the workplace with some degree of safety awareness.
- Details of what the participant should expect from their host with respect to managing their safety; for example the expected levels of host oversight and safety training, expectations around the provision of personal protective equipment, access to safety documentation, *etc.*

The aim of providing students and researchers with safety awareness training, information and knowledge before they go on placement is twofold; the first reason is to help them safeguard their own safety in the workplace; the second reason is to give them the necessary knowledge to be able to identify an unsafe workplace. In the latter scenario it is important that participants have the option to revert back to their HEI/RI in the event that they consider that their placement environment is unsafe and they wish to change. Once again HEIs and RIs are reminded that academic progression and research objectives must not be allowed to take precedence over safety.

8.2.4 A Note on Learning and Research Objectives

For every PWE placement there must exist a clear set of learning or research objectives as applicable. Clear and unambiguous learning or research objectives allow for an accurate assessment of the type of workplace tasks that participants are likely to undertake when on placement and consequently allow for an understanding of the type of hazards they are likely to be exposed to. A lack of clear objectives may create difficulties in identifying potentially high risk placements and the level of safety awareness training and information that students and researchers will need in advance of same.

8.2.5 Lines of Communication

For all PWE placements there should exist regular contact between the participant and their placement coordinator, academic supervisor or equivalent. Such contact will allow the participants' progress to be monitored and will also provide them with a regular opportunity to raise concerns with their HEI or RI about any aspects of the placement they are worried about.

Similarly a formal line of communication between the host and the participants' home institution will allow for a regular review of student or researcher progress and will allow for either party to raise any concerns that it may have with respect to participant progression, working conditions, safety, *etc.*

8.2.6 Placement Risk Assessment

Other than in respect of the high risk placements as outlined in Section 8.2.7, for many HEIs and RIs it may not be practicable to undertake a risk assessment of every individual PWE placement. It is suggested that a more practical approach may be to risk assess cohorts of placements on a programme by programme or module by module basis. Risk assessing the type of placement associated with a programme or module at a high level will inform the HEI/RI as to the extent and nature of the safety awareness training required by the participant in advance of their placement, and will also inform them as to the level of host vetting required. Linking the learning and research objectives of the work placement to the likely activities that this will require the participant to undertake is a key part of this risk assessment process.

For example, the placement of law students with legal firms is at the lower end of the workplace safety risk spectrum. For such placements a HEI may vet potential hosts *via* confirmation of their membership of a relevant professional body and/or the use of a host vetting checklist, whilst providing students with a basic level of safety awareness information in advance of the placement. In contrast placing forestry students with an arborist where they may need to use chainsaws, operate quad bikes, climb trees, *etc.* would be at the higher end of the workplace safety risk spectrum. In this case the

HEI may decide to vet hosts by reviewing their safety documentation and perhaps by undertaking a site visit; whilst providing students with detailed safety training in advance of the placement given the nature of the hazards involved with this type or work.

8.2.7 High Risk Placements

In my experience there are a small cohort of work placements that have the potential to expose participants to unusual or complex hazards which may be difficult to manage. These are hazards of a nature over and above the hazards encountered in typical workplaces.

One example I have encountered are placements for students engaged in humanitarian action programmes. These students may be placed for example with non-governmental organisations in areas with extreme societal challenges where there exists a very real risk of violence, civil unrest or crime. For these types of placements it is suggested that a much more thorough host selection process be undertaken. As part of placing students with hosts in these extreme environments a robust risk assessment process must be undertaken to ensure that the student's safety is protected in so far is possible during the placement. Depending on the nature of the potentially high risk environment the student is entering the HEI may need to seek the assistance of trusted entities 'on the ground' who can advise the HEI on the current state of the environment and emerging hazards. In some cases a HEI may decide that the risk to a student's safety is too great and may prohibit such placements. Managing a student's safety on work placements of this type can be challenging and in many ways is similar to managing safety in high risk fieldwork operations, with consideration of issues such as emergency response and planning, communications, emergency repatriation, *etc.* required.

8.2.8 Assignment of Responsibilities

Ideally there should be some clarity for hosts from the HEI/RI on where responsibility lies in respect of a PWE participant's safety. This may be done *via* an information note issued by the HEI/RI to the host in advance of the placement, or it may be done *via* the use of a formal agreement.

When considering a participant's safety it is important that no ambiguity exists between the host and the HEI/RI on who is responsible for issues such as:

- The tasking of the participant.
- The ongoing day to day workplace supervision of the student or researcher.
- The provision of safety training for the participant and the extent of same.

- The issuing of personal protective equipment and associated training in its use.
- Workplace safety induction.
- Accident reporting and investigation.
- The overall learning and research objectives of the placement.

Both parties must also be clear as to the level and nature of the relevant competencies held by the student or researcher in advance of the placement.

Participants must also be clear as to their own responsibilities in the workplace in respect of their own safety and that of others. The expectation from the HEI or RI that they work in a safe manner, adhere to safety instructions given by the host, and work within their own competencies must be communicated to the them as part of their preplacement training/information.

8.2.9 Case Study – The University College Dublin Agricultural Science PWE Programme

Traditionally Agricultural Science students in University College Dublin (UCD) have undertaken PWE placements as part of their course of study. Given the broad nature of the Agricultural Science programme in UCD placements can be with farms, forestry operations, laboratories, manufacturing facilities or office based organisations.

Historically placements were somewhat informal in nature and it was recognised some time ago that a more formal structure was required in respect of the placements, particularly as the hazards associated with agricultural work can be significant.

In developing a suitable safety management structure for these work placements the University concentrated on the key areas of host selection, assignment of responsibilities and the concept of the 'safe student'.

Having evolved over a number of years the current system operates as follows:

- Hosts may be selected by a student or sourced by the University.
- If selected by the student details of the host must be passed to a University module coordinator for approval.
- Initial approval of a potential host is made on the basis that it can be shown that the learning objectives of the placement can be met by placing a student with that host and that no obvious reason exists to exclude the host from the programme.
- The University then issues a formal work placement contract to the host that covers the following:

  - It makes the host responsible for the student's overall safety.
  - It makes the host responsible for the student's supervision, the provision of safety training, the provision of personal protective equipment and the reporting of accidents involving the student to the University.
  - It requires the host to confirm it adheres to all relevant health and safety legislation.
  - It also details the learning objectives of the placement and what the University expects from the host regarding the work that will be undertaken by the student.
  - It is also used where relevant to require the host to have appropriate insurance cover in place for hosting students.
- In the event that a potential host declines to sign the contract then that host will not be used for student work placements.
- The use of such contracts also allows the University to track all PWE placements.
- The University also issues a formal contract to the student involved in the placement. This contract requires the student to agree to:
  - Adhere to any instructions given by the host regarding workplace health and safety including wearing personal protective equipment, attending safety training, *etc.*
  - Work in a safe manner at all times.
  - Behave in an appropriate manner when on placement.
  - Inform their placement coordinator if they consider that their safety is at risk on their placement for any reason.
- Students are also provided with safety training in advance of their placement. Initially this consisted of a one off safety session with information provided by both faculty and the University's central health and safety function. However given the recognition that Agricultural Science programmes can give rise to placements in potentially hazardous environments, students are now required to complete a mandatory module on workplace health and safety which is delivered over a number of weeks by experts in various fields relevant to their placement programme. Subjects such as behavioural safety, accident causation, zoonoses, agricultural chemical safety, machinery safety, *etc.* are covered. Like other parts of their curriculum, students are formally examined on this module and it must be completed before a student can begin a work placement.

UCD considers that this approach has led to significant improvements in the safety of Agricultural Science students on professional work placements. This approach has now been adopted across the University, with the use of the 'University – Host' and 'University – Student' contracts now mandatory for most PWE placements, with students receiving safety training/information in advance of their placement which is commensurate with the hazards likely to be encountered whilst on same.

8.3 Fieldwork Safety

8.3.1 Introduction

For many academic disciplines, fieldwork is an essential part of teaching, learning and research activities. In this context fieldwork is defined as any work carried out by staff or students for the purposes of teaching or research whilst away from a fixed Higher Education Institute (HEI) or Research Institute (RI) managed premises. This definition does not extend to include work placements or internships.

For many scientists fieldwork may be the most dangerous activity that they engage in. Fieldwork can expose the individual or the group to hazards and environments over which they have little or no control and which may be difficult or impossible to manage. The online resource, Memorial To Fallen Naturalists, lists fieldworker fatalities due to plane crashes, vehicular accidents, animal attacks, disease, murder, drowning and equipment failure.[10] Consequently, an appropriate level of preplanning and risk assessment is required for all fieldwork, commensurate with the hazards posed by the work.

Due to the variety and complexity of fieldwork it is beyond the scope of this paper to provide an exhaustive guide to fieldwork safety. Rather this chapter will, at a high level, outline the key considerations that I consider are essential to managing fieldwork safety based on my own experiences. Detailed fieldwork safety advice can be obtained from other sources, such as the Universities Safety and Health Association.[11]

8.3.2 Preplanning

All fieldwork must be subjected to some form of preplanning. Fieldwork should not be undertaken without a review and consideration of the work in advance. At a minimum preplanning should include the preparation of a fieldwork risk assessment and an associated emergency plan. For some types of fieldwork a HEI or RI may decide that a formal approval process is also required.

8.3.2.1 Fieldwork Risk Assessment

The ultimate aim of a fieldwork risk assessment is to assess the risk from the planned fieldwork to the health and safety of persons carrying out the fieldwork and to identify control measures designed to reduce the risk to as low a level as possible. It is important that each risk assessment is completed by a person or persons who have the necessary knowledge and experience so as to be able to identify potential hazards and the control measures required to reduce the risks to participants from these hazards. Such persons must consider the activity being undertaken, the location in which the work is being carried out and any other relevant issues, such as security considerations. As a result, many fieldwork risk assessments are complex multifaceted documents, hence the need for their completion by competent persons or teams of persons.

Persons completing a fieldwork risk assessment should:

- Identify the hazards associated with the fieldwork, taking into consideration for example the location, the planned work, travel arrangements, *etc.*
- Consider where relevant the level of physical fitness required to undertake the fieldwork.
- Evaluate the associated risks and consider who might be harmed and how, paying particular consideration to those who may be at greater risk, such as solo fieldworkers or those lacking experience.
- Identify control measures that can be taken to reduce the risk. Risk control measures may be procedural in nature, could involve the use of specialist equipment, be based around participant training, *etc.*
- Identify minimum training requirements for participants. Depending on the nature of the fieldwork this could extend to include training in self-care and wellness techniques, especially if a fieldworker is to operate away from home in an isolated setting for an extended period of time.
- Assess the residual risk rating after the implementation of all risk control measures and decide if it is acceptable.

The same risk assessment may be used to manage safety for repeated routine fieldwork as long as the nature of the work and the environment remain broadly static. However, in many instances the variability between differing fieldwork expeditions will require a unique fieldwork risk assessment for every trip.

Where deemed appropriate by a risk assessment it may be useful for a fieldworker to nominate a responsible person with whom they can make regular contact at prearranged times during their fieldwork in order to confirm that they are safe and well. This may be useful, for example, when persons are working in remote locations, when solo fieldwork is taking place, or when fieldworkers are dealing with members of the public or research subjects in the field and there is a concern about their personal safety. If such a protocol is to be used as a risk control measure, then the responsible person must be clear as to how a failure to 'check in' by the fieldworker should be responded to. The responsible person must also have clear instructions from the organisation as to how to respond if they are concerned for the fieldworker's welfare, for example if the fieldworker is experiencing mental health challenges or has reported a crime against their person.

Whilst risk assessments lay down steps to minimise risk to fieldwork participants, they should not restrict the initiative of individual fieldworkers or group leaders 'on the ground' in responding to the prevailing conditions or circumstances. In particular when operating in rapidly changing, potentially high risk environments, participants should have the necessary experience and training such that they can respond to changing conditions or circumstances by engaging in dynamic risk assessment. In this context dynamic risk assessment is defined as the continuous process of monitoring and identifying emerging hazards and taking action to eliminate or reduce risk in a changing environment. For example, a researcher plans a trip abroad and arranges with their in-country collaborators to have access to a driver and an off-road vehicle. When they arrive the off-road vehicle is in a poor state with broken lights, no seat belt and obvious damage. Do they still use the vehicle? What if their project requires them to work with this group and this is all they have available? Fieldworkers need to be able to respond to and dynamically risk assess emerging scenarios.

8.3.2.2 Emergency Planning

All fieldworkers should prepare an emergency plan for all reasonably foreseeable emergencies in advance of undertaking fieldwork. Emergencies may include:

- personal injury
- assault
- theft
- loss of way
- isolation
- first aid requirement or medical emergency
- vehicle accident
- engine or equipment failure
- loss of group member
- bad weather
- civil unrest

The extent of any emergency plan should reflect the nature of the work being undertaken, but also the location where the fieldwork is being carried out. In some cases, the actual fieldwork tasks could be low risk but could, for example, be undertaken in an area where crime or civil unrest is a threat.

At a minimum an emergency plan should include at least the following:

- Details of foreseeable emergencies and the actions that must be taken in the event that such an emergency arises. For some fieldwork expeditions a basic emergency response plan can be developed, which can then form the basis for more detailed case specific responses, *e.g.* a general emergency response to crimes against fieldworkers can be developed, with more detailed information added for specific incidents such as kidnap and ransom demands.

- Details of the emergency response equipment that must be carried in the field and when and how it should be used.
- Contact details for external emergency services and other supports available locally and how to contact them.
- Details of the means of summoning help that must be available when on fieldwork. For remote locations the use of a mobile telephone may not be sufficient, consideration to the use of satellite phone, VHF radio, distress flares, *etc.* may be required. The use of 'travel safety' Apps may also be useful. For some scenarios something as simple as a whistle can be useful in summoning assistance.
- A protocol for contacting the home HEI/RI and the next of kin of participants where relevant.
- A full expedition itinerary.

Ideally the emergency plan should be lodged with an appropriate person/office in the home HEI/RI with an agreed protocol of what steps must be taken in the event that an emergency occurs, or the fieldworker fails to make contact or return from the field.

Before travelling abroad to undertake fieldwork persons should familiarise themselves with their destination country. They must be aware of local cultural considerations and customs and be comfortable with same. An assessment of the risk of crime, civil or political unrest and disease outbreak will also be required for some locations. The information collected should then be used to inform the fieldwork risk assessment, minimum training requirements, predeparture health checks, the emergency plan, *etc.*

Together the fieldwork risk assessment and the expedition emergency plan form the basis for the management of safety on a fieldwork expedition. In order to assist fieldworkers a HEI/RI may decide to issue a template for both of these documents, which will allow the organisation to ensure that all of the issues that it considers pertinent with respect to fieldwork safety are considered and documented. For example, the hazard identification and risk assessment section of the fieldwork risk assessment template issued by University College Dublin (the author's home HEI) requires persons to consider physical hazards, biological hazards, chemical hazards, manmade hazards, personal safety, environmental impact of the fieldwork and any other hazards that may be relevant.

8.3.2.3 Formal Preapproval of Fieldwork

Notwithstanding the need to prepare risk assessments and an emergency plan for all fieldwork, a HEI or RI should implement a formal approval process for at least some types of fieldwork. Fieldwork in certain locations/of a certain type (*e.g.* offshore marine fieldwork) should always require an additional formal approval of some nature before commencing. Certainly, lone or sole fieldwork should always be subjected to a robust hazard identification and risk assessment process and be approved by the highest level of management within an institution.

This formal approval could be given by a centralised health and safety function, by a departmental safety committee/group, or by the head of the relevant academic or research department. Any such review and approval process should be based upon a review of the fieldwork risk assessments and emergency plan and the quality and efficacy of same in the context of the nature of the fieldwork being carried out.

8.3.3 The Participants

Fieldwork participants should have a minimum level of training and information so as to allow them to undertake their tasks in a safe manner and also to be able to recognise the types of hazards that they are likely to encounter in the field. The level of training and competency required by an individual fieldworker will depend on the nature and complexity of their working environment, the task at hand, and the types of equipment in use. In addition, those persons that lead groups of persons such as undergraduate students in fieldwork or who engage in solo fieldwork must be competent to do so safely.

8.3.3.1 The Fieldwork Leader

When fieldwork is undertaken in groups each expedition should have an appointed leader(s). It is the responsibility of this person(s) to ensure that the fieldwork is undertaken with due regard paid to the safety of all who participate. Fieldwork leaders taking inexperienced persons into the field have a significant duty of care to these individuals.

In group fieldwork the leader-participant ratio can be important. How many persons can one group leader manage? Is a deputy required? Is the ratio affected at all by the nature of the fieldwork and the experience of the fieldworkers? Selecting the right leader to lead the right size group of fieldworkers can become very important if an emergency situation were to arise and the group were to rely upon their leader's competency and experience to protect their safety.

The fieldwork leader has the task of overall field supervision. They must ensure that there is a detailed knowledge and understanding of safety measures amongst the group. Fieldwork leaders must be competent and experienced in the relevant area of fieldwork activities, must ensure that a risk assessment of the fieldwork has been carried out, an emergency plan has been prepared, a safe system of work has been established, and that roles and responsibilities are clearly understood.

8.3.3.2 Fieldwork Participants

All fieldwork participants must take steps to ensure their own ongoing safety and that of those who may be affected by their actions. They must adhere to any direction given by a group leader and have access to the necessary information and training so as to allow the work in a safe manner.

All persons planning to engage in fieldwork must be physically capable of coping with the conditions likely to be encountered. Fieldwork participants must inform the expedition leader of any medical condition that might affect their ability to undertake fieldwork.

8.3.3.3 Solo Fieldworkers

Lone working during fieldwork should be the exception rather than the norm, and ideally should only take place after being approved by the highest level of management within an institution following a very robust hazard identification and risk assessment process. The default position and policy should be that lone work is only permitted in exceptional circumstances.

Solo or independent fieldworkers are responsible for ensuring their own safety in the field, and as such require a much greater level of training and knowledge than a participant in a group expedition. Given the nature of solo fieldwork it is imperative that no activities take place that may place the worker at risk of incapacitation, as in the event that this happens there may be a significant delay in summoning assistance.

8.3.4 Fieldwork Risk Assessment and Emergency Planning – Key Safety Considerations

The following listing details what I deem to be key safety considerations when planning and risk assessing fieldwork. This listing is not intended to be an exhaustive checklist of fieldwork hazards or considerations, but rather reflects the type of issues and questions that a person completing a fieldwork risk assessment and emergency plan should consider.

8.3.4.1 Transportation Issues

- How does the fieldworker travel to their destination?
- Do they use public or private transport?
- Do they use their own vehicle?
- Does the HEI or RI provide a vehicle?
- Is specialist equipment such as a tracked vehicle, quad bike or a boat used?

- Will the travel involve driving abroad?
- Will third parties be hired to drive?
- Will there be a need to tow a trailer?
- Is there bulky or high value equipment to be transported?

8.3.4.2 Communications

- How does the worker communicate with others involved in the fieldwork and those back 'at base'?
- Is communication *via* a cell phone or radio for example or is it *via* a third party?
- Will all communication devices function in the fieldwork location?
- How does a participant or a leader summon help in an emergency?
- Are backup communication devices required?

8.3.4.3 Fieldwork Tasks

- Is the fieldwork physically arduous?
- Do participants require specialist equipment?
- Do participants require specialist training?
- Will persons be working alone?
- Will workers be vulnerable?
- Will workers be in contact with potentially unstable or unpredictable individuals?
- Will persons be working in an unpredictable environment or one prone to rapid changes in conditions?

8.3.4.4 Equipment Issues

- Are there any safety critical pieces of equipment that are needed?
- Does safety critical equipment need regular inspection and maintenance?
- Will electricity be used and if so, how is it sourced?
- Will petrol/diesel fuels or other chemicals be used?
- Can all equipment be manually handled easily?
- Does it require significant effort to move and use the equipment?
- Is the equipment being left unattended at any time?
- What is the response if a piece of equipment fails or malfunctions?

8.3.4.5 Clothing Considerations

- Is every participant's clothing suitable for the terrain and the conditions likely to be encountered?
- Do workers need specialist or waterproof footwear?
- Is clothing likely to get wet and if so, is quick drying clothing required?
- Is spare clothing required?

- Do workers need to consider weather extremes?
- Do participants need to keep their hands and head warm?
- Is participant clothing sufficient to spend a night in the field if an emergency arises?
- Is all outer clothing of a high visibility?
- Do participants need specialist 'clothing' such as a dry suit or a life jacket?

8.3.4.6 Health Hazards

- Will workers be exposed to extremes of temperature?
- Is sunscreen required?
- Is thermal clothing required?
- Do participants need to consider biological agents such as Lyme disease, Leptospirosis, *etc.*?
- Will workers come into contact with any dangerous flora or fauna, *e.g.* agricultural animals, blue green algae, *etc.*
- Are vaccinations or immunisations required (*i.e.* are they mandatory) or are they recommended (*i.e.* considered to be good practice)?
- Is prophylaxis medication required for specific diseases such as malaria?

8.3.4.7 Fieldwork Abroad

- Do participants need travel insurance and/or visas?
- Will participants have easy access to medication and healthcare when abroad?
- Is a medical or dental check-up required before travelling?
- Will there be a need to drive when abroad?
- Are all participants aware of local laws and customs?
- Can persons eat and drink freely or do they need to avoid certain sources of water and food types?
- Is the region stable?
- Is the region currently subject to a disease outbreak of any sort?
- Are there seasonal weather or other phenomena that could increase the risk to fieldworkers?

8.3.4.8 Does the Fieldwork Involve Dealing or Interacting with Members of the Public?

- Are fieldworkers dealing with persons who may be unstable or unpredictable?
- Have suitable meeting venues in which to meet subjects been identified?
- Is there a protocol and a process in place for tracking fieldworkers' movements *via* the use of regular 'check ins' with a responsible person?

- If a worker doesn't 'check in' on time does the responsible person know what to do?
- Are workers familiar with the location of the meetings/interactions, such as exit routes, sources of assistance, *etc.*
- Do persons have ease of exit at all times?
- Will a fieldworker ever be alone with a subject/person?
- Is crime or antisocial behaviour an issue?

8.3.4.9 The Location

- Is it at home or abroad?
- Does the working location itself pose a risk?
- How is it accessed?
- Is it managed by a third party?
- Is it dynamic in nature?
- How will a fieldworker communicate from the location?
- Is it secluded?
- Is crime an issue?
- Is it prone to bad weather or other natural phenomena?
- Will workers have limited control over adjacent activities?
- Are there risks from the activities of third parties?

Table 8.3 gives examples of three different fieldwork environments and associated safety precautions (extracted from the University College Dublin Fieldwork Safety Guidelines).

Table 8.3 Sample fieldwork locations and safety considerations – extract from the University College Dublin Fieldwork Safety Guidelines.

Elevated Land – Safety Considerations

- Avoid solo work in mountainous areas.
- Move carefully over wet grass especially on slopes as it can be very slippery.
- Move carefully over rough or rocky terrain avoiding loose boulders, burrows *etc.*
- Never run down screes or steep hills and take care not to dislodge loose rocks or other objects.
- Protective headgear should be worn where there is a danger of falling stones. If necessary shout a warning to those below.
- Take particular care in areas where uneven compaction may lead to instability *e.g.* tips, spoil heaps, *etc.*
- Avoid becoming tired, cold or wet.
- Observe weather conditions continuously as deterioration can be rapid.
- In conditions of poor visibility, constantly refer to a compass and make progress cautiously. Stop frequently and each time take a sight on an object ahead.
- If immobilised for any reason, put on all spare clothing and summon help.

Table 8.3 (*continued*)

Woods and Forests – Safety Considerations

- Work in woods and forests is tiring and there may be difficulty in movement and there may also be limited visibility. Fieldwork plans should take account of this.
- If lost in a woodland or forest, backtrack.
- Avoid areas where growth is dense and the nature of the ground and any obstructions or holes are obscured.
- Watch for whiplash of branches.
- Rocks and boulders and fallen trees in forests frequently bear a covering of moss and are slippery when wet.
- Avoid screes in forests whenever practicable.
- If climbing slopes, take care not to rely too heavily on vegetation for support. It may not be firmly anchored.
- Do not smoke or light camp fires at times of high fire risk, or as decreed by local regulations.
- If applicable, do not park your vehicle in tracks or fire breaks.
- Do not climb trees.

Prisons – Safety Considerations

- Always book visits well in advance.
- Fieldworkers should be aware that they may be subjected to a personal search before entering restricted areas.
- Always ensure that you are accompanied by a member of staff when dealing directly with a prisoner or moving in a restricted area.
- Bring as little as possible into the prison.
- Leave wallets, jewellery, phones, *etc.* outside the prison/restricted areas.
- Do not bring anything into or out of a prison for a third party.
- Wear suitable clothing – seek advice if unclear.
- Wear any issued ID at all times.
- Treat all persons you encounter with respect.
- Adhere to all instructions given by prison staff.
- Do not enter into any disputes with a prisoner.
- If feeling unsafe at any time remove yourself from the area.

Table 8.4 A non-exhaustive list of fieldwork safety equipment.

Map & compass	Personal alarm	VHF radio	Whistle	Tide tables
Dry suits/Wet suits	Camping knife or similar	Mobile telephone	Safety glasses/face shield	High visibility clothing
Lifejackets/buoyancy aids	Waterproof watch	Hard hat/bump caps	Safety harnesses	Safety boots
Torch & spare batteries	First aid kit/Thermal blankets	Wading poles	Ropes	Sunscreen
EPIRB (Emergency Position Indication Radio Beacon – marine field work only)	Thermos flask filled with a hot drink/cold drink	Emergency overnight rations (*e.g.* chocolate, glucose tablets)	Hand held/airborne distress flares	Tents/Ponchos

8.3.4.10 *Safety Equipment*

In addition to the equipment required to carry out the actual fieldwork it may be necessary to carry further equipment for safety purposes. Table 8.4 lists the various types of safety equipment that a fieldworker may wish to consider.

8.3.5 Final Checks

Before departing for fieldwork operations there are a number of final checks that can be made as required. These include:

- checking weather forecasts
- ensuring you have landowner permission
- completing final equipment checks
- ensuring all persons are well enough to travel
- reviewing maps and work plans
- charging equipment including mobile telephones
- reviewing risk assessments
- rechecking emergency plans
- ensuring your contact person is informed
- undertaking final participant briefing

8.4 Endnote

Whilst for HEIs and RIs managing safety in laboratory and classroom settings can be challenging, the institution to a significant degree retains control over the generation of hazards within such environments and can take steps to control and curtail their occurrence. However, when persons leave the classroom environment and enter the wider world they have less control over their environment, and consequently difficulties in managing their safety can arise.

In this chapter I have considered three areas outside of the classroom that pose challenges for HEIs and RIs in managing safety. I chose sporting activity because in the almost twenty years I have been involved in managing and advising on safety in a university it is the only activity that I have first-hand experience of that has led directly to the death of a student; I chose work placements/internships as a subject because when students and researchers are engaged in such activities the HEI/RI loses control to a significant degree over their working environment and working activities and consequently their safety; and finally I chose fieldwork because I consider it to be one of the most dangerous activities that any academic, student or researcher can engage in.

I have not set out to provide detailed granular advice on addressing the hazards associated with the aforementioned activities, rather I have drawn on my own experiences to highlight what I consider to be the key aspects of managing safety in these areas, which I hope will be of use to safety practitioners and other stakeholders who are tasked with ensuring student and researcher safety outside of the classroom.

References

1. The Mirror Newspaper Online (UK), *Hockey Player and 'perfect Son' Dies after Being Hit on Head with a Stick in Freak Accident during Training.* Available from https://www.mirror.co.uk/news/uk-news/hockey-player-perfect-son-died-6994717, last accessed on 39 March 2020.
2. The Daily Mail Newspaper Online (UK), *Following the Death of Lizzie Watkins, Why Does Hockey Refuse to Protect its Players?* Available from https://www.dailymail.co.uk/sport/othersports/article-2873726/Following-death-Lizzie-Watkins-does-hockey-refuse-protect-players.html, last accessed on 30 March 2020.
3. The Independent Newspaper Online (UK), *Fourth Rugby Player Dies in France in Space of Eight Months after Losing Consciousness Following Tackle.* Available from https://www.independent.co.uk/sport/rugby/rugby-union/news-comment/nathan-soyeux-dead-fourth-french-rugby-player-dies-nicolas-chauvin-france-a8721256.html, last accessed on 30 March 2020.
4. Atlanta News Now Online (USA), *Student Boxer Dies in Amateur Charity Event, after Collapsing during Fight.* Available from https://www.ajc.com/news/national/student-boxer-dies-amateur-charity-event-after-collapsing-during-fight/E8oiPRVOpHTePwGxWJBxYO/, last accessed on 30 March 2020.
5. BBC News Online (UK), *Phillip Hughes Inquest Says Death was 'Inevitable' After Hit.* Available from https://www.bbc.com/news/world-australia-37604252, last accessed on 30 March 2020.
6. Sky News Online (UK), *'Extremely Popular' Teen Footballer Dies in 'freak Accident' on Pitch.* Available from https://news.sky.com/story/extremely-popular-teen-footballer-dies-in-freak-accident-on-pitch-11578072, last accessed on 30 March 2020.
7. Health and Safety Executive UK, *Amateur Sports Clubs: Guidance on Running a Safe Sports Club.* Available from https://www.hse.gov.uk/entertainment/leisure/amateur-sports-club.htm, last accessed on March 30th 2020.
8. Royal Society For The Prevention of Accidents UK, *Sport, Leisure and Water Safety.* Available from https://www.rospa.com/School-College-Safety/Teaching-Safely/Sport-Leisure-Water, last accessed on 30 March 2020.
9. Irish Marine Casualty Investigation Board, *Report of the Investigation into the Death of Kayaker on the River Gaddagh, Co. Kerry on 28th December 2007,* Marine Casualty Investigation Board, REPORT No. MCIB/155. Available from https://www.mcib.ie/_fileupload/Documents/reports/River_Gaddagh_revised_verson_for_website.pdf, last accessed on 30 March 2020.

10. Strange Behaviours Website, *The Wall of the Dead: A Memorial to Fallen Naturalists*. Available from https://strangebehaviors.wordpress.com/2011/01/14/the-wall-of-the-dead/, last accessed on 30 March 2020.
11. Universities Safety and Health Association, *Higher Education Sector Guidance on Health and Safety in Fieldwork and Travel*. Available from https://www.usha.org.uk/images/stories/files/guidance-documents/MASTERUSHA_Safety-in-Fieldwork-Guide.pdf, last accessed on 30 March 2020.

CHAPTER 9

How to Do Health and Safety Right from the Beginning

M. ULUSOY*[a], J. PRESLAND[a], J. BROWNLEES[a], E. HOWLEY[a], S. GILLOTIN[a], T. LEPKO[a], V. JAGANNATH[a] AND F. H. MARSHALL[a]

[a]MSD, London Bioscience Innovation Centre, 2 Royal College Street, London NW1 0NH, UK
*E-mail: mehriban.ulusoy@msd.com

9.1 Introduction

Merck & Co., Inc., Kenilworth, NJ, USA, known as MSD outside the USA and Canada, is an American multinational pharmaceutical company. The MSD UK Discovery Centre is an expansion of its cross-disciplinary drug discovery model into Europe with its interim facilities located in central London. The concept of the London Discovery Centre was to establish a nimble research group built on the premise of a flexible and exciting biotech company while being part of a multinational pharma company with very established and regulated practices. The Discovery Centre consists of a multidisciplinary team of cellular biologists, pharmacologists and medicinal chemists with a shared purpose and commitment to highly principled and rigorous early bioscience research to drive medical advances against some of the most challenging diseases facing ageing populations. Starting up the Discovery Centre with new research facilities and scientific and support teams has been a great challenge, in particular seeking to balance the benefits of a biotech company within the legal and health and safety framework of a large pharma. In this

Challenges for Health and Safety in Higher Education and Research Organisations
Edited by Olga Kuzmina and Stefan Hoyle

Published by the Royal Society of Chemistry, www.rsc.org

book chapter, we have focussed on the challenges we have experienced from a health and safety perspective, how we tried to overcome them, and the key lessons learned.

Multinational organisations generally govern environment, health and safety (EHS) by a centralised EHS department which establishes companies' global EHS strategies and practices and monitors site level compliance against those standards. The global standards provide a certain level of flexibility for the incorporation of specific national and local legislation. However, the expectation is that all sites will meet company standards as well as national requirements. This is key to achieve and maintain the highest standards of EHS with equal importance and consistency globally across all operational units from research and development, clinical studies, manufacturing, supply chain, sales, and patient care.

9.2 Challenges of Getting EHS Right from the Beginning

EHS is a discipline and speciality that works to protect the environment and the health and safety of personnel at work. Safety assessment of working practices and the introduction of safety policies and guidelines are of the utmost importance in achieving these goals not only from a moral obligation but also to ensure adherence to legislation. The five pillars of occupational safety and the advantages to both employee and employer include (a) moral and humanitarian obligation, (b) legal and regulatory, (c) economic and financial, (d) risk management and corporate governance and (e) occupational safety as a human right.[1] The consequences of not implementing appropriate safety control measures can be very significant, including fines for the organisations and in extreme cases the potential of imprisonment of company management should staff be exposed to harm or a more serious incident. Therefore, getting EHS right from the beginning is essential.

What is required from a safety perspective to enable laboratory-based research at a new research site and where should you begin? These are the questions our first staff were faced with when establishing the site and we believe they will be strongly relevant to any new research sites. It is crucial to get support from corporate safety professionals with in-depth knowledge of global EHS standards to help the start-up site from the time the lease on the premises is agreed. This will help the introduction of company EHS standards relevant to the business in order of priority and help identify the level of support needed from local/national EHS professionals for concurrent introduction of national regulations. For instance, the UK's Health and Safety Executive (HSE) requires new research sites to notify them of the intended use of biological agents including genetically modified organisms (GMOs) from the inception of the new premises. Support from a biosafety expert at this point is very important to help the site to make the notification to the HSE in the most compliant and effective manner, the nature of the notification being dependent on the vision and intended use of the site. In our case, the intended use of the laboratory suite was to house a team of cell and molecular biologists and associated specialist equipment to study the underlying biological mechanisms of ageing, with particular reference to neurodegenerative diseases, and to engage in early stage target identification and drug discovery to feed into the MSD Neuroscience portfolio. There was a strategic aim to focus on human biology informed by genetics as well as human samples including cell lines, induced pluripotent stem cells (iPSCs) and tissues when possible. Cell biology often involves the use of transfected cells and so there was a need to form an EHS committee in order to appropriately assess the risk level of the proposed work with GMOs and advice on suitable containment measures.

For the EHS committee to perform the function of assigning containment levels, biological risk assessments had to be carried out initially for planned experimental procedures such as culturing cells and running analytical techniques, for example, Western blots, with information on suitable waste streams (including disinfectants) and safe compound handling practices which falls under the Control of Substances Hazardous to Health Regulations 2002 (COSHH). The corporate biosafety manual was adapted to UK specific requirements with support and advice from corporate biosafety experts and site EHS professionals. Relevant corporate risk assessments covering basic biological experimental procedures were utilised as a starting point, which later were revised and improved to meet local and site-specific needs. For instance, the selection and use of disinfectants, one may assume is fairly straightforward, however, we found this particularly taxing as we had to adhere to the landlord regulations in terms of effluent waste requirements and at the same time to comply with corporate industrial hygiene standards. We managed to overcome this challenge by collaborating with landlord and company subject matter experts in identifying and sourcing disinfectants that were approved by both parties and compliant with national environmental regulations. After we managed to source an approved disinfectant it became possible to develop our waste management procedures.

Once the EHS committee was established and reviewed the proposed GMO work, the corporate biosafety manual was adapted to local rules and waste streams were implemented according to landlord requirements, we were able to notify the HSE for first contained use of the site. Without HSE notification, the majority of the cell biology work could not have taken place.

Once the necessary regulatory notifications and safety documentation are in place training of staff is of vital importance in order to allow work to begin and this should be approached in a structured manner starting with the most critical processes such as emergency procedures, waste streams, risk assessments for experimental procedures, and safe use of chemicals, biological agents and laboratory work equipment. Most of the core safety training courses are currently provided online but some are still required to be delivered in person such as use of centrifuges and blood-borne pathogen safety awareness training. It is very important to have a structured training programme that is tailored to the permitted work and staff are trained in a timely manner so that newly acquired knowledge can be used when it is most needed. Allocating laboratory supervisors, super-users of equipment and including scientist representatives on the EHS committee helped in balancing the workload of training and ensuring good working practices were adhered to. However, since the site was built from nothing this could only be done as suitable staff were recruited. This naturally led to a delay between recruiting staff and getting lab work up and running. In the long term, building the team in parallel with safe working practices helped to create and maintain the positive safety culture throughout from the outset.

Support from EHS professionals with in-depth knowledge of company standards as well as national legislation and regulatory requirements applicable to a research environment was critical at all stages of setting up the site EHS management systems. Recruitment of EHS professionals with the right skill sets and in-depth understanding of the research and development environment can take quite a lot of time for new sites. As an interim solution we advise recruitment of local consultants and/or subject matter experts who can liaise with corporate EHS professionals to bring quick and effective solutions to high priority tasks ensuring a safe and compliant start to experimentation. At the very start, it was critical to bring together corporate biosafety subject matter experts with the scientists to progress basic Containment Level 1 (CL1) cell work. Later on, more advances were made once the site-based EHS professional was recruited. This individual issued further HSE notifications for Class 2 GMO use, developed and streamlined risk assessment processes, set up a more advanced EHS committee with wider representation from all disciplines to increase employee engagement at the start, and devised structured document storage, record keeping and safety communication platforms. A significant part of the role also included the introduction and adaption of corporate EHS standards to the UK site and inclusion of global procedures for reporting incidents and accidents and sharing associated lessons learned.

The time and effort required to put in place safety procedures and documentation for a new site should not be underestimated and should be initiated as soon as possible to avoid delays in starting scientific operations. The London site is now seen throughout the company as an excellent example of how to get laboratory work up and running rapidly in the safest manner. This was quite an achievement and is the result of the efforts and commitment of all those involved in the very early days. Getting EHS right from the beginning is instrumental in instilling a safety culture which becomes a way of work life and leads to a very safe and productive working environment.

9.3 Challenges Arising from the International Nature of Corporate Organisations

In common with many industries, the pharmaceutical industry has undergone dramatic globalisation over the last twenty years. As well as modifying global footprints many companies have also expanded and diversified their networks of, for example, collaborations and contracted-out activities. These changes have impacted all stages of the drug discovery and development process from initial screening and compound synthesis to clinical trials and active pharmaceutical ingredient (API) production. It is within this complex and fluid environment that effective workplace safety processes and procedures need to function.

When thinking about enacting EHS programmes internationally it is important to consider not just the legislative requirements but also cultural factors too. Organisations establish EHS processes specifically designed to address the statutory requirements of the authorities in whose jurisdiction they operate. In theory, expanding to new locations just increases the administrative burden, however, in reality, this is an oversimplification as it does not account for the 'human component' present in drafting and implementing EHS legislation and procedures.

In this section, we focus on the factors that impact the implementation of EHS processes within a new research organisation, but many concepts are applicable to those people and procedures involved in drafting policies and implementing the EHS laws that regulate how companies operate.

The HSE produced a report[2] that showed that internationally health and safety is enshrined in specific safety legislation rather than in corporate law, but that unsurprisingly these regulations were significantly different between countries. This study, focussing on the legislative burden on company directors and executives found stark differences but that these did not correlate with differences in their safety records. While this study was very specific in its remit it is a good example demonstrating that while different nations agree that EHS needs specific standalone legislation, each country enacts rules and regulations that suit their own particular needs.

Just comparing the EU and US systems it is clear to see that they both have multiple layers of bureaucracy, co-ordinating differently. The EU-OSHA is an overarching organisation that empowers and educates national safety regimes, but each member nation is responsible for enforcement, having national regulators akin to the UK's HSE. Conversely, in the US the principle authority, OSHA, can create regulations and administer fines nationally. In addition, states and cities have their own powers to implement local safety rules and associated penalties. All these nuances are important to understand in establishing a new site in a multinational organisation. More specifically, there are clear differences in how different authorities approach similar issues, a good example is the approach of UK and US authorities to GMOs where the regulations placed on genetically modified organisms in the UK are much stricter than those in the US with more robust notification requirements to the competent authorities to protect human and animal health and release of GMOs to the environment from contained use.

A strong safety culture within an organisation, that is the safety-related component of organisational culture,[3] is critical for the successful implementation of EHS policies. This safety culture is a set of shared norms and values that directly impact safe working practices. With the assumption that an organisation should only have a single, unified, global safety culture, it is important to ask, 'how do you transfer and then establish this culture in a new site that may be geographically distant, culturally distinct and under different legislative regulation?'. It may be useful at this point to separate safety culture from safety climate, which is the 'surface feature of a safety culture at a given point in time'[4] or in this case a given geography as well. So why is this distinction necessary? Safety climate embodies the practical implementation of EHS procedures so is a more focussed and dynamic parameter than safety culture.[5,6] It is possible therefore to have multiple equally successful safety climates driven by site-based needs, derived from a single safety culture. These distinct safety climates may be driven by cultural as well as practical considerations.

From a practical perspective, a large established organisation will have well-developed processes and established EHS colleagues able to carry out much of the EHS work for the site. In a newly formed location, there is very often a lot of upfront work to be done requiring intensive administrative tasks, such as risk assessments, facilities checks, induction and training *etc.*, and with small locations having small teams, these roles and responsibilities may be shared, under the supervision of EHS experts. In small biotech startups, EHS policies and documentation are usually established through the engagement of external consultants as well as individuals within the company bringing documentation from previous roles or shared across other biotech companies. While these may be initially 'fit for purpose' they usually do not meet the rigour set by larger pharma companies and have to be continually updated and modified as the company develops. Biotech companies spun out of university labs that frequently have less robust EHS policies and procedures, *e.g.* with regard to personal protective equipment (PPE) due to handling hazardous materials at a lower scale, have a more difficult task than those emerging from established companies.

Influential work by Geert Hofstede[7] defined six cultural characteristics that could be measured, through surveys and questionnaires *etc.*, that provide a framework for identifying cultural differences across countries. Figure 9.1 shows these dimensions and the spectrum of behaviours associated with them.

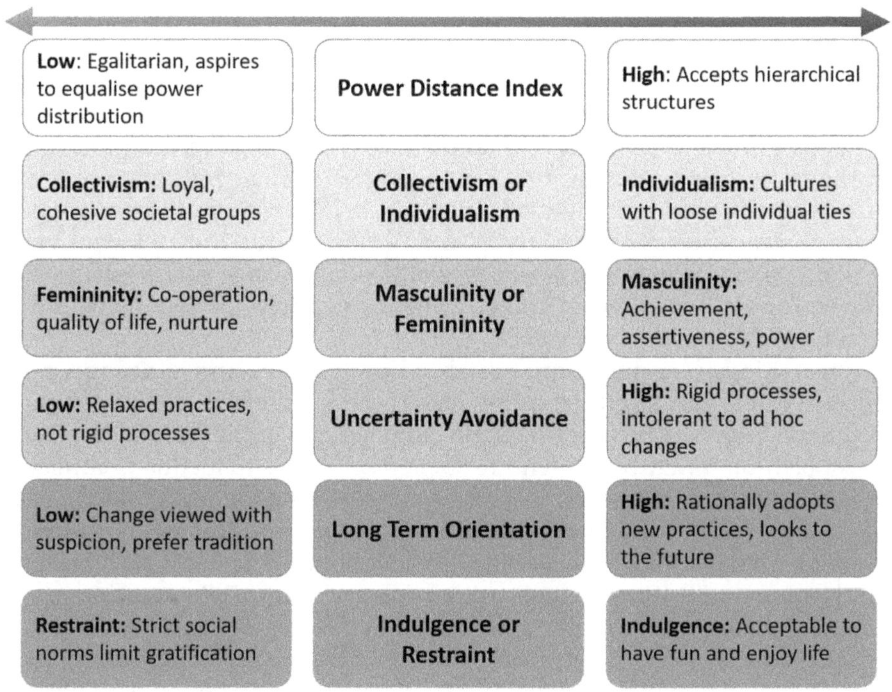

Figure 9.1 The six dimensions defined in Hofstede's model of national culture.[7] Adapted from ref. 8 with permission from CFI™ Education Inc. Copyright, 2020.

If we take a superficial view of four of the leader nations for pharmaceutical research and development, China, France, the UK and US, there are clear differences in cultural dimensions (Figure 9.2). In this example, there are two dimensions where a single nation diverges from the others. France scores high on the uncertainty avoidance index which suggests that in that culture they accept uncertainty and risk taking. Conversely, the other three countries have a propensity to minimise the unknown (*i.e.* risk) through rules and regulation. Secondly, while the western nations scored high for individualism, China demonstrated more collectivist traits. This may suggest that there are differences in the importance of personal attainment, group goals and well-being. Also, China and France share a high power distance rating indicating that they tolerate power differences, this means that they respect rank and accept bureaucracy so may find it more difficult to speak up and raise issues when they occur. The UK and US have lower scores for this parameter indicating a flat, decentralised organisational structure with shared decision making and responsibility, and while this may be associated with designing and implementing novel ideas, often it could be also linked with a lower rate of compliance.

While the concepts put forward by Hofstede and other researchers are acknowledged as having relevant and useful parameters, many consider the foundational data to be less relevant in the modern workplace. While the original data and subsequent analysis was based on a relatively homogeneous workforce at International Business Machines Corporation (IBM) in the 1970s, today's innovation-led industries such as pharma, are much more diverse and multicultural. Diversity brings with it many positives: increases in innovation, creativity and flexibility, but it also brings challenges. These challenges, such as conflict and miscommunication can impact the implementation and compliance required for EHS policies and standards.

What this means in practice is that how policies and procedures are implemented and administered needs to be tailored to each institution and each country. In places, where policies are implemented in a top-down fashion it may be difficult for bench scientists to feedback concerns or areas for improvement, but compliance would be anticipated to be high. Conversely, in a flat organisation or a newly set up biotech company, multiple perspectives may adversely affect compliance and how policies are implemented. In multicultural environments, simple solutions such as using pictograms can be a powerful way to overcome communication or language barriers. Even these, however, can be interpreted differently by different groups of people. This type of communication should be considered as an aid which supplements strong training schemes, and these should be designed to cover the range of personality types and cultural styles so should include classroom learning, role playing and interactive teaching methods.

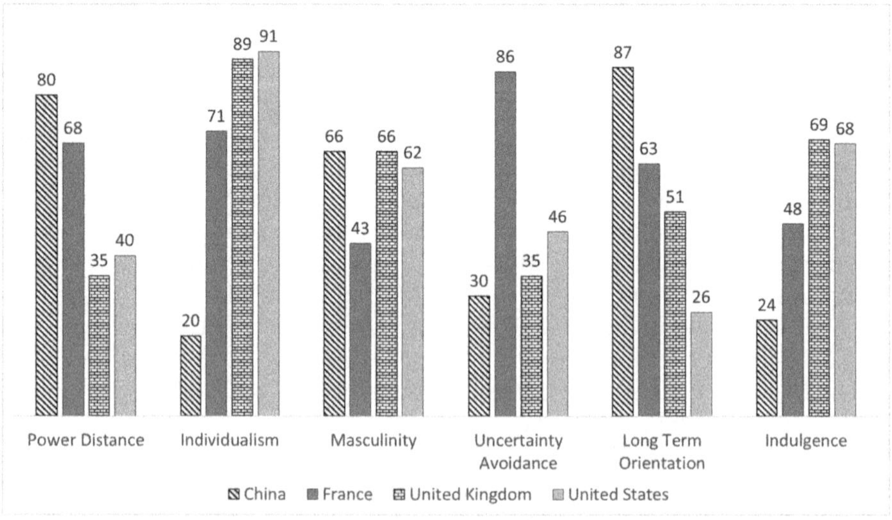

Figure 9.2 Hofstede Cultural Dimensions comparison for four major R&D intensive nations. Data from ref. 9 and 10.

Within diverse teams, increasing inclusivity through cross-cultural training can help improve group identity and it is this type of environment, built on respect and recognition that maintains a healthy safety climate. Leadership style is also an important driver of efficacy within a diverse workforce. While most managers employ a transactional style on a day to day basis, which may suit many team members, a transformational style may be more suitable in establishing a strong EHS environment in a new site with multicultural setting. This is because a transformational leader acts as a role model, a leader and mentor. In effect, a transformational leader coaches colleagues and encourages everyone to be proactive and take ownership for EHS compliance.

In summary, establishing a new site is not just a case of replicating an existing location in, usually, smaller scale but requires a bespoke solution. This solution should be a holistic combination of policies that adhere to the necessary laws, procedures, and company standards that enable compliance and an atmosphere that engenders inclusivity and respect.

We recommend gap analyses are carried out between corporate EHS standards and national regulatory requirements to identify how best bespoke solutions can be created to fill the gaps to ensure compliance while at the same time not generating unnecessary burden to the new sites, but instead, provide practicable solutions to enable scientific work. Gap assessment tools specifically developed for each applicable EHS standard such as industrial hygiene, contractor management, waste management, PPE *etc.* can provide a structured and consistent way of identifying the gaps. Based on our experience effective communication with corporate EHS policy experts, site management and site EHS representatives is very helpful to plan out a road map where a baseline is drawn, gaps are identified and prioritised based on the site's scientific strategy drives, and action plans are generated with innovative solutions.

9.4 Importance of Understanding the Science When Devising EHS Management Systems

9.4.1 Scientific Needs of an EHS Management System

The scientific needs of our new drug discovery research site can be broadly divided into two areas: chemical and biological. The needs of medicinal and synthetic chemists are directly related to the synthesis being performed, the chemical building blocks, solvents and catalysts being used, and the compounds being made; they are not related to the therapeutic areas or specific projects. Conversely, the biological needs are determined by the therapeutic areas under investigation, the pharmacological tools being used, and the biological techniques being employed.

For example, the risks associated with activities to develop antiviral therapies will be different from those associated with exploring chemotherapeutics. In the first case, using virally infected patient samples is an essential aspect of the research, in the latter case, the cytotoxic nature of the compounds is a more paramount safety consideration. It is also possible to parse the risks based on the stage of the drug discovery process, for example, scientists working to elucidate the fundamental mechanisms associated with very novel biology may utilise an array of well characterised tool compounds across a broad range of techniques. These techniques may use patient-derived samples, cutting edge molecular manipulations such as CRISPR (Clustered Regularly Interspaced Short Palindromic Repeats) or viral-mediated transduction methods, all with their own inherent risks. These techniques may be used at short notice and for brief durations (days to weeks). Equally, scientists working on later stage, chemical optimisation approaches use more established and routine assays and processes, potentially over many months rather than days, often still utilising patient samples for example, but with often hundreds of newly synthesised compounds of unknown hazard.

An EHS management system that can enable compliance without hindering project progression considering both the rapidly changing nature of research at the early stages and more iterative changes later is essential. If we consider an EHS management system not to be just a back-end repository of documents tracking policies, procedures and audits but a more holistic system, then another key component has to be the EHS professionals liaising with the ultimate end-users, the scientists. Regular communication between scientists and their EHS colleagues is key to the successful implementation of, and compliance with, EHS policies and this effective teamwork is essential to effectively assess the risks associated with new experiments as they arise.

9.4.2 Fundamental Knowledge of Scientists and EHS Personnel

It must be assumed that scientists and EHS counterparts possess the experience, skills, knowledge, and abilities that are necessary to carry out their responsibilities. However, when working within a new site neither group will be aware of the pre-existing scientific or safety culture within the larger organisation. Layered on top of this is the disconnect within the organisation when there are any legislative differences brought about if a new site is in a different jurisdiction. For scientists new to an organisation there is very often another divide, that between academic and industrial EHS. It is commonly considered that EHS in an industrial setting is more rigorous in assessing risk, implementing policies and procedures, and documentation. Trying to implement scientific procedures as both parties learn about their new organisation requires patience, empathy and good communication.

It is during this phase, when scientists are new, that they might not identify and assess hazards accurately, especially when using new methods and laboratory equipment they have not used before. New EHS staff need to become familiar with their functional area before they can effectively perform their role. Furthermore, EHS personnel with a relevant scientific background are an asset because they understand the scientific principles and very often the needs of the scientists and can try wherever possible to marry this to the EHS policies being implemented. EHS personnel without a relevant subject matter of expertise may implement EHS policies focussing on regulatory interpretations, without understanding the practicalities and feasibility of enacting these within a laboratory environment, such as implementing a chemical hygiene programme without sufficient knowledge of chemicals and their exposure routes. Therefore, collaboration between scientists and EHS personnel are important for the successful implementation of EHS policies and programmes.

9.4.3 Interaction Between Scientists and EHS Personnel

Clear and frequent communication between the scientific and EHS functions is vital to achieve and maintain compliance with EHS policies. Very often the overarching research goals are shared across the organisation in order to gain cross functional alignment. However, scientists need to be able to communicate to their EHS colleagues the practical needs within their function. Conversely, the EHS staff need to communicate the legislative and policy requirements. This interaction is enhanced if all involved can provide context and justification for their needs. Scientists are, after all, driven by data and facts, so providing them with the background to EHS policies most often enhances compliance. Similarly, EHS professionals focus on implementing safety and environmental procedures which stem from a detailed understanding of upstream processes *i.e.* experiments. In our experience, an EHS management system that permits and even encourages this bi-directional communication is fundamentally more successful than a dictatorial system, as it empowers scientists and EHS counterparts to work together rather than providing a top-down system where the EHS staff enforce and police rules and regulations. When the EHS personnel have relevant experience in supporting research operations, it is easier for them to understand the scientific needs and expectations. Then, EHS personnel are more capable of identifying EHS concerns and seek clarity using the same lexicon as scientists. Scientists are then able to respond without having to translate into non-scientific terms. Therefore, the interaction between scientists and EHS personnel is an ongoing process aimed at mutual trust and respect such that scientific experimentation can continue in the most efficient way while maintaining the highest EHS standards (Figure 9.3).

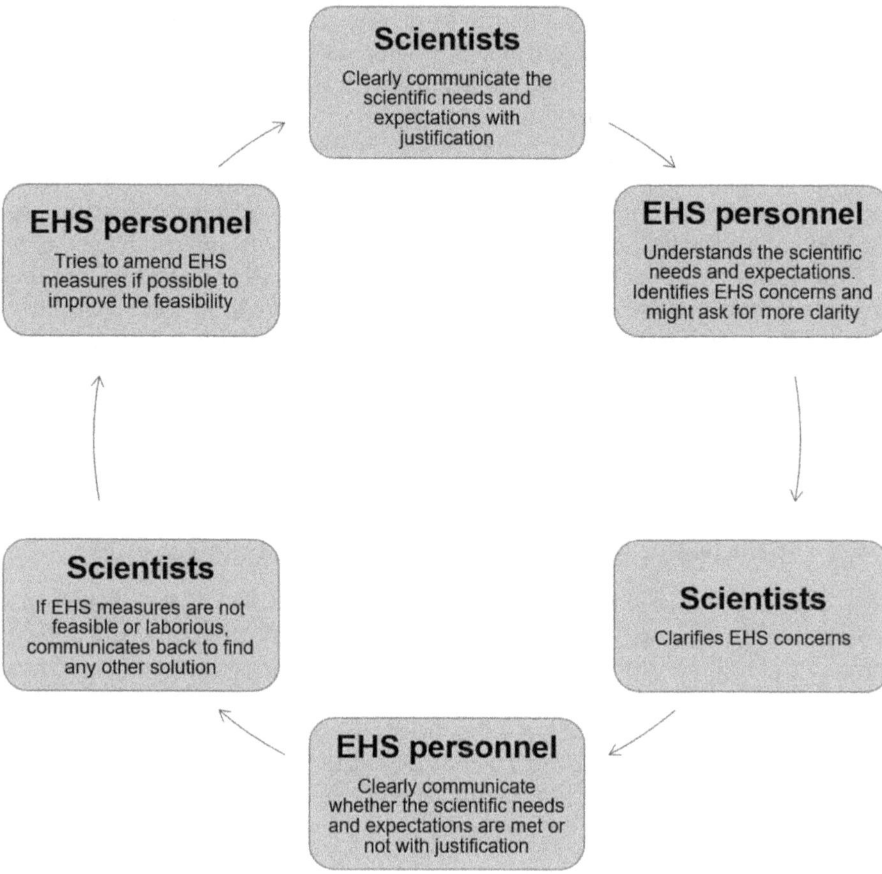

Figure 9.3 Interaction cycle between scientists and EHS personnel.

9.4.4 Compliance of Scientists with EHS Policies and Programmes

It is the most fundamental requirement of any organisation to achieve compliance with EHS legislation through the implementation of appropriate policies and procedures. However, if these policies are not clear they can be open to misinterpretation or incorrect implementation. If they are not relatively concise, they may also be ignored or skim-read. If the processes put in place as a result of EHS policies are not feasible or are very labour-intensive, productivity can be severely impacted, and compliance may also be compromised. In these instances, explaining the reasons or providing justifications often improves compliance and identifies strategies to maintain research productivity. To establish and maintain EHS compliance, it is important to implement procedures that are clear, feasible and justified. The EHS policies/programmes need to focus on end-users and stakeholders so that they can be adapted to their needs of carrying out experimental work while maintaining compliance. The EHS policies/training programmes need to be current, which should take into account changing regulations and changing scientific needs.

If the EHS personnel advise a complex or arduous safety measure to scientists without giving a clear justification, there is a higher risk of scientists not being compliant than if they understand the reasoning and justification for the additional work. Let us look at a few scenarios showing how explaining the reasons and providing justifications improve compliance and maintain research productivity with some common examples of working with potentially hazardous chemicals. Often scientists need to work with potentially hazardous chemicals which are typically purchased in powder form. From an EHS perspective the first response is to advise substituting the chemical with an alternative with a lower hazard or a liquid form consistent with the COSHH hierarchy of controls to reduce risk of exposure *via* inhalation. As often happens, these tool compounds cannot be replaced as they have a specific activity and they are not available as solutions, which often leads EHS personnel and scientists to consider the next step in the COSHH hierarchy of control measures which is to reduce the quantity. Once the residual risk is determined to be negligible or acceptable by using the powder in small quantities, such a chemical will be suitable to be used with adequate control measures.

Synthesis of novel chemical entities with potential therapeutic activities can require using highly hazardous substances including pyrophoric reagents, sensitisers such as amide coupling reagents, and toxic compounds. Understanding the mechanism of action by which these substances can pose health hazards is key to determine whether the reaction conditions desired by scientists can be adequately controlled. Hence to increase compliance with sound and effective chemical hygiene standards, it is imperative that EHS personnel and scientists work together to evaluate the risks accounting for experimental conditions that can change the risk score dramatically, such as chemical combinations, elevated temperatures and pressures, and sealed conditions. This approach will help to determine hazards arising from the experiment rather than from the individual chemicals being used. This is particularly important when unique combinations of chemicals can generate hazards that the individual chemical compounds would not produce. If those hazards are not identified correctly, it may place researchers at risk of exposure to hazardous chemicals, for example due to lack of appropriate engineering controls in place or proper personal protective equipment (PPE) to adequately protect themselves. A recent example of this is the reported case of a hydrofluoric acid (HF) burns researchers suffered at Northwestern University in Chicago, USA.[11] HF alone doesn't permeate nitrile gloves easily, however, a carbonate mixture carrying HF can rapidly pass through nitrile gloves, hence leaving nitrile gloves an inadequate PPE for that experiment.[11] Proper risk assessment is crucial to prevent such accidents from recurring. Therefore, when EHS personnel advise and support researchers on the risk assessment process, understanding the science is essential to identify the hazards and risks correctly and to implement control measures required to minimise those risks. A clear interaction between scientists and EHS personnel resulting in customisable EHS policies/programmes can improve the EHS standards to a great extent making compliance by scientists easily achievable. Large companies like MSD may have a process for reviewing hazardous reactions with their management or a group of specially trained scientists that review and approve higher risk reactions before they may be run.

9.5 Establishing a Robust Risk Assessment Process

The risk assessment (RA) process is the key step to recognise hazards of a given work activity and deliver solutions to maintain health and safety of employees and comply with health and safety regulations. We can divide the RA process into four main elements:

(a) Identifying workplace activities based on scientific needs.
(b) Carrying out risk assessment for those identified workplace activities that have inherent hazards.
(c) Communicating and sharing risk assessment outcomes including control measures required to reduce the risks to acceptable levels.
(d) Maintaining the risk assessment to ensure it remains relevant and effective.

Involvement of different stakeholders is necessary to make the RA process effective and straightforward and to deliver a clear document that will successfully identify all possible hazards and risks associated with the activities being assessed. When starting up a new site, project leaders will need to prioritise scientific projects that are crucial to initiate scientific work in the laboratories. Once those activities are decided, project leaders and scientists will need to start the RA process. To carry out a robust RA, it is important that all stakeholders have a clear idea of the whole procedure and think collectively about all the associated risks, such as use of hazardous chemicals, biologicals and work equipment. Resources from EHS representatives, corporate safety documents and national regulatory bodies (*e.g.* HSE) will be needed to identify applicable regulatory requirements to the site, which will then help to determine how those requirements translate into control measures needed. For example, work with GMOs is subject to Genetically Modified Organisms (Contained Use) Regulations 2014 in the UK. Hence, when assessing risks and identifying control measures for work with GMOs, the GMO regulations will need to be interpreted carefully to classify GMOs correctly, identify and implement corresponding engineering and administrative control measures, security measures, and notification requirements before work can begin. The fact that scientific staff likely won't be fully familiar with the corporate standards or the local regulatory requirements presents a critical challenge for the site management in avoiding gaps and potentially facing non-compliance since the site EHS management system won't be fully implemented at the start. To prevent this from happening, we have worked very closely with site and corporate EHS professionals at every stage of project planning and prioritisation, developing and implementing robust risk assessments for those projects so that compliance measures could be sufficiently devised. At our new site, the initiation of risk assessments has been an iterative exercise resulting in a streamlined process. At first, RAs would be simple and clear with all the essential regulations and safety measures to guarantee a safe start for scientific work. Corporate risk management policies and existing risk assessments for companywide common research activities were used as the starting point. We then gradually tailored them to a more thorough and comprehensive RA template to match the increasing scientific demand for new procedures and to adapt to changing hazards and risks associated with the specificity of the site, such as the risk assessment for liquid handling can initially be for manual pipetting and gradually be tailored for robotic liquid handling units as the need for high throughput experiments increases.

Having a flexible template that covers all possible and relevant safety measures is a great tool to have. For instance, the template we have devised for the use of biological agents contains eleven compulsory sections, namely project details, control measures, waste, access and storage, PPE, maintenance of equipment, emergency procedures, lone working, containment level/GM Class, training, occupational health, and three facultative sections for biological agents, namely microorganisms (non-genetically modified), GMOs, and tissues/cells/body fluids or excreta. Each section contains subsections that directs scientists to question what their procedure would need to be safely executed and allows them to consider the entire process or task *e.g.* from delivery of materials until their safe disposal. This also guides scientists on how to mitigate the risks by suggesting preventive and protective measures. As a simple example, if scientists identify a risk of exposure to hazardous aerosols during manual pipetting with cultured cells, the risk assessment template directs the assessor to the appropriate selection and use of biosafety cabinets as an engineering control measure. Addition of hyperlinks to gain access to references and available training material could be implemented to improve efficiency. Moreover, interactive tools such as qualitative risk rating tools to identify the hazard and toxicity of the chemicals would simplify this process and make it more consistent and time effective.

We recommend a review of the initial draft by the EHS professional for gap analysis. This will trigger a series of discussions to make sure the procedure is fully thought through with all the implementation of practical aspects in place for the required preventive measures. In our experience, a well-designed template expedites bi-directional communication between the scientists and EHS professionals and eases the review process, for example by EHS committee if needed for further expertise and advice.

Once the risk assessment is fully reviewed by project leaders and EHS representatives, the next step is to communicate it to all relevant end-users. The communication and implementation of risk assessments is mostly the responsibility of group leaders. However, studies have shown that the involvement of end-users and lab managers in the crafting of risk assessments makes the implementation a more effective process.[12] The ultimate goals of hazard communication are to produce an informed and involved workforce and to positively influence safety behaviour. Scientists should be able to express their knowledge, concerns and opinions that should be considered in the risk assessment. Therefore, the communication of risk assessments should be a fluid process with continual feedback from end-users. The latter is particularly important at newly established sites in order to improve the process but also for maintaining the responsiveness of a company to an ever changing operating environment. To reinforce agility, we have appointed equipment super-users as first point of contact for streamlining the needs of new RAs. Together with adjustable safety documents and clear leadership vision this structure allows every stakeholder to be proactive for new safety measures. An additional challenge that appears for new sites is the overflow of information with the expectation for scientists to read and understand multiple documents at the same time prior to starting their work. Therefore, different communicating methods tailored to the nature of the risks and to the target audience become critical to focus attention on the major points and the preventive measures. In addition to publishing risk assessment outcomes on easily accessible online platforms, there are other ways of communicating them effectively, such as participative exercises, safety posters, group discussions, and animated media.

Visual communication can be particularly effective when presenting possible risks and consequences by putting a particular risk into context. In addition to being a very direct way of showing what could happen, it could help scientists to connect with the consequences of unsafe practices. Showing images of the damage caused to and by unbalanced centrifuge rotors is a common example. Another good example of visual communication is showing good and bad safety practices in the laboratories with very clear and informative illustrations such as in the electrical safety poster designed very recently by Northwestern University as part of The Research Safety Student Initiative programme (Figure 9.4). Oral communication on the other hand has the advantage of getting immediate audience feedback, promotes discussion and ensures messages are understood effectively. Scientists communicating the risk assessment should not only be able to interpret, clarify and simplify the RA in order to make it easily understandable for any scientific community, they should also be able to influence their colleagues to change practices and their habits to effectively follow the safety standards. However, it is for the end-users themselves to make sure that the safe practices will be followed. The implementation of risk assessments also sometimes requires some adjustments to laboratory settings and equipment selection, installation and use. For example, selection of a correct type of ventilated balance

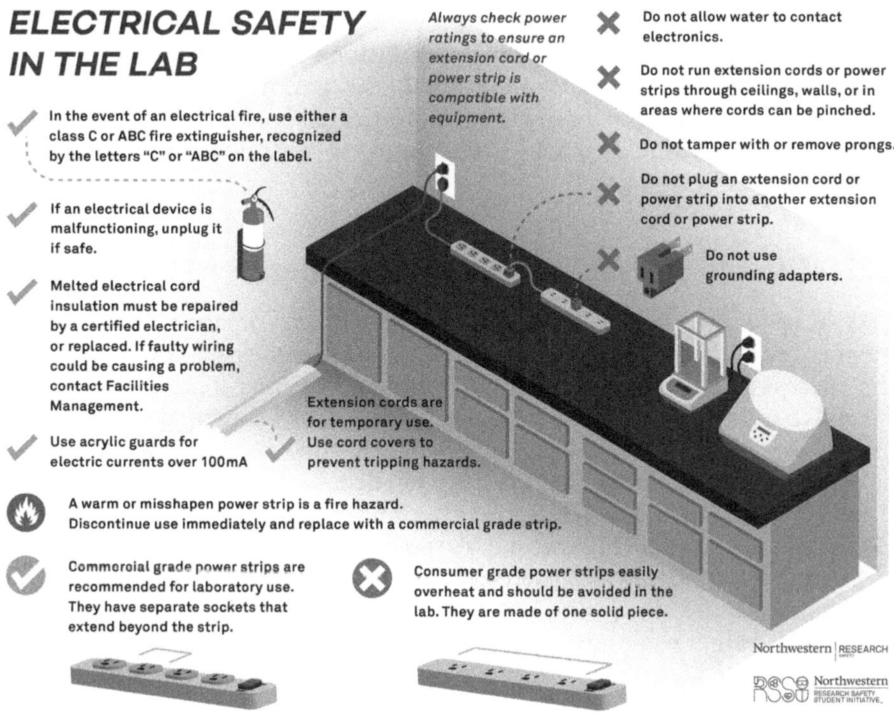

Figure 9.4 Electrical safety in the lab poster. Reproduced from ref. 13 with permission from Anna Stasek, The Research Safety Student Initiative, Northwestern University. Copyright, 2020.

enclosure (VBE) should be made based on type and quantity of solid materials to be used. Furthermore, factors such as dustiness and toxicity of the compounds will define local rules around use of PPE, waste disposal, and signage when working at VBE. Use of novel chemical entities with a high occupational exposure band (OEB) rating may require use of double nitrile gloves, disposable sleeves and lab coats and special hazard signage. Whereas use of compounds with lower OEB rating may not require as stringent administrative and PPE controls. Involving end-users to this phase of selection and use of equipment and especially engineering controls based on risk assessment outcome makes it easier to get commitment from the end-users to implementing and maintaining established safe handling practices.

The effectiveness of RA communication can be monitored by the number of near-misses and incidents. Regular laboratory inspections by line management can also be an opportunity to review effectiveness of risk assessment process since a poorly carried out RA would result in unsafe conditions due to hazards not being identified correctly and/or control measures not being suitable and sufficient. It is important to review the procedures and risk factors on a regular basis. Risk assessment updates should ideally be carried out by project leaders and scientists so that evolving scientific strategies requiring new research activities and use of state-of-the art equipment to bring innovative solutions to drug discovery can be integrated into the site EHS management system. Any changes resulting from the review process should be communicated back to the end-users to ensure they are aware of those changes and the resulting safety measures they need to follow.

9.6 Effective Stakeholder Engagement

EHS is becoming increasingly intertwined with an organisation's overall business strategy, hence the success of EHS programmes will directly impact the success of the business. Since stakeholder buy-in and participation are crucial to moving efforts forward, there is a need for a solid and smart stakeholder engagement strategy, especially when starting up the site from scratch as this is when stakeholder help is most needed.

Success in EHS is business imperative and although it is heavily rooted in compliance, its most important purpose is to ensure health, safety and wellbeing of employees and protection of the environment. In addition, for the business to be successful, EHS procedures must be correctly positioned to enhance rather than hinder research productivity. Hence EHS is unquestionably a collective effort and requires all employees and stakeholders to play their part. Safety touches a variety of different stakeholder groups, each possessing unique insights and influence and a participative model is necessary to generate support from these stakeholders.

During the start-up stage, the first step for us was identifying the stakeholder groups which played critical roles in facilitating the science. Once identified, we then established a stakeholder engagement strategy based on effective communication and mutual trust. We experienced challenges in finding the correct stakeholders at the beginning due to the complex nature of the multinational corporate structure. The physical location of stakeholders being in different countries (*e.g.* the US and the UK) added an additional layer of difficulty in communicating the business needs and developing collaborative working methods. Clear communication, prioritisation of business-critical deliverables, allocating responsible project leaders with clear roles and responsibilities to ensure successful achievement of targets have been among the tactics we have deployed.

Figure 9.5 lists major stakeholder groups we have worked intensively with when designing and implementing EHS management systems in order to initiate the research operations safely.

Each stakeholder plays an important role and one's contribution cannot be considered without the others. In order to define the appropriate nature and level of support required from each stakeholder it is necessary to firstly identify different requirements needed from national regulations, corporate and landlord standards, and how these requirements will apply to the research operations. For example, emergency showers and eyewash stations are required as part of the emergency response programme in the event of

Figure 9.5 Stakeholder engagement in implementing site EHS management systems when starting up the site from scratch.

chemical exposure. Facility management was one of the key stakeholders we collaborated with for the selection, installation, and maintenance of the emergency showers. Another key stakeholder is the HSE for notification of first contained use of GMOs. Before we notified the HSE, we had to create an EHS safety committee, carry out project biorisk assessments, and establish site procedures for working safely with GMOs. To successfully complete these tasks, we engaged with a variety of stakeholders including subject matter experts on genetic modification, corporate biosafety, facility management, and the HSE. Information technology has also been one of the key stakeholders we have engaged with when creating an online data management system on Microsoft SharePoint, to enable the provision of safety information tailored to each research department including risk assessment review processes, training records, incident reporting, and safety onboarding for new starters, *etc.*

One rather challenging area has been contractor management due to the nature of laboratory space as we operate in two different sites leased from different landlords. It has required a great deal of engagement and coordination between the landlords, facility management and the contractors managed by each to reach alignment in EHS requirements and to understand the shared roles and responsibilities. Miscommunication between these stakeholders can create significant gaps in health and safety systems, such as fire control, emergency management, environmental safety, hazardous waste control, and water management. Routine governance meetings between key management groups can help to ensure all parties are on the same page in terms of expectations and are on track to achieving common goals while maintaining clear communication and trust. Once EHS policies are put in place in organisations it is difficult to operate outside their framework. This proved a problem when significant differences occurred in the global corporate policies compared to those of the landlords governing our rented space. A very high standard was set in all cases, however the standard, while being equivalent, was not necessarily the same, such as waste management and refrigerant management requirements.

It should be expected that there will be some situations in which there is a difference in alignment between the landlord and the corporate safety standards, and in these situations, we should remember that the highest priority is ensuring the health and safety of employees and protecting the environment. Hence a common ground must be met when deviating from stakeholder standards to ensure this priority is never compromised.

9.7 Effective EHS Induction Process for New Employees

A new employee EHS onboarding process constitutes the very first step that introduces new employees to the company and prepares them for their new roles. It ensures employees are fully informed about the EHS organisational structure, where the employees are placed within that structure, their roles and responsibilities and essential health and safety information they need at the start. It also serves as a starting point to introduce a positive culture of safety in the workplace.

Therefore, we strongly believe that a fluid workflow between various stakeholders including the EHS department, security, occupational health, and human resources (HR) is essential for line/site management to communicate and promote a positive safety culture from the very first day. Introducing new starters to different parts of an organisation accelerates the process of acquainting each other in the workplace but also promotes the idea that everyone is responsible for their own safety and that of others.

Dividing health and safety inductions into general and specific categories can help line managers to communicate required EHS information effectively. General induction provides site and area related EHS and security information whereas specific induction provides role/responsibility related EHS information based on the new employee's anticipated tasks and associated hazard and risks (see Figure 9.6). This can help site management to identify stakeholder groups who will then need to develop and deliver the general and role specific inductions so that the onboarding process can be tailored effectively to the new employee's job specifications.

Onboarding new employees on a new site creates additional challenges. These mostly arise from the fact that the majority of the site EHS management systems and procedures are still in the design and implementation stage. This situation compels line managers to be even more proactive in identifying the most essential EHS information and arrangements needed by working together with the other relevant stakeholders. Based on our experience, this situation can create a convoluted process leaving the new employee confused about what is needed before they can start working safely in the laboratory. Also, the recruitment of scientists and site management at the same time adds another layer of challenges since the implementation of EHS policies and the onboarding process occur simultaneously. Line managers will require extra time and training to become fully familiar with their EHS responsibilities and what is required from them to onboard their employees safely to their new roles. We have utilised support from corporate EHS to overcome this particular challenge and provided a corporate developed training programme for senior leaders and managers to introduce to them their EHS responsibilities.

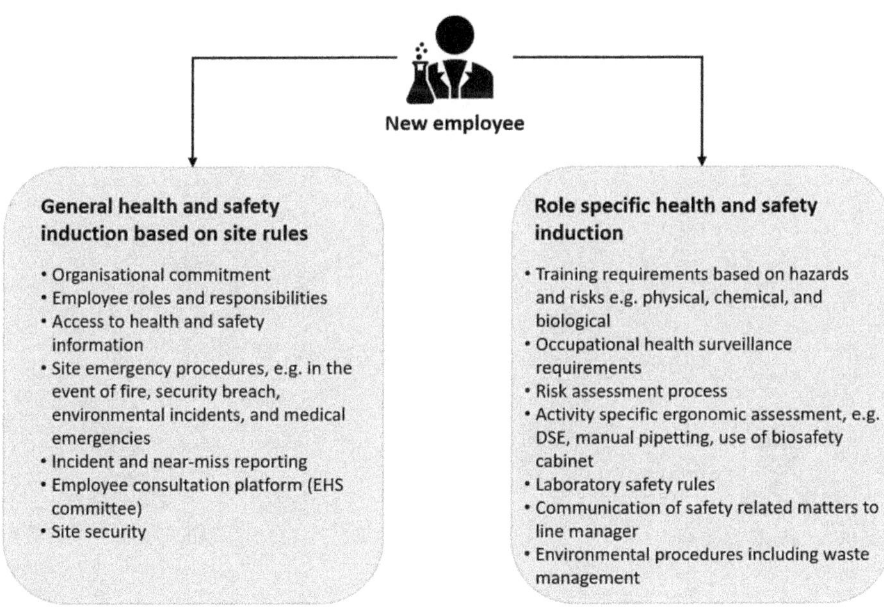

Figure 9.6 Content of new employee EHS induction process divided into general and role specific induction.

The time pressure to get a new site up and running often conflicts with the complexity of developing EHS procedures that would enable basic scientific operations in chemistry and biology laboratories. We have tried to gain additional time from the infrastructural set up of laboratories and installation of laboratory equipment to develop the new employee EHS onboarding process. However, to efficiently start, we have found that having a clear vision for the overall project that will be carried out at the new site allows managers to anticipate safety requirements while planning experimental work and associated risk assessments for new employees.

Most of the initial hurdles can be overcome with clear communication and, after a few trials and errors, we have managed to optimise the process and developed a workflow as outlined in Figure 9.7. To facilitate the onboarding process, we recommend devising an EHS induction checklist that clearly indicates which inductions are required for a specific role and clearly identifies the responsible stakeholder for each induction. Following this checklist, the new employee is first directed to the EHS representative to be inducted with area and site-specific general rules and EHS arrangements (step 1). This leads to the next step between the line manager and the new employee to outline required safety training and ensure that specific risk assessments for the role are well understood and implemented before starting (step 2). Concomitantly, the line manager coordinates with occupational health services to identify health surveillance requirements based on hazards and associated risks and arranges required medical surveillance programmes as needed (step 3). This step further ensures identification of any additional preventive and protective measures required for the new employee so that line management

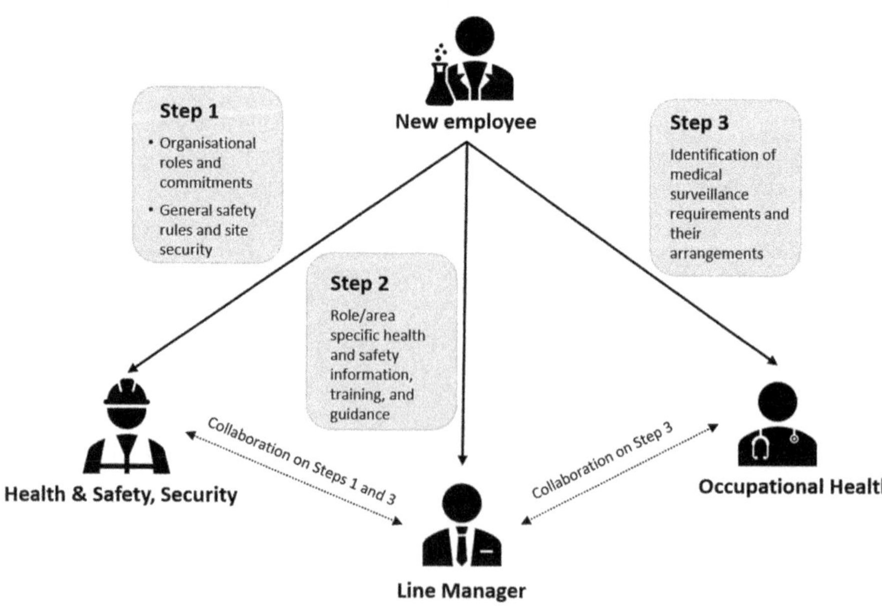

Figure 9.7 New employee EHS induction workflow with relevant stakeholders.

can put these in place to facilitate the employee's start in the safest manner. At first glance this workflow might look like bi-directional communication between the new employee and the relevant stakeholder. However, it is more of a collaborative work between other stakeholders. For example, the EHS representative also provides support and guidance to the line manager on identification of medical surveillance requirements based on applicable regulations, such as COSHH (Regulation 11) and corporate requirements so that the line manager and occupational health services can make provisions for the required surveillance programme in a timely manner. For instance, if a scientist's project involves use of skin and respiratory sensitisers, the developed checklist identifies the need for enrolment to annual skin and respiratory questionnaires. Overall, we find this tool very efficient for line managers to guide their new employees through a more streamlined trajectory and to remove any ambiguity before their new employees can safely and successfully start their roles.

9.8 Creating a Positive Safety Culture from Scratch

In April 1912, Captain EJ Smith is believed to have said ‘I will say that I cannot imagine any condition which could cause a ship to flounder. I cannot conceive of any vital disaster happening to this vessel. Modern shipbuilding has gone beyond that’, right before the first and last voyage of the RMS Titanic.[14]

When faced with starting a state-of-the art Discovery Centre, trying to build a culture of safety while also maintaining the zeal that will drive growth is a difficult balance. Building an infrastructure that will ensure that there never will be that valley between foresight and hindsight, is our task at hand.

Creating a positive safety culture from scratch has been a very exciting and at the same time very challenging experience. Employees with varying seniority, experience and backgrounds, all new recruits to the company, have contributed towards building this positive safety culture.

The term “safety culture” was introduced in literature by the International Atomic Energy Agency in its 1986 Chernobyl Accident Summary Report to describe the thinking and behaviours of people in the organisation who were responsible for safety in that nuclear plant accident.[15] In 1993, the Advisory Committee on Safety of Nuclear Installation (ACSNI) investigated disasters such as the Chernobyl meltdown (1986), the Kings Cross fire (1987), the Piper Alpha explosion (1988), and the train crash at Clapham Junction (1988). They concluded that these breakdowns were not caused by deficits in safety management methods, but were a consequence of poor safety culture.[16] The investigation concluded that creating a corporate atmosphere or culture in which safety is understood to be and is accepted as the number one priority is essential.[16,17] According to the ACSNI Study Group on Human Factors report, safety culture is defined as “the product of individual and group values, attitudes, perceptions, competencies, and patterns of behaviour that determine the commitment to, and the style and proficiency of an organisation's health and safety management. Organisations with a positive safety culture are characterised by communications founded on mutual trust, by shared perceptions of the importance of safety and by confidence in the efficacy of preventive measures".[16]

Responsible organisations commit to achieve the highest EHS standards and all their sites are expected to reach and maintain those standards. It has been a critical step for us to assess and establish the initial safety culture on site so that a path forward could be determined in order to achieve the highest EHS goals as set out by the company as well as compliance with national regulations. However, it has been challenging to assess and establish the initial safety culture due to the ongoing recruitment process.

One of the learning points has been that incorporating safety assessment into the recruitment process can have an impact on building a strong safety culture. A few questions can assess candidates' existing safety values and experiences they have developed in their past employment. The interview process can be tailored based on the seniority of the role and associated safety responsibilities in a way that candidates can discuss their vision on health and safety and the kind of management strategy they can bring to create a positive safety culture with the same core values as the organisation. Understanding candidates' existing safety values will help the organisation to draw the baseline and set targets accordingly in order to achieve the desired standard. This will include identification of resources and training they will need to allocate to the candidates based on their role and responsibility. Especially when recruiting managers and senior leaders, emphasis should be given to their health and safety values as this will influence behaviour, development and performance of their employees. Senior management needs to pay attention to setting productivity goals which do not compromise health and safety and always aim at finding the balance right between safety and productivity while placing safety as the highest priority.[18,19]

It is well known that making changes in organisational safety culture can take many years, meaning it will require immense effort, resource, and time. Therefore, organisations should focus on shaping the safety culture at a very early stage starting from recruitment. Unfortunately, it is not common practice in academia and research organisations to discuss a candidate's safety experience in an interview unless the candidate is being considered for a specific health and safety role. Sadly, safety-related questions are not among common interview questions and even if they are, they don't have the same priority as the role specific subjects which makes it very tricky to depict a clear picture of a candidate's safety vision in a limited time. This practice forces organisations to share and evaluate their organisational safety culture at a later stage and try to make improvements based on the outcome of the evaluation.

Various online tools have been developed to measure safety culture and provide organisations with a starting point to identify gaps and to develop solutions to improve their standards. For example, the online Safety Climate Tool (SCT) developed by HSE is being utilised by over 300 organisations both in the UK and overseas to measure and improve safety culture.[20] The Human Factors Toolkit is another example that helps companies gauge their relationship between human action and safety performance.[21] The Rail Safety and Standards Board's (RSSB) Safety Culture Toolkit includes multiple questions to assess employee attitudes, values and perceptions towards safety management systems across various safety culture factors.[22] iAuditor by Safety Culture also provides customisable online behaviour-based safety checklists aiming to achieve a total safety culture by observing routine tasks, providing timely feedback, and analysing the root cause of unsafe acts.[23] These online tools can be used by individuals, teams, departments, sites, and whole organisations while maintaining anonymity to gather honest feedback and prompt constructive discussions with an aim to devise better practices and policies. Although these tools are developed for existing cultures, they can be a good starting point in order to understand and get a picture of initial safety culture once candidates are recruited and the organisational structure is created.

The Just Culture has been widely accepted to be one of the best behavioural safety approaches used by organisations to improve their overall organisational safety performance.[24,25] The model works on the assumption that mistakes and unsafe acts are generally the product of faulty organisational culture. 90% of the time the reason for an unsafe act is environmental, not personal. Hence organisations should spend 90% of their time and resources improving the environment in which they want behaviours to be safe, and just 10% analysing the individuals concerned.[18] We have adapted this approach when building the working environment from scratch in which we want a positive safety culture to flourish. Organisational commitment, visible leadership, a focus on proactive and preventive measures rather than reactive measures, employee incentives programmes and engagement activities, commitment to organisational learning, and promoting safety first with near-miss reporting are the main strategies we have promoted to create a positive safety culture. Let's expand on those strategies a little bit below.

Health and safety require engagement across all levels of seniority from top-to-bottom and bottom-to-top. EHS management systems can be effective only when they are designed together with the whole workforce. This results in a more user-centric infrastructure, which takes into account not just organisational and management responsibility but also feasibility, specific user needs and is ultimately more effective. To achieve the level of engagement required for this input into safety practices, we encourage participation *via* activities and employee incentives. The consequent participation is accepted by management to be an important part of development and progress, and ultimately, results in safety practices that will benefit business and represent employee values. For example, we utilise a point-based employee recognition programme, which provides immediate rewards to help promote an organisational culture with effective ways of working. The modifications to our working practices we encourage are encapsulated by the following concepts: knowing to learning, withholding to sharing, controlling to empowering, planning to experimentation, silos to network, build a diverse and inclusive culture, demonstrate ethics and integrity, and drive results. Almost every employee played an active role in designing and implementing the site EHS management systems, such as helping with development of the risk assessment process, forming the EHS committee, devising the waste management process, helping with notification applications to HSE, providing safety information and labelling in the laboratories for chemicals and biological agents, and keeping a critical view and eye in identifying the shortfalls and gaps in the procedures so that they could be fixed at an early stage. Overall, health and safety has become a team activity where every member of the team has taken ownership and responsibility. Acknowledgement of all those employee involvements by senior management *via* our recognition programme has kept the momentum and motivation ongoing. This also has enabled a safety culture where mistakes are not punished but instead approached in a positive manner so that everyone can speak up freely as they all have experienced the benefits of openness and team work to create a safety system that promotes rather than competes with productivity.

An effective near-miss reporting programme can also help create a positive safety culture by removing fear of sanctions. It enables employees to actively identify and control hazards, reduce risks and protect themselves and others from harm. Near-miss reporting should be very easy, quick, straightforward and preferably anonymous to remove barriers from employees reporting near-misses. There are many off-the-shelf health and safety software packages with user-friendly forms offering drop-down menu functionality, the ability to attach photographs and tailored questions based upon answers given. Use of online reporting tools also enables data collection for analysis. Management should persistently reinforce near-miss reporting and reporters should be acknowledged for their commitment and active participation in developing safety defence mechanisms to prevent accidents and to learn how to avoid incidents with more serious consequences in the future.

Safety quizzes are good activities to gamify employee interest towards health and safety. In its monthly membership letters the Royal Society of Chemistry (RSC) asks its members 'What would you do?' for a given health and safety incident, such as chemical spill, hazardous gas leak, chemical fire *etc.* where the winner gets a prize. With some further creativity this can be developed into a regular safety quiz with scenarios not only about how to respond to emergencies but also about developing reactive and preventive measures, such as 'which engineering control will you need if you want to work with infectious biological agents to prevent exposure?'. This will guide scientists to think about preventive measures hence use their effort and resources on proactive measures rather than reactive measures.

9.9 Conclusion

Working with innovators is exciting. The central aim for pharma companies is developing novel medicines to treat patients with unmet medical needs. The same drive that turns young scientists into great inventors can occasionally be difficult to constrain within a rule book. Our approach is to share our vision. We are just as keen that the team achieve their scientific goals as they are. Our purpose is to remove barriers that compromise safety, and not place barriers in the way of scientific achievements. Good communication and integrating with the team are crucial in delivering this message. The time invested in generating relationships with the team, and focussing on good communication, can go a long way in generating the safety culture. The best safety culture can only come from a collaborative team spirit; one that is likely to last, and scale as the site grows.

Acknowledgements

We would like to acknowledge Radha Desai (Scientist, MSD UK) for her highly valuable input to this chapter.

References

1. A. Waring, *Saf. Sci.*, 2019, **117**, 152–163.
2. D. Bergman, C. Davis and B. Rigby, *RR535 – International Comparison of Health and Safety Responsibilities of Company Directors*, HSE Books, Norwich, 2007.
3. F. W. Guldenmund, *Saf. Sci.*, 2000, **34**, 215–257.
4. R. Flin, K. Mearns, P. O'Connor and R. Bryden, *Saf. Sci.*, 2000, **34**, 177–192.
5. D. M. DeJoy, *Saf. Sci.*, 2005, **43**, 105–129.
6. A. Neal and M. A. Griffin, *Aust. J. Manag.*, 2002, **27**, 67–75.
7. G. Hofstede, *Culture's Consequences: International Differences in Work-related Values*, SAGE Publications, USA, 1984.
8. CFI Education Inc, https://corporatefinanceinstitute.com/resources/knowledge/other/hofstedes-cultural-dimensions-theory/, last accessed March 2020.
9. G. Hofstede, *Culture's Consequences: Comparing Values, Behaviors, Institutions, and Organizations across Nations*. SAGE Publications, USA, 2001.
10. G. Hofstede, G. J. Hofstede and M. Minkov, *Cultures and Organizations: Software of the Mind*, McGraw-Hill Education, New York, 3rd edn, 2010, p. 3.
11. G. S. Svenningsen, B. R. Williams, M. B. Blayney, E. Czornyj, I. Schröder and C. A. Merlic, *ACS Chem. Health Saf.*, 2020, **27**, 40–42.
12. K. L. Ng and D. M. Hamby, *Health Phys.*, 1997, **73**, 473–482.
13. A. H. Stasek, *Electrical Safety in the Lab*, 2020, M. Ulusoy, personal communication.

14. Disaster at last befalls Capt. Smith, New York Times, Tuesday 16th April 1912, https://www.encyclopedia-titanica.org/disaster-at-last-befalls-capt-smith.html, last accessed March 2020.
15. International Nuclear Safety Advisory Group, *Summary Report on the Post-accident Review Meeting on the Chernobyl Accident*, International Atomic Energy Agency, Vienna, 1986.
16. Health and Safety Executive, *Third Report of ACSNI Study Group on Human Factors – Organising for Safety*, HSE Books, Norwich, 1994, p. 1.
17. Y. Kim, J. Park and M. Park, *Saf. Health Work*, 2016, **7**, 89–96.
18. T. Marsh, *A Definitive Guide to Behavioural Safety*, Routledge, Oxon, 2017, p. 1.
19. HSG48, *Reducing Error and Influencing Behaviour*, Health and Safety Executive, Norwich, 1999, p. 2.
20. Safety Climate Tool, https://www.hsl.gov.uk/media/78921/sctbrochure-forweb.pdf, last accessed January 2020.
21. RSSB's Human Factor Tool Kit, https://www.rssb.co.uk/Standards-and-Safety/Tools--Resources/Human-Factors-Toolkit, accessed January 2020.
22. RSSB's Safety Culture Tool Kit, http://safetyculturetoolkit.rssb.co.uk/home.aspx, last accessed January 2020.
23. Safety Culture iAuditor, https://safetyculture.com/iauditor/, last accessed January 2020.
24. A. S. Frankel, M. W. Leonard and C. R. Denham, *BMC Health Serv. Res.*, 2006, **41**, 1690–1709.
25. P. G. Boysen 2nd, *Ochsner J.*, 2013, **13**, 400–406.

CHAPTER 10

The Role of Technical Staff in Maintaining Good Health and Safety

J. S. HUBBLE*[a], H. SHEELEY[b] AND A. GALLAGHER[c]

[a]Cranfield University, Cranfield, Bedford, MK43 0AL, UK; [b]Public Health England, Porton, Salisbury SP4 0JG, UK; [c]University of Glasgow, Glasgow, G12 8QQ, UK
*E-mail: J.S.Hubble@cranfield.ac.uk

10.1 Introduction

Technicians play a key role in maintaining good health and safety culture and standards in the higher education and research sectors. The technical role and the ways in which technicians work and support higher education and research promotes good health and safety practices. There are several challenges facing technicians including a lack of understanding of the technical role which has led to an ageing population profile and anticipated loss of technical skills and institutional memory. Initiatives to increase awareness of the duties (including hidden duties) associated with the role may help increase the status of technicians and preserve the often-underestimated contribution that they provide.

Challenges for Health and Safety in Higher Education and Research Organisations
Edited by Olga Kuzmina and Stefan Hoyle

Published by the Royal Society of Chemistry, www.rsc.org

10.2 What is a Technician?

Currently, there are around 33 000 technicians working in English universities. This number rises to greater than 65 000 when Scottish universities and research council institutes are included.[1] The technical community traditionally works "behind the scenes" and as such has largely been overlooked. Technical roles have perhaps been perceived as a lower status than academic and other research staff and there is a lack of understanding and recognition of the technical role and the way it has changed over the years. The traditional image of a science technician, particularly when first encountered in the school environment, is of someone pushing a trolley full of glassware or equipment that will be used by others. When faced with increased financial pressure and further pressure to demonstrate value for money, it appears to have been an easy, and perhaps cost-saving option to allow technician numbers to decline through natural loss. The decline of suitable apprenticeships from the 1980s has reduced the number of entry points into the technical career. Career progression routes for technical staff have not been updated in many institutions, with many technicians progressing only through ad hoc opportunity, or by fulfilling the rare vacancies made by retiring technicians. This has led to an ageing population profile amongst technical staff. There are risks to the higher education and research sectors from the loss of technical staff and the skills, knowledge and institutional memory that they hold, through retirement. Additionally, technicians are highly sought after by industry, who will tempt technicians with higher salaries and perhaps better career prospects. The technical role in higher education and research institutions is also complicated by its diversity, large number of job titles, and line management structures thereby making it very difficult to clearly identify all technicians and take a strategic approach to address this trend.

There are many definitions and descriptions of technicians. One useful definition that has been more frequently used is:

> "A technician is a person who is skilled in the use of particular techniques and procedures to solve practical problems, often in ways that require considerable ingenuity and creativity. Technicians typically work with complex instruments and equipment, and require specialised training as well as considerable practical experience, to do their job effectively.[2]"

Technicians work at the interface between the material and symbolic worlds, supporting academics to realise theory into tangible outputs.[2]

Regardless of definition, technicians play an essential part in maintaining health and safety within their institutions. The technicians' influence on health and safety may well be underestimated given the poor understanding of the technical role within the higher education sector. There is an ongoing study by the National Technician Development Centre which helps institutions to understand their technical staff through a skills, roles and responsibilities survey. Across a sample of five universities it is reported that 99%

of technicians who responded to their institutions survey have experience in carrying out health and safety activities as part of their role.[3]

10.2.1 The Technical Role

Technicians can work in a wide variety of departments and undertake an even wider range of tasks. Regardless of where they work, their role is likely to fit into one of the following role types or will combine aspects of several types.

- Specialist Support for Research

  This role involves carrying out experiments, analysing data from those experiments and supporting the operation of the research environment through training of new researchers, keeping accounts, and organising rotas. As part of this role the technician is expected to work to excellent health and safety standards and to encourage others to do so too. The technicians may be involved in setting these standards through the establishment of written laboratory rules and standard operating procedures.

- Support for Teaching

  This role involves the preparation of material for practical teaching sessions and attendance at those sessions to assist the teaching and ensure health and safety for staff, students and themselves when these activities are conducted. The technicians' presence will contribute to safe staff:student ratios and they will provide expert assistance should an emergency arise, such as first aid provision and fire evacuations. Through direct involvement with practical sessions the technicians can impart safe skills and knowledge to students, setting them up with good practical skills for their future career. Technicians may demonstrate apparatus as part of these sessions or separately. In some cases, technicians may participate in the teaching process by delivering lectures, devising and leading the practical sessions and marking any related assignments.

- Infrastructure Support

  This role involves supporting the infrastructure where research and teaching take place. This can be quite diverse such as maintaining the upkeep of buildings and performing odd jobs like installation of shelving. Alternatively, it can refer to the performance of core tasks such as washing glassware, dealing with laboratory wastes, supply of gases *etc.* In more senior roles technicians may organise activities, schedule work, develop and implement policies and procedures and allocate space in shared facilities. In this role technicians will ensure the safe upkeep of the working environment and the equipment and plant contained within, for example by arranging and overseeing statutory inspections of fume cupboards, and by arranging prompt repairs as necessary. They will also ensure the correct disposal route for a variety of laboratory wastes to prevent accident or contamination of waste handlers and the environment.

- Analytical Service Technicians

  This role provides specialist support for specific instrumentation that is managed as a cross-departmental resource such as electron microscopy suites. They generally run samples and maintain the equipment building up specialist expertise in the operation of the instrumentation. As part of this role they are expected to use and maintain the equipment safely and ensure the safe use and operation of the equipment by others through training and inductions.

- Workshop Technicians

  Engineering workshop environments may be research facilities in their own right or may be managed as a central resource serving several areas. The technicians working within will either operate as research support or technical service technicians, or combinations of those. The technicians within will be highly skilled in their discipline and will work to develop, design and construct, maintain and repair experimental equipment and instrumentation. They are expected to have an excellent working knowledge of how to use the equipment safely and to ensure any people working in the workshop environment adhere to safe working practices, for example correct use of guards and covers, not covering emergency stop buttons *etc.*

Technicians cannot always be easily categorised into the above roles, with many technicians carrying out different roles, and consequently health and safety duties, at different times. Where teaching loads are small then practical sessions may be supported by research technicians. Likewise, they may carry out infrastructure support tasks such as hazardous waste disposal if there is no dedicated support for this. Some specialist practicals may be supported by more appropriate technicians who may usually otherwise work in a central workshop or core facility environment. It is likely that all technicians to some degree will have a role in the maintenance, fault diagnosis and repair of equipment and the environment they work in.

Technicians may also contribute to proposal writing and the publication of research papers. It is likely that some will attend various committees and working groups at their institution such as health, safety and environment committees, quality assurance working groups *etc.* Some may go on to represent their institution at national or international events. Technicians may work at the institutions' premises or may take their skills off site, sometimes working in the external environment, or at a client's premises.

10.2.2 Technical Management Structures

Technicians work under a variety of management structures. The traditional research set up has typically been that a technician would work closely with a single or small group of academic staff who would be considered the leader of the laboratory. In this scenario, the academic and the technician would be the relative constants within changes of research students and postdoctoral researchers.

The pressures on the management of universities and research institutes have resulted in a continuing trend for technical staff to be increasingly shared amongst larger groups of academics and to be working in shared facilities. Increasingly universities are required to demonstrate value for money and that they are sharing research council funded equipment. It can be significantly cheaper to operate using shared facilities and a pooled technical resource. Cross-disciplinary collaborations may also be more successful at winning funding, leading to a breakdown of disciplinary boundaries. It is therefore increasingly common for new research facilities and buildings to be shared between research groups and the technical staff to support this shared environment. Within a shared environment it is less appropriate for any one academic leader to manage the technical staff. Consequently, technicians may be managed as part of a technical team that is led by a technical manager or chief technician.

It is not unusual for both management structures to be present within the same research institution. This can have arisen organically through historic changes within research departments or may be a purposeful arrangement if there are shared core resources such as electron microscopy supporting smaller research groups.

10.2.3 Key Qualities of Technical Staff

The technical role requires technicians to possess and utilise some key characteristics if they are to be successful in their role. They will also possess specialist skills and experience that will have been built up through practical work, over several years. It is likely that technicians will also have an academic qualification that may have been obtained through full or part time study. Technicians may also be members of relevant professional bodies such as the Royal Society of Chemistry, Royal Society of Biology or the Institute of Physics, or be recognised with professional designations like RSci or RSciTech by the Science Council. Similarly, engineering based technicians may be members of the Institute of Engineering Technology or other more specialist organisation and have professional designations such as EngTech. Technicians will also have a wider knowledge such as the application of relevant health and safety legislation related to their area of expertise. Technicians typically attend specialist training to ensure the health and safety of their operations, *e.g.* Control of Substances Hazardous to Health (COSHH) training. They may hold health and safety qualifications such as the National General Occupational Health and Safety Certificate or Managing Safely Certificate from the National Examination Board in Occupational Safety and Health (NEBOSH) or Institution of Occupational Safety and Health (IOSH) respectively. They may also hold relevant certification and perform specific roles in biological and radiological safety such as Biological Safety Advisor, or deal with licencing such as for duty free spirit or import of controlled materials. Crucially, the technician will apply this knowledge practically in their day to day activities.

Technicians must be well organised to cope with a variety of demands made on them. A degree of flexibility is also required. The technical role, by its nature, has a reactive component and the need to frequently reprioritise is essential. It is not unusual for technicians to end up dealing with something unexpected that arises from the research process or perhaps an emergency. In the event of an emergency it is usually the technical staff that can make something safe or assertively manage the situation to ensure everyone's safety.

A further essential technician quality arises from the implication that they make tangible real-world activities happen to support the theoretical world of academic staff. A technician must be creative and able to problem solve. It is not unusual for the requirements to be described in brief terms and roughly drawn sketches, rather than following detailed technical drawings or specifications. A skilled technician will draw on their knowledge skills and experience to develop the ideas into safe functional equipment and experimental systems.

Technicians are frequently in the middle of the laboratory activities and need to keep up with the progress of the various projects happening in their laboratory. It is important that they are effective communicators and able to communicate using a variety of styles to suit any situation. Technicians are also providing key communication channels between themselves and researchers with other parts of the institution such as facilities or estates who may have responsibility for maintenance of large plant like air handling units, air conditioning or local exhaust ventilation systems, and who may arrange for statutory tests and inspections to be carried out.

Technicians work at the frontline and are student-facing roles in many institutions. Technicians often have no formal line management or supervisory relationships with the people working in their areas. Consequently, they are required to be personable and be able to effectively influence people to set standards to ensure good working practices and conditions are established in the research environment. This is done through providing inductions, demonstrating good laboratory practice and establishing good discipline amongst the researchers working in their environment.

10.2.4 The Technical Role in Setting Standards and Providing Inductions

The initiation into the laboratory and related facilities may be taken over a period because of the complex environment and need for timely introduction in a progressive manner. Where the facility is shared it is often the technical staff who have an overview of the whole facility and are best placed to provide information on the hazards that a researcher may encounter and the control measures that have been established. The supervisory lecturer or principal researcher may not have time or be available to undertake this and the technician is often involved to some degree. The induction is not just a run through of rules and arrangements but should include an assessment of how much the new person has absorbed the information and their attitude to following the systems. Technicians have been known to devise and implement a best practice induction consisting of a written safety manual, short open book competency test and laboratory safety tour to ensure that new lab users are fully aware of the work going on in a large shared facility. High achieving academic ability is not a predictor of rule understanding and compliance. This in turn may inform the type of ongoing support a new person may require. Technicians will adapt the induction delivery to suit the person. Unfortunately, this may also be the point at which it is discovered that a new lab user may be unsuitable for the work or need further support. In our experience a research student attended a safety induction tour but could not understand what was being said. Upon investigation it was found the student had falsified their English qualifications. It is crucial that the induction process includes assessment, informal or otherwise, of understanding. The technician may have delivered inductions many times and have a set of topics to include in the induction which will include the lessons learned from previous adverse events and examples to reinforce the message especially for prevention of future health and safety incidents.

10.2.5 Good Laboratory Practice

The laboratory and science environment require a specific approach through a set of behaviours that respects the scientific method and to ensure safety of oneself and those in the vicinity. This must be learned. Where this has not been done, and staff and students "make their own methods" it has resulted in injury such as repetitive strain from poor pipetting techniques.[4] The technician makes time to demonstrate these key skills that will be carried through a person's career. The lecturer supervising may not be available or that skilled themselves to adequately and patiently demonstrate and explain why these techniques are important. Examples include:

- pipetting to prevent splashing and aerosol generation, also repetitive strain injury[5]
- the hierarchy of volume accuracy in glassware measures

- avoiding exothermic reactions in making dilutions
- types of mixing and separation
- benchtop workflows to avoid cross contamination
- prevention of cross contamination
- use and disposal of scalpels and other sharps
- use of water baths or other thermal systems
- use of fume cupboards and other local ventilation systems
- speed and density calculations for high speed centrifugation
- importance of sterility and cleanliness of vessels and implements

Confirmation of these is best when observation is made of technique and any adjustments made early and understood to be consistently used. It is unfortunately common for less experienced lab users to adopt poor technique and practices. This is usually not malicious in nature, but rather down to inexperience and pressure to complete practical work to tight, and sometimes unrealistic, deadlines. Examples witnessed by us include:

- removal of safety glasses or positioning them on top of their heads rather than over their eyes
- students misusing fume cupboards by working with the sash at high level and their head inside the fume cupboard with a hazardous substance
- researchers pipetting too close to their face, rather than at arm's length on the benchtop
- less experienced researchers attempting to filter solutions through a fluted filter paper without the aid of a supporting funnel
- people failing to label containers of clear fluid

If these are not picked up quickly it can lead to accidents. Additionally, people react differently when accidents have happened. If they are visiting from overseas they may adopt their cultural norms rather than the laboratory practice that they have been inducted to. An example of this was the breakage of a large piece of glassware which was found disposed of in the regular laboratory bin instead of the glass bin. Despite having undergone an induction process where the bins were described and shown, the researcher was upset and embarrassed by the breakage, and disposed of the broken glass in the regular bin as was normal for their home country. Technicians in the lab are therefore sometimes required to take on the role of the trainer and advisor, and sometimes enforcer, on an ongoing basis. A health and safety officer cannot generally do this as they may be based away from the laboratory. Academic staff may also not be in the laboratory as frequently and may not observe the poor practice. In a shared laboratory environment there may be a poorer understanding of the hazards as each research group will focus on their own work, rather than that of other groups.

10.3 Quality Management Systems

The scientific environment is increasingly expected to be the bedrock of future products and innovations. The need for reproducibility and repetition of studies has increasingly become a normal part of research and teaching laboratories. Thus, quality systems have been adopted to meet the needs of customers and other stakeholders whilst meeting statutory and regulatory requirements related to a product or service. This applies to both the scientific processes and in the selection and use of key equipment.

A quality system sets out what is to be achieved, and:

- how it is to be achieved
- how it was assured as done
- how the operatives are competent
- how change is to be managed
- how errors, complaints or deviations are discovered and addressed

Implementation of quality management has considerable advantages for laboratory operations as this aids the control of risk and records deviations. Whilst this comes with costs, it enables the technician and researchers to operate in a controlled environment producing reliable results and maintaining high standards of research. The need to monitor low temperature systems for integrity of samples is a commonly encountered example in many laboratories. Early detection and appropriate action of rising temperatures alerts to potential failures and prevents leaks from thawed contents which may cause slip and trip accidents.

It is often a joint role of the technician to complete the quality system requirements such as recording, checking and monitoring. The keeping of records assists building a history of equipment failures, due dates for calibration, servicing and replacement. The technician should be invited as a competent expert to audits and inspections against accredited standards such as ISO 9000, 9001, ISO 17025, ISO 15189, ISO 14001, GxP (including GLP), JCoPr and OHSAS18001 (replaced in 2018 with ISO 45001). In reality, this may not always be the case which can sometimes lead to assumptions and misinterpretations by the auditing team or other complications.

10.3.1 Laboratory Quality Management Systems

The laboratory quality management systems (LQM) include developing, implementing, and sustaining relevant policies, processes, and procedures for each quality system essential. Health and safety-related essentials are listed below:

- organisation
- process management
- document and records

- facilities and safety
- information management
- personnel
- nonconforming event management
- purchasing and inventory
- assessments
- equipment
- continual improvement

Each of these requirements enhances the health and safety as well as the quality of the research but must be maintained by appropriate record keeping and diligent application of policies and practices. For example, anyone responsible for producing risk or COSHH assessments should have received appropriate training which should be documented. Any new chemical purchased must be COSHH assessed and recorded. Any new equipment purchased must be risk assessed before use and its documentation such as instruction manuals and certificates of compliance filed appropriately in case required by audit, or accident investigation. As a centre-point of the laboratory it is usual for technical staff to be involved in many aspects of quality compliance which assists with health and safety compliance. The organisation and research grants are dependent on producing safe and high-quality outputs to attract students, grants and funding as well as students. Technicians are a key part of that team effort.

10.3.2 Discipline

The technician with responsibilities to support a laboratory or function will set the standards for that area of responsibility be it a physical space such as a laboratory or a service function such as electron microscopy or histology. This will include:

- cleaning and hygiene arrangements and practices
- personal protective equipment use and availability
- storage and dispensing of hazardous substances
- waste management
- techniques and operating practice
- documentation and record keeping
- access and security
- lone working arrangements and requirements
- emergency and adverse event procedures

The rules and set procedures are not only for the smooth running of the laboratory or service but also act as best practice examples for early career scientists to take into their own work and future standards. For example, at the beginning of a new student intake the technician demonstrates the correct donning, use and removal of gloves or other types of PPE. This not only ensures correct procedures and sets expectations but is part of the learning processes for students and trainees to take through to their future careers.

The technician then reinforces these practices by example, encouragement and rapprochement during the frequent interactions. These behaviours may not be consistently demonstrated or reinforced by supervisors, colleagues or peers without the technician's endorsement.

10.4 The Technicians Role in Training, Teaching and Instruction in Research and Teaching Laboratories

Training, teaching and instruction in research and teaching laboratories is fundamental to enable staff and students to perform tasks and operate equipment in a safe, productive and efficient manner. A good training experience by highly competent staff familiar with the task or experienced in the use of key equipment can ensure that competencies are maintained to the standard operating procedures. This enables both staff and students to have a good learning experience to fuel the future generations in higher education establishments and non-academic environments. Whilst academics or senior management staff may have gained competencies in various processes and tasks over the years the demands on their time in a demanding role may mean that they are not available to help with hands on training at the 'front line'. Indeed, whilst they maintain the accountability for what is undertaken in their area of management the responsibility for training is usually delegated to a senior team member or available technicians. This frontline training of junior staff and students or inexperienced personnel is often led by the technicians or laboratory managers within an area, who play a critical role in maintaining standards of training and ensuring that there is effective oversight of activities on a day to day basis. It is highly likely that any new person entering the laboratory will have been made aware of the obligation to be trained either through senior colleagues who have accountability, or by technical staff who provide the training and uphold working standards.

When technical expertise provided by managers or technicians is not present within an area, the detrimental effect to the overall safety culture and consistent competencies becomes apparent and variability and complacency can appear. Inappropriately trained staff or students could begin to train others, and without the knowledge and skills input of key personnel unintentionally introduce poor practices. This type of scenario can lead to an unsafe culture, affect productivity and reproducibility and can be an 'accident waiting to happen'. Reputational damage from such scenarios can be immense in addition to financial loss and, more importantly, possible injury to the lab users and others.

Ensuring that the technical presence and expertise is maintained across an organisation for both teaching and research activities will not only improve the teaching and working experience and an organisation's reputation, but also ensure that the organisation has a frontline means of staying safe, compliant and productive.

10.4.1 Recognising Inappropriate Behaviours

Ensuring that processes, training, instruction and supervision are of a consistent standard and that the relevant safety management systems are in place can forge a safe and productive working environment. This can not only help ensure the safety of personnel but also prevent wastage of materials and equipment through failures in protocols and procedures. In research environments funding and time can be limited therefore getting things right at the start can ensure that smooth processes and that output is maximised. Additionally, research staff are often focussed on the outputs of their work and may overlook legislative requirements such as notification of the appropriate regulatory body when importing research materials. Fundamental to a safe working environment is the establishment of a good health and safety culture where working correctly and dealing with situations appropriately should be the norm rather than the exception. A good safety culture is driven by strong and active leadership and engagement at the local level by key personnel.[6] In our experience over many years in research laboratory environments, areas with a strong positive technical presence are often better resourced to recognise early inappropriate behaviours and act accordingly before this becomes the cultural norm, such as the accepted level of tidiness and good housekeeping. Some behaviours can be individualistic and isolated in nature such as personnel thinking it may be better or faster to do something their way rather than the way they have been shown. One such incident happened in our laboratory, discovered when a student's experiment continually did not work after previously working well. When going through exactly what had been done with the student it became apparent that they did not follow the protocol but preferred a faster method that made more sense to them. On this occasion adding reagents in the wrong order resulted in a loss of data, time and a waste of expensive reagents and consumable products. In other circumstances, breaking protocols may have safety implications if the wrong reagents or plastic wear are used. One student decided to use a similar looking tube to the one that they had been told to use. They should have used polypropylene but instead used plastic. The chloroform solvent being used came through the bottom of the tube spilling the contents into the fume cabinet.

Additional inappropriate behaviours can include:

- A lack of/inappropriate personal protective equipment (PPE) such as wearing lightweight and disposable visitors lab coats which offer less protection than the regular versions.

- Ignoring warning signs, particularly if the lab has been closed for maintenance.
- Ignoring standard operating procedures.
- Covering up mistakes.
- Use of personal phones/eating and drinking in restricted areas.
- Misuse of equipment.
- Untidy working areas.
- Not following set up and exit procedures.
- Lone working at unusual hours.
- Inaccurate recording of data.
- Failing to evacuate when fire alarms have sounded.

When a technical leader is present within an area and has an authoritative laboratory managerial style then the likelihood of sustained inappropriate behaviours is unlikely. That technical presence may also be key to recognising changes in peoples' behaviours such as carelessness, noticeable tiredness and longer hours of working due to wellbeing issues.[7] In our facility the technical staff noticed a student had amassed an unusually large number of samples which were poorly labelled and stored. It was clear that the student was feeling the pressure of an impending deadline. Through notifying their supervisor and offering support the technical staff were able to provide assistance with the analysis and correct disposal of the samples. Communication between key technical staff within an area and management can ensure that there is a commitment to sustained good working practices. Robust everyday laboratory management can help ensure that a good safety culture is maintained, and that poor safety culture is frowned upon and not accepted either locally or at the organisational level.

10.4.2 Laboratory Practical Classes

The role of key technical personnel in the laboratory practical class setting is well established and recognised; however, that role goes well beyond making the class itself run smoothly for the student and teacher/demonstrator. Technical staff are often at the front end of undertaking the risk assessment process of activities and considering and preparing what is required to enable laboratory practical classes to run smoothly and safely. Additionally, they may prepare reagents from stock solutions for use within the classes and must risk assess their own activities in addition to what is undertaken by the students subsequently. Practical classes can involve many students and ensuring that all equipment and processes have been thoroughly checked before allowing inexperienced personnel to attempt practical work is an enormous task in undergraduate laboratories. Technical personnel must have a thorough understanding of equipment operations and laboratory processes in advance of all practical classes to facilitate troubleshooting and be able to recognise behaviours that impact either the success of the class itself or the safety of both staff and students. The importance of technical personnel in the laboratory practical class should never be underestimated or taken for granted. Laboratory classes that run smoothly and safely are only achieved by the interaction of the academics/management leading the teaching and the technicians fully supporting activities before, during and after practical classes. Emergency response such as first aid and fire warden duties can additionally be held by technicians within research and teaching laboratories ensuring that emergency processes are *in situ* and consistent.

10.4.3 Field Operations

The laboratory is not the only place where technicians contribute to maintaining health and safety, many institutes operate field study and study centres from where activities take place. This may involve temporarily moving key equipment to a non-standard location. All equipment, substances and devices need to be prepared, packaged for transport and for such use as intended. This can range from spades and buckets to complex sensitive electronic equipment. There will also be vehicles, some specially fitted to accommodate equipment, and possibly researchers, in a safe manner. Previous experience of when this has gone wrong or improvement opportunities can be included in future planning and thus prevent re-occurrence of adverse events.

The technical role comes to the fore when the equipment is shared across several courses and research groups. One example from our experience is the servicing and testing of an inflatable boat and life jackets that is cared for by a dedicated field equipment technician. The technician will ensure everything is returned in a clean state and inspect the equipment for damage and to ensure no faults have developed. Where minor damage has occurred, the technician is able to repair it, such as small punctures found in waders. The technician can also check that equipment is properly prepared before returning it to storage, such as the correct storage of pH probes, and removal of batteries from handheld instrumentation.

10.4.4 Expert Equipment Users

In the research environment particularly, the equipment usage may vary from the intended use envisaged by the manufacturer or supplier. It may be a different environment for the intended use, such as in a wet laboratory, in an animal house or associated with other equipment such as in robotic set ups. The familiarity and careful understanding of the equipment performance, operational parameters and limitations can be gained by an individual who regularly sets up such equipment and undertakes calibration or adjusts set points. Thus, the relevant instrumental technician becomes an expert user of the equipment, beyond that of a researcher or student and is well placed to instruct new users and contribute to safe use of the equipment for that purpose in that setting. This knowledge and experience prolongs equipment life and ensures good science as the equipment is set up consistently providing reproducibility and reliability. The technician can also suggest better ways to gain valuable information from the equipment due to this intimate knowledge. They may also be able to do small adjustments and alterations without affecting safety, an example may be ensuring guards are replaced or adding splash protection to control panels. Some of this knowledge is built up through frequent contact with calibration and servicing contractors. The technician can keep up to date with developments in technology and techniques through contact with distributors and other users to enhance investigation and analysis. This can be through specialist formal training, or broader activities such as attendance at technical conferences and demonstrations. User focus groups may also exist such as the Midlands Innovation mass spectrometry[8] and flow cytometry groups, which bring together researchers and technicians from across the Midlands Innovation member universities.

Laboratories are complex environments with specialist equipment, sensitive instruments and demand constant attention to operate optimally and safely under ever changing demands of research and constant changes of occupants. Knowledge of the routines around operation, maintenance and servicing can take a prolonged period to learn and to be intuitive. Researchers and students have short term goals and tend to care less about longevity and optimal lifespan of instruments and equipment. The technicians longer term presence provides continuity of knowledge and can help ensure equipment is used correctly. It is here that technicians can directly influence students and researchers by providing training and instruction in the use of instruments and equipment that can be complex, cutting edge and even prototype in its use or application.

10.4.5 Potentially Hazardous and Specialist Equipment

Whilst it is recognised in core facilities that there may be key technical personnel overseeing cutting edge pieces of equipment, there are many pieces of routinely used general equipment that require induction and training before use. General inductions and safety inductions can often cover generic topics with targeted inductions and training being delivered locally for fundamental processes and equipment use. Areas having a technical presence will have the means to train staff and students in a consistent and uniform manner with local oversight of ongoing activities.

Areas without this technical presence can find that untrained staff and students may attempt to use equipment that is accessible in general laboratories such as centrifuges or microbiological safety cabinets routinely, without considering the appropriate operating procedures beforehand or understanding the correct purpose and function of the equipment. The inappropriate use of equipment may not only mean the loss of a very expensive piece of laboratory hardware and potential damage to infrastructure, it may result in serious injury/illness to operators and others within the facilities as a consequence of activities undertaken. Ensuring that personnel have a training checklist for all the key pieces of equipment that are in use within a given area can form part of a training portfolio for staff and students that may be used to demonstrate compliance. Technicians can and do help facilitate such processes.

Equipment is a massive expense, from small commonly used pieces such as pipettes, microfuges and whirl mixers through to specialist equipment that has been procured for key tasks with expensive consumables used for certain processes. Untrained personnel can generate spurious results, ruin equipment, affect productivity and potentially put the safety of themselves and others at risk. In our unfortunate experience we have had to deal with grossly contaminated equipment such as liquid chromatographs contaminated with oil, mass spectrometers contaminated with sugars and oil and cleaning out autoclaves and centrifuges where the contents leaked from their containers. In each case there was significant downtime on the equipment and diversion of technician time away from normal duties. Maintaining a technical presence and expertise in the use and operation of potentially hazardous equipment is essential in the teaching and research environments in higher education establishments and other organisations.

10.4.6 Maintenance and Care of Equipment

Basics of care of accessories and components of large equipment contributes to safety and optimal performance, this can include cleaning routines, subtleties of operation and fault identification to prevent expensive and catastrophic damage or injury. An illustrative case witnessed by one of the authors is a visiting post doctorate researcher assuming to know how to use an ultra-high-speed centrifuge with fixed angle rotor head. The centrifuge has potential to be very dangerous and correct use and care of the rotor is essential to preserve its integrity and lifespan. These types of equipment are expensive. For whatever reason the post doc did not ask for help from local technical staff or the supervisor and in the course of a week, three rotor heads were damaged and rendered useless other than for scrap metal. It was lucky that this incompetency did not result in centrifuge malfunction and harming himself or others. A ten minute demonstration and instruction would have avoided this dangerous and expensive oversight.

10.5 Regulatory and Statutory Requirements

The Provision and Use of Work Equipment Regulations, 1995, (PUWER) covers the use of work equipment. It applies to any machinery, apparatus or tool, or installation. "Use" is widely interpreted and covers any activity involving work equipment. The key parts to these regulations include:

- That equipment must be fit for purpose.
- Equipment must be maintained in good working order, with any maintenance logs kept up to date.

- Appropriate inspections are carried out, such as after installation and at periodic intervals to check for deterioration.
- Suitable information, training and supervision is provided including restriction of use only to those trained and authorised where necessary.
- Taking measures to prevent people being injured through access to dangerous parts, ejection of materials or exposure to excessive heat or cold.
- Ensuring machinery has clearly identified controls and means of isolation.
- Ensuring new equipment purchased conforms to appropriate laws, *e.g.* CE marking and Declaration of Conformity.

Whilst it is commonly thought that work equipment, and application of the regulations applies to large plant and machinery, the wider definition of work equipment means that the requirements apply to all work equipment, including that found in laboratories and other research environments. It also applies to mobile equipment. Additionally, there may be other, more specific regulations such as those for lifting equipment. The regulations refer to the use of equipment by employees and other affected people which would include students. The research environment is populated by researchers and students who may not have worked in the environment for very long, and in the case of students may be young and inexperienced. Through provision of training and instruction, as well as any work carried out to ensure good condition of the equipment the technicians' contribution to their institutions compliance with these regulations can be significant.

Technicians tend to remain in post for a longer period and build up a body of experience and knowledge which is vital to the support of the studies and the health and safety. This includes the regulatory environment such as permits, licences and safety standards expected by the regulatory bodies for example electrical, chemical and biological safety in use, storage and shipping. The technician also builds a knowledge of the facility, equipment and infrastructure of the laboratory so is well placed to work with facilities and maintenance teams to anticipate problems, failures and to plan preventative maintenance. In addition they may, in conjunction with facilities and/or the laboratory manager or chief technician, maintain databases on equipment held, condition and status. Some facility and infrastructure items are subject to statutory requirements and planning for these can be assisted by technical staff that know when they are due and to plan for their availability. Examples include statutory steam, electrical and power blackout arrangements.

10.6 Management Systems

Any effective Health and Safety Management Systems (HSMS) requires leadership and drive from the top of the organisation and cooperation and engagement from all stakeholders to ensure the effective delivery and maintenance of that system.[9] Integral to a good HSMS is the role that technical staff play in ensuring consistency and expertise across an organisation for a range of safety critical activities, procedures and protocols, emergency response and monitoring/review systems.

In higher education establishments traditional academic scientists and researchers form a crucial part of the safety management structure overall. Active safety involvement of the principal investigator (PI) or laboratory supervisor also correlates significantly with lower numbers of accidents reported by students and postdoctoral fellows in academic labs.[10] Technical staff and area managers are those who additionally help drive forward the safety policy of an organisation at the frontline ensuring that safety management systems are properly implemented and that the operational goals are adopted in real working practices.

Research breakthroughs and innovations will not happen without highly trained and properly supported technicians who are the people who design and build crucial equipment, make possible sophisticated experiments and form a vital part of the specialist teams. They are skilled people who use their science, engineering or technology knowledge to identify and solve practical problems. They are also crucial to the success of many of the United Kingdom's future-growth areas, including the aerospace, chemical, digital, engineering and manufacturing industries.[12]

10.7 Risk of Loss of Key Technical Staff

Over many years it has been recognised that technicians play a crucial role in higher education, including supporting safety management systems and organisational functions. The realisation that technical staff may start to decline with a detrimental impact on organisations in years to come is now fully recognised, and actions are underway to intervene before the situation declines further such as the Technician Commitment initiative.[12]

10.7.1 Retirement

Over 1.5 million technicians are employed in the UK, which includes all research institutions and those in universities. Engineering roles are the major areas of employment for technicians however science, health and technology disciplines also employ significant numbers. "An ageing workforce means that 50 000 of our best technicians are retiring every year, with forecasts showing that as many as 700 000 more technicians may be required to meet demand over the next 10 years or so.[11]" Retirement coupled with the decline of apprenticeships through the 1980s and 1990s has led to a shortage of skilled technicians to replace those leaving through retirement.

10.7.2 Impact of Fixed Term Contracts

Sustainability is perhaps easier to justify and facilitate within the teaching side of an organisation *versus* the research environment to ensure the consistency and delivery of classes for undergraduates in the future. As experienced technicians leave research areas the replacement and justification of such posts with identified funding can be difficult where the technician can be placed amongst various groups with individual research funding. Technicians who are specifically grant funded are often tied to a specific group or research project with little scope to move to other core tasks because of the funding constraints and time pressures.

For many years the norm within research environments was to have fixed term contracts of limited duration because they mirrored research project funding and provided 'flexibility'. Those on fixed term contracts are more likely to be focussed on a particular project or less able or willing to contribute to other core activities within an organisation conventionally provided by core funded technical staff.

Fixed term contracts have many negative effects. Staff can feel marginalised or undervalued and this can lead to loss of key staff members. Resources are wasted as trained staff look for more secure employment and ultimately projects can be jeopardised if staff leave before the end of a contract to take up a more secure post.[13,14]

The introduction of the fixed term employee regulations in 2002 did bring changes such as entitlement to redundancy payments for fixed term staff and equal treatment to permanent staff; however, this has not led to the end of successive short term contracts as originally hoped.

It is important that an organisation thinks about its use of fixed term contracts to make sure not only that they comply with the law but that the best use of staff and resources is adopted. This can affect overall productivity and continuity within the organisation. Key highly trained technical staff are required that can provide consistency and continuity within areas across the organisation over the years rather than continually retraining.

This remains a difficult situation for those on fixed term contracts but also a key concern in research areas where permanent core funded technical staff are required to enable the expertise and business continuity to be maintained. It is particularly within the research environments that this will impact most heavily as existing core staff retire or leave and are not replaced. This is less evident in teaching and other areas where a more permanent solution and permanent posts and continuity is perhaps easier to justify and maintain.

The politics and financial constraints at the organisational level will determine the future sustainability as technical staff retire. Outcomes should and can be positive if communications across organisations are improved and consultations between senior management, stakeholders and technical representatives try to address matters arising in terms of future sustainability.

10.7.3 Attracting New Recruits for the Future

Working as a technician can be a varied and rewarding role with good levels of pay and opportunities for career progression in some areas. One difficulty is that the best technician jobs are not always visible on the radar of most young people and large numbers of young people leaving school are unaware of the qualifications that employers look for when recruiting technicians. A previous lack of good career guidance in schools and colleges has not helped and is perhaps indicative of a wider cultural perception of the technical role. From personal experience in the 1990s, the career advisor did not mention technical roles and instead attempted to direct those with a science interest into a nursing career. For decades, government and society at large have overlooked how important technicians are to the economy.[11]

Modern apprenticeships are available and successful across primarily teaching and engineering areas with many universities and research institutions starting to employ apprentices through use of the apprenticeship levy. Some have devised a structured pathway through the apprenticeship to ensure a wide range of experience is gained.[15] Additionally, the National Technician Development Centre has collaborated to submit an apprenticeship standard for a Level 3 Higher Education Assistant Technician.[16] The difficulties with funding of 'core' technicians supporting the research environments can be challenging. Decisions at senior management level are made over finance and recoup of salaries when everyone strives to grab a piece of the technician's time. Their expertise and time is an extremely valuable resource but not one that can easily be pulled in different directions without loss of stability.

It appears that if the posts exist then there is a willingness of a new generation of key technical staff to pursue such a career. If programmes are varied and job roles diversified, then technicians can be multidisciplinary and move to different parts of the organisation if required. Ensuring that technical career pathways are clearly visible, accessible and stable will enable future generations of technicians to thrive.

10.8 Technician Commitment

More than 83 universities and research institutions from across the UK have backed a pledge to support their technicians. The Technician Commitment is a sector-wide initiative led by the Science Council, supported by the Gatsby Foundation to help address key challenges facing technical staff working in research.[11] Four target areas have been identified, which universities and institutions will work to improve to safeguard vital technical skills. The commitment will ensure greater visibility, recognition, career development and sustainability for technicians across all disciplines.

The commitment comes at a key time as the demand for technicians is increasing. Over 1.5 million technicians currently work in the UK, which is expected to rise by around 70 000 each year. Vice chancellors and directors of research institutes from leading institutions across the UK have recognised the need to ensure sustainability by safeguarding technical skills across their organisations by utilising and developing expertise.

Whilst many educational establishments have embraced the Technician Commitment ethos how that is progressed can vary across organisations depending on the size and nature of the business and the whole establishment. Typically, within a large higher educational organisation, such as a university, then the representation and commitment to such an undertaking requires a carefully coordinated approach led by the organisation and 'buy in' from interested parties across the relevant areas.

Signing up to the Technician Commitment is only the first step in the process by evaluating the impact through self-assessment and future action planning. Maintaining the technical cohort and ensuring business continuity for the future is not an easy task to address and appropriately sustain.

Whilst large organisations may already have several programmes and initiatives in place to support staff generally, prior to embarking on the Technician Commitment programme, these may not have previously specifically involved a focus on the technical staff only. Thereby addressing the specific considerations required for these roles aside from other staff members can allow a more targeted approach moving forward.

10.8.1 Institutional Action Plans

A 'One-year in', report from 2018 [17] across the sector showcases some of the initial impact of the Technician Commitment and illustrates key themes/approaches that are emerging from signatory institutions. A year after pledging their support to the Technician Commitment, signatories undertake a self-assessment exercise and submit a two-year action plan to the Technician Commitment Steering Board. Members of the Steering Board reviewed the first wave of self-assessments and action plans and the main findings were included in the report.

Institutional action plans vary but are reported and available on each signatory organisations website alongside the Technician Commitment statement. In addition to ongoing initiatives an organisation can plan for the future, with support available to scope the nature and extent of support from the National Technician Development Centre.

Examples of themes from institutional action plans include:

- Clear pathways for succession planning for specialist technical roles.
- Workforce planning – training and development for existing and new staff.
- Support for the induction and training of new staff and apprentices.
- Up-skilling of current staff through clear career pathways.
- Signposting towards training from a range of sources for individuals.
- Route to professional registration – guides and mentors for technical staff.
- Provision of an agile and flexible workforce.

- The development and retention of talent through improved induction and faster progression.
- Greater job flexibility and mobility to meet the challenges of the future.
- Increased links with industry.
- The need to improve the student experience.

Of key importance is that the submissions showed that the Technician Commitment has driven an increased awareness of technical staff numbers and structures across institutions and prompted efforts to understand the scope of technical expertise. This Technician Commitment activity is aligning to wider sector initiatives, such as the Teaching Excellence Framework, the Research Excellence Framework and the Athena SWAN Charter.

Professor Sir Anton Muscatelli, Chair of the Russell Group stated that, "We need to think carefully on how we maintain and sustain the pipeline of technical skills essential to our success as over 9000 technicians work within the Russell Group. The Technician Commitment stands as a call to recognise the very real contribution our technicians make to the life, success and achievement of our universities.[17]"

10.8.2 Increasing Recognition

Integral to the Technician Commitment is to provide a range of mechanisms where the contribution and role of technicians can be recognised, celebrated and rewarded. This varies across organisations with several initiatives and processes to increase the recognition of technicians which may include:

- Reward and recognition schemes for one-off bonuses or consolidated pay progression.
- Celebrating excellent service to the organisation or wider community recognising outstanding and sustained service through annual awards.
- Recognition rewards voucher scheme through which excellent service can be recognised and rewarded through a small gift and/or 'thank you' throughout the year.
- Long service awards through which staff attaining long service are recognised and celebrated.
- Organisational newsletters to celebrate and recognize the contribution of technicians individually and collectively.
- Steering and leadership groups.
- Formation of technical websites for information and training opportunities.
- Annual technician conferences.
- Definition of 'researchers' to include all staff including technicians playing vital research roles.
- Leadership development programmes/masterclasses to improve CPD.
- New apprenticeship programmes.
- Programmes of technical events.

The Technician Commitment is fully underway with a strong push apparent to sustain and enhance the technical workforce within the higher education sector. The Technician Commitment is also being embraced by professional bodies and funding bodies with the Institute of Physics launching its own action plan in 2019[18] and UK Research and Innovation signing up to the commitment and providing funding towards technicians' development in 2020.[19]

Ensuring that technicians have the opportunity and financial support to progress and attain professional qualifications and registrations that are recognised across the sector will demonstrate the commitment to the recognition of the technical roles. The Higher Education and Technician's Educational Development (HEaTED) is a leading provider for professional development of the technical workforce.[20]

Various levels of recognised registration such as RSciTech and RSci can also be achieved through the Science Council's initiatives.[21] A few discipline specific professional registrations exist such as the professional registration for IT technicians.[22]

The Recognising Excellence in Teaching (RET) scheme is the University of Glasgow's CPD framework based on the UK Professional Standards Framework for Education (UKPSF) and has four levels including Associate Fellow, Fellow, Senior Fellow, and Principle Fellow, based on the four descriptors within the UKPSF. Technical staff who "support learning or teaching" are eligible to apply for the Associate Fellowship (AFRET). One of the big advantages of this scheme is that it is accredited to a national framework and recognised across the entire higher education sector. Within teaching, The Postgraduate Certificate in Academic Practice (PgCAP) is a formal qualification awarded by the University and forms the first part of the integrated MEd in Academic Practice. Academic staff at the University are contractually obliged to complete the PgCAP as part of the Early Career Development Programme, however technicians who have enough teaching workload can also apply. Technicians are not contractually obliged to complete PgCAP but can choose to do the full programme or pick up specific modules as CPD.[23]

Professional accreditations and qualifications that are recognised across the sector are vital for the recognition and sustainability of the technical workforce moving forward.

10.8.3 Technician Sustainability

There has been significant effort and investment over several decades to improve the number of science, technology, engineering and maths (STEM) graduates, whilst successive governments have turned a blind eye to an equally critical shortage of technicians in these exact disciplines. Following the first year of the Technician Commitment there has been an undeniable shift in culture in terms of how technician roles across higher education and research are viewed, valued, recognised, developed and sustained. The Steering Board reported that they "were delighted with the progress to date and look forward to continuing their work with all signatories to ensure that technical roles, skills and careers are afforded the esteem they so richly deserve.[17]"

In years to come the Commitment may not be necessary as practices put in place become the accepted norm and become embedded into organisational business as standard. A critical mass of signatories developing and enacting their action plans is required in the meantime before that vision can be confidently realised by the steering group.

10.9 Conclusion

The technician role within the higher education and research sector is varied and diverse. However, this role is key to ensuring good levels of health and safety within the sector. Technicians set and enforce good working standards, ensuring through their teaching that students and researchers develop safe skills that will last them throughout their career. Additionally, they are crucial to ensuring the safe use of equipment and instrumentation, and its maintenance and repair. The technical input if reduced, can result in unsafe working practices that in turn can lead to accidents and non-compliance with health and safety regulations. Whilst it is recognised that more technicians are needed for the future, many institutions are facing the loss of experienced technical staff through retirement and the constraints of fixed term employment contracts.[14] Initiatives such as the Technician Commitment and the work of the National Technician Development Centre are increasing awareness of this situation, within institutions and in the wider sector. Through raising awareness and increasing the recognition for technicians it is hoped that the technical role will be secured and that safe standards of working within the research sector can be maintained.

Abbreviations

CE mark	Conformité Européenne mark, a certification mark that indicates conformity with health, safety, and environmental protection standards for products sold within the European Economic Area (EEA).
COSHH	Control of Substances Hazardous to Health Regulations (2002)
CPD	Continuing Professional Development
GLP	Good Laboratory Practice, managerial quality control system covering the organisational process and the conditions under which non-clinical health and environmental studies are planned, performed, monitored, recorded, reported and retained (or archived).
G*x*P	is a general abbreviation for the "good practice" quality guidelines and regulations. The "*x*" stands for the various fields, including the pharmaceutical and food industries.

IOSH	Institution of Occupational Safety and Health
ISO 15189	an international standard that specifies the quality management system requirements to medical laboratories.
ISO 9000, 9001	an international standard that specifies requirements for a quality management system.
ISO 17025	an international standard that specifies the general requirements for the competence, impartiality and consistent operation of laboratories.
ISO 14001	an international standard that sets out the criteria for an environmental management system and can be certified to.
ISO 45001	an international standard for occupational health and safety management systems.
JCoPR	Joint Code of Practice for Research standards for the quality of science and research processes for contractors who carry out research on behalf of the UK Department for Environment, Food and Rural Affairs and other signatory organisations.
LQM	Laboratory Quality Management
NEBOSH	National Examination Board in Occupational Safety and Health
OHSAS 18001	a British standard for occupational health and safety management systems.
PgCAP	Postgraduate Certificate in Academic Practice
PPE	Personal Protective Equipment
RET	Recognised Excellence in Teaching
RSciTech	Registered Science Technician
RSci	Registered Scientist
UKPSF	UK Professional Standards Framework for Education

Acknowledgements

The authors wish to acknowledge the input of the technicians at Cranfield University, Public Health England and University of Glasgow in the creation of this chapter.

Additionally, Jared Carnie at National Technician Development Centre.

References

1. National Technician Development Centre, *Background Information*, 2018, https://nationaltechnicianscentre.ac.uk/background/, last accessed 29 December 2019.
2. P. A. Lewis and H. Gospel, *Technicians under the Microscope*, The Gatsby Foundation, London, 2013.
3. J. Carnie, *National Technician Development Centre*, personal communication.

4. M. G. Björksten, B. Almby and E. S. Jansson, *Appl. Ergon.*, 1994, **25**(2), 88.
5. *Smart Pipetting: Using Ergonomics to Prevent Injury*, http://www.anachem-docs.co.uk/Laboratory%20Ergonomics/Smart_Pipetting_Using_Ergonomics_to_Prevent_RSI.pdf, last accessed January 2020.
6. Health and Safety Executive, *Leading Health and Safety at Work. INDG417 (Rev1)*, 2013, https://www.hse.gov.uk/pubns/indg417.pdf, last accessed March 2020.
7. Science Council, *Technicians: Providing Frontline and Vital Support for Student Mental Health and Wellbeing*, www.technicians.org.uk/assets/technician-commitment/pdfs/report-technicians-student-well-being.pdf, last accessed January 2020.
8. Midlands Innovation, Mass Spectrometry Group, https://midlandsinnovation.org.uk/Midlands-Mass-Spectrometry, accessed February 2020.
9. USHA, *Leadership and Management in Health and Safety in Higher Education Institutions*, 2015, http://usha.org.uk/images/stories/files/guidance-documents/USHA-Leadership-and-Management-in-Health-Safety-in-Higher-Education-Institutions.pdf, last accessed January 2020.
10. I. Schroder, D. Yan Qun Huang, O. Ellis, J. Gibson and N. Wayne, *J. Chem. Health Saf.*, 2016, 12.
11. The Gatsby Charitable Foundation, *Stem Skills in the Workplace*, https://www.gatsby.org.uk/education/focus-areas/stem-skills-in-the-workforce, last accessed January 2020.
12. The Gatsby Charitable Foundation, *Technicians Make it Happen*, https://www.technicians.org.uk/technician-commitment, last accessed January 2020.
13. A. Webber, *Job Insecurity Fuelling University Staff Hardship*, https://www.personneltoday.com/hr/university-job-insecurity-fuelling-staff-hardship/, last accessed January 2020.
14. UCU, *Research Managers Employing Fixed Term Staff*, https://www.ucu.org.uk/media/5242/Advice-for-research-managers-employing-fixed-term-research-staff/pdf/ACC_res_managers_briefing.pdf, last accessed January 2020.
15. Imperial College London, *Apprenticeship Structure*, https://www.imperial.ac.uk/jobs/career-programmes/apprenticeships/faqs/technician-apprenticeship-scheme/structure/, accessed February 2020.
16. National Technician Development Centre, *HE Assistant Technician Standard*, https://nationaltechnicianscentre.ac.uk/he-assistant-technician-standard/, last accessed February 2020.
17. Technician Commitment, One-year in, https://www.technicians.org.uk/assets/technician-commitment/pdfs/technician-commitment-one-year-in.pdf, last accessed January 2020.
18. Institute of Physics, *IOP Celebrates Launch of Technician Commitment Action Plan*, https://beta.iop.org/iop-celebrates-launch-of-technician-commitment-action-plan#gref, last accessed February 2020.
19. UKRI, *Technicians Receive Support from the UK's Largest Research Funder*. https://www.ukri.org/news/technicians-receive-support-from-the-uks-largest-research-funder1/, last accessed March 2020.

20. STEM Learning Ltd, *Higher Education and Technician's Educational Development (HEaTED)*, https://www.stem.org.uk/heated, last accessed January 2020.
21. Science Council, *Scientists and Science Technicians*, https://sciencecouncil.org/scientists-science-technicians/, last accessed January 2020.
22. Chartered Institute for IT, *Professional Registration for IT Technicians (RITTech)*, https://www.bcs.org/membership/get-registered/professional-registration-for-it-technicians-rittech/, last accessed January 2020.
23. University of Glasgow, *AFRET & PGCAP for Technicians*, https://www.gla.ac.uk/myglasgow/staff/technicians/news/headline_651446_en.html, last accessed January 2020.

CHAPTER 11

Liaising with the Emergency Services

J. BLUNT*

Formerly Safety Officer at The Department of Physics, University of Cambridge, Cambridge, UK
*E-mail: fjb27@cam.ac.uk

11.1 Risks to Emergency Personnel

There are two sorts of emergency personnel who may attend your institution or department – there are the external emergency services such as the Fire and Rescue Service, Police and Ambulance service, and the services internal to the organisation, such as the Security or Estate Management services.

During emergency situations, it is frequently the case that the identified control measures have failed in some way, and a dynamic risk assessment may be necessary in order to decide on the safest course of action. It will frequently be necessary to pause and repeat this process in the light of developments during the handling of the emergency. Much of the information required to conduct this assessment will be possessed by one or more of the department researchers, maintenance personnel or safety personnel. Whoever is taking charge of the department's emergency response may need to draw on the expertise of many other people and liaise with the emergency services in order to ensure their safety.

In most departments there will be a control point, which is frequently close to the master fire panel, where any designated emergency team will be collected and where the Fire and Rescue Service will report if they are called to a situation.

Challenges for Health and Safety in Higher Education and Research Organisations
Edited by Olga Kuzmina and Stefan Hoyle

Published by the Royal Society of Chemistry, www.rsc.org

Science research institutions can contain many hazards that are unfamiliar to emergency personnel. These can include (not an exhaustive list) biological hazards, chemical hazards, asphyxiants, gas cylinders containing a variety of hazardous gases, explosives, high magnetic fields, optical radiation, ionising radiation and high voltages. It is the nature of research that these hazards are not always present in a readily recognisable form to the emergency personnel and, in consequence, they may not recognise the risks, or may misinterpret the risks, and not know how to avoid harm.

In general, the emergency services will adopt a cautious approach and this may severely delay the resolution of the problem. It may lead to medical help for personnel being delayed, or increased damage to the premises. Nevertheless, if they do not have knowledge of an unseen risk (such as asphyxiant gases, magnetic fields, *etc.*) they can still risk injury, however cautious they are. Departmental personnel play a key role in preventing this from happening.

During the normal working day, the research personnel and support staff will almost certainly be available to guide and assist the emergency personnel. The training regime should encourage research personnel to come forward and help in the event of an emergency that involves their own laboratories; their input is crucial to the safe and speedy handling of emergencies in their laboratories.

During the night and at weekends the emergency services may enter the building before there is anyone available with the knowledge and authority to assist them.

By providing information to the emergency services, and fitting suitable alarms and signage, the risks that they encounter can be minimised. While there may be 24 hours security provision in your buildings, the personnel who run these services are unlikely to be fully aware of the risks in all of the laboratories and they will also need information and guidance.

In the author's experience the Fire and Rescue Service pay fairly frequent familiarisation visits to the premises and this is an ideal opportunity to show them the information you have compiled, discuss the risks with them and ask for their advice and recommendations. They may be keeping an information card in their vehicles, which gives them an overview of the significant risks in your premises. These familiarisation visits are a good opportunity to reinforce understanding of some of the less obvious risks to firefighters.

11.2 Effective Internal Procedures

11.2.1 Effective Evacuation Protocols

In the event of a fire occurring in the building, the Fire and Rescue Service's first consideration is the preservation of human life. They will want to know with certainty that the buildings are empty and that there is no one to rescue. This takes priority over the fighting of the fire. In order to avoid unnecessary delay in tackling the fire, or exposing the Fire and Rescue personnel to unnecessary risk in entering the building to conduct a search, the department needs to train its personnel to evacuate the building when the alarm is raised and it needs a protocol for ensuring the building is empty. In large and complex buildings, with occupants ranging from undergraduates to postgraduates, visitors, members of other departments, and overseas visitors, it is a challenge to be sure that the buildings are empty.

Obviously, everyone will have been trained that they should leave a building by the nearest exit when the fire alarm sounds and gather at the nearest assembly point. However, human nature dictates that some people will be reluctant to leave, some will deliberately walk through a building in which the alarm is sounding in order to reach another building, and others will hover around the exit door rather than go to the assembly point. These behaviours need to be tackled at every opportunity (*e.g.* when conducting evacuation exercises) and the dangers that they pose need to be highlighted. Unfortunately, a few personnel may become used to the idea that when the fire alarm sounds it is only a practice, a test, or a fault and may stay in their rooms. There is also considerable evidence that most people have a strong preference for leaving the building by their preferred entry point, and they need encouragement to explore a wider range of options, so that they can identify their nearest fire exits and alternatives.

Because of the diverse and changing occupancy of higher education and research buildings, it is not normally possible to have a roll call or any other means of positively identifying who has entered and who has left. In these circumstances one solution to ensuring the building is empty is to have fire wardens or appointed emergency response teams do a sweep of the buildings. In the institutions in which the author has worked, to allow each fire warden to complete their task in a short time, the buildings were divided into search areas containing up to about twelve rooms each. Each volunteer fire warden was given a search area – which included their own office or laboratory. They were given laminated plans showing exactly which rooms and areas were to be searched. The instructions were printed on the reverse. These instructions included checking door handles for heat, what to do if a fire was encountered, how to check the toilets (of both genders) and what to do if a casualty was found. The area to be searched was shaded in colour on their plan and a master plan was situated beside the fire master panel (the control point). Since the coloured plans resembled the pieces of a jigsaw it was very easy to see at a glance whether every area had been checked (Figure 11.1). Fire wardens were asked to bring their plan to the control point when they had finished, and report anything important that they had encountered (such as a person refusing to leave) prior to going to the assembly area.

This method of checking is not guaranteed success. The fire warden may be absent from their area at the time of the alarm. Having nominated deputies can help to reduce this risk, but this doubles the number of people needing training and the maintenance of an up to date list of trained fire wardens can become a problem. In the author's department there were over

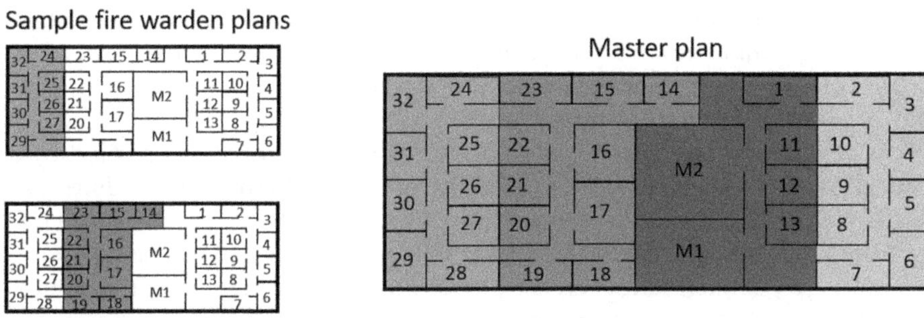

Figure 11.1 An example of organising search areas for fire evacuation.

fifty search areas due to the sheer size of the department. The fire warden may also forget to bring their plan to the control point, so that confirmation that their area is empty is lacking.

If evacuated personnel remain close to the exit doors this poses a challenge if there is a real emergency. In the event of a fire involving gas cylinders, the evacuated personnel will need to be moved to a place of safety hundreds of metres away. Clearly if everyone is at the assembly areas it is a lot easier to relay this information to them than if they are scattered around the periphery of the buildings. Staff meetings and training courses can be used to reinforce the importance that people should go to the assembly areas. It needs to be emphasised that assembly areas are not necessarily places of safety, but are points at which people should gather to receive further instructions.

At night and during the weekends it is even more of a challenge to establish that the building is empty. In the event that the department has only a single night entry point it is relatively simple to have a signing-in and signing-out book. If there are multiple entry points, and the buildings are large and interconnected, it is not feasible to do this efficiently. One solution is to have a computerised system, with terminals at the entry doors. The software can be tailored to ask the person to sign in and give details of their likely whereabouts. A terminal can be placed at the control point where the emergency team congregate to handle emergencies, with a printer to print off a list of who is in the buildings. Unfortunately, this system does rely on personnel remembering to sign in and out. Spot checks can help to improve the compliance with this system. They may need reassurance that the data is wiped daily so that no one can scrutinise their working patterns for any purpose.

Such a signing-in system can be more efficient than using the door entry system; in the author's experience the latter just lists people going in and out but does not carry a running record of who is inside. The door entry system also fails if two or more people leave or enter using a single card and does not lend itself to indicating the location of the person if there are several interconnected buildings.

If there is an emergency evacuation at night people should be trained either to sign themselves out on leaving or, if they are unable to do so, to go to the control point where the Fire and Rescue Service will report, so that they can confirm that they are safe.

It is suggested that the department emergency response is handled by a team of people. There may be a need for one or more people to distribute instructions to evacuated personnel, and there needs to be someone to liaise with the emergency services. The team may include the safety officer, facilities manager, department secretary, chief maintenance technician and others. A team of this nature combines a good mixture of authority, knowledge and skill.

11.2.2 Effective Casualty Management

First aiders, First Responders and Ambulance crews are all taught to assess the situation for danger before approaching a casualty. Therefore, wherever possible, the area should be made safe before the First Responders and/or Ambulance crew arrive. First Aiders should be encouraged to take the lead in making the area safe, as they will often be on scene before the emergency services arrive. It is their task to try to prevent the casualty's condition from worsening and they should be given the authority to request isolation of supplies, clean-up of spills, ventilation, or removal of a casualty to a safer place.

It is advised that the training courses in risk assessment should mention that it would greatly help emergency response if the principal investigators and research workers would plan how to make the situation safe, or plan how to remove the casualty from the danger area, while at the same time summoning help. Moving the casualty will be in their best interests in the case of fumes or spillages; provided there is no possibility of a spinal injury. They should remain on hand to give any specialist information, such as the safety data for a substance to which a casualty has been exposed.

11.3 Specific Risks in Scientific Institutions

The following are discussed, in alphabetical order, in the following sub-sections:

- asphyxiating atmospheres
- biological hazards
- bottled gases
- chemical hazards
- explosive atmospheres
- explosive substances
- ionising radiation
- lasers
- magnetic fields

11.3.1 Asphyxiating Atmospheres

Asphyxiating atmospheres are a particular challenge, as they have no warning properties and people can inadvertently walk into danger if they are not wearing a working oxygen monitor. There have been numerous fatal accidents associated with asphyxiating atmospheres. Many are traced to poor ventilation and oxygen monitors not being switched on.[1,2]

In the author's experience, training in using oxygen monitoring is not always effective. People can be issued with a personal oxygen monitor, trained about the dangers of asphyxiating atmospheres and shown how to check that the monitor is working properly before using it in an emergency. However, their understanding may nevertheless be poor. The author once asked a security guard what he would do if, when walking along the corridor, his oxygen monitor went into its alarm mode. His response was that he would head for the nearest fire exit. Unfortunately, this might be on the far side of the oxygen deficient area. The training course was adapted in the light of this misunderstanding and guards were explicitly told to turn around and retrace their steps into the area from which they had come, and that they knew to be safe.

On another occasion a personal oxygen monitor worn by a security guard had alarmed in an atmosphere where there had been a large helium release. On discussing this with the security guard after the incident the author was surprised that he was unaware that his oxygen monitor could have saved his life. On further discussion he revealed that he felt he had not been in any danger because helium was merely the gas that made your voice sound like Donald Duck. The author suspects this might be a fairly common belief! The training course for security personnel was amended in the light of this revelation, to stress the dangers.

Significant sources of asphyxiating gases include dewars which have failed or whose valves have stuck in the open position, quenching events in superconducting magnets and leaks from piped supplies.

A dewar has an internal vessel which is suspended from a neck. Transporting the dewar over rough surfaces can cause the internal vessel to vibrate and this can, over a period of years, cause a fatigue failure of the neck. When this happens the liquid nitrogen leaks into the vacuum jacket and causes the pressure to rise. The plug will then blow out and because the thermal insulation is now absent the dewar will begin to boil off and vent gas at an increased rate. The author encountered a dewar that was about 35 years old, which failed in this way. If the failing dewar can be approached safely (using an oxygen monitor), removed to a place of safety such as out of doors or a space with a very large volume of air then it can be left to vent until it is empty. If there is a casualty near the failing equipment, an attempt may be made to retrieve them by the emergency services. However, in attempting to remove a casualty, it is imperative that an oxygen monitor is used to check that the would-be rescuers are not entering an oxygen deficient atmosphere. Unfortunately, if the atmosphere is still oxygen deficient, and ventilation cannot be improved to make the area safe, a casualty can only be retrieved by emergency responders with breathing apparatus.

The department may have pipelines containing liquid nitrogen, piped supplies of nitrogen gas from bulk containers (*e.g.* nitrogen tanks containing more than 1000 Litres) or piped supplies from a nitrogen generator. Some research departments also have helium or argon gas pipes. If any of these are present the risk assessment for activities in the rooms that are fed these supplies should include a consideration of the possibility that the gas might escape into the room continuously. For example, the hoses attached to the supply could become accidentally disconnected. Where the risk assessment indicates that this could happen, it is advised that some calculation is undertaken to decide whether an oxygen deficient atmosphere can develop. There are some sample calculations in the British Compressed Gases Association code of practice CP27.[3] If there is a reasonably foreseeable risk that an oxygen deficient atmosphere can develop oxygen monitoring for the room is strongly recommended. The alarm should be audible, and the percent oxygen in the room should be visible from outside the room where practicable. In new buildings it is possible to incorporate proper monitoring and automatic shut-off for the supplies at the construction and fit-out stages of the laboratory. It is more difficult to fit retrospectively.

There should be labels on the door telling people what they should do in the event that the alarm is going off. This label should warn them not to go into the room without checking the oxygen level is safe. Therefore, there should be some monitors available for emergency use, possibly at the fire alarm control point, and the notice should give the user instructions as to how to check that the monitor works properly. The check with this second monitor may well show that it is a false alarm, in which case the faulty alarm should be disabled and the person responsible for the room should be informed of what has happened.

The procedure should also include how to make a safe check for a casualty, what to do if a casualty is present, and contact details for the person responsible for the room.

The people responsible for the room should be aware of the layout of the supplies and the cut-off points. However, in the event that they cannot be contacted, plans should be available which show the layout of any pipes containing cryogens and asphyxiating gases, showing the location of all the valves. This will enable any incident involving a breach of the pipework to be made safe as quickly as possible.

Carbon dioxide is also supplied to some laboratories in large quantities, either *via* pipes or from stores of dry ice. Many people are of the mistaken belief that carbon dioxide is an asphyxiant, but it is also toxic and therefore it is appropriate to have a carbon dioxide monitor, and not an oxygen monitor in these laboratories. The short term exposure limit for carbon dioxide, according to EH40,[4] is 15 000 ppm, and if carbon dioxide is released to this concentration in a laboratory it will only result in a fall of 0.5% in the oxygen level. Oxygen monitors are routinely set to alarm at 19.5% oxygen, which is significantly lower, so they fail to give adequate warning of rising carbon dioxide levels.

The risk assessment for areas that can become oxygen deficient should include a review for the need for a ventilation alarm. This alarm indicates if the ventilation has failed, making the build-up of hazardous gases a possibility. The lab users can then arrange for the ventilation to be fixed as soon as possible, can suspend work if necessary and are aware that they need to use gas monitors to check the atmosphere before entry.

11.3.2 Biological Hazards

If the institution works with biological agents or has biosecurity areas in which there are items listed in ATCSA 2001 Schedule 5[5] they will have classified the biohazards according to the risk to the workers and public, and there will be appropriate containment levels (CL1 to 4) and where appropriate, security measures instituted. There are approximately 600 CL3 laboratories in the UK, and approximately nine that are CL4.

If the laboratories affected are only Containment Levels 1 and 2, there is little to report to the emergency services if they are called.

However, if there are Containment Level 3 and/or 4 laboratories it is advised that the handling of emergencies such as fire or flood is discussed beforehand with the Fire and Rescue Service. For Containment Level 4 laboratories, the Fire and Rescue Service are likely to adopt a policy of allowing the facility to burn down in the event of a fire. It is important to develop a procedure for dealing with a casualty in such a laboratory. CL3 and CL4 laboratories will have specific emergency procedures and are required to practice emergency responses regularly.

11.3.3 Bottled Gases

Bottled gases can present a variety of hazards, particularly in a fire. Standard gas bottles/cylinders, such as nitrogen, argon, oxygen *etc.* may explode if exposed to fire while the main valve is closed. If the main valve is open, the gas will vent its contents into the room at some time during the fire.

It is preferred that gases are piped into the building, with the means to shut off the supply out of doors, but it is recognised that few laboratories have the infrastructure to allow them to keep all the gas bottles out of doors. Ethyne (which the emergency services will know as acetylene) is particularly dangerous as it can polymerise inside the gas bottle and subsequently explode. Gases such as propane are stored in containers which have pressure relief to prevent explosion. However, the leaking gas adds to the fire risks.

You need to be able to indicate the approximate locations of gas bottles in the event of fire, so that the fire fighters can take suitable precautions. They will definitely want to know the location of any acetylene cylinders on site. It is appropriate to include gas bottles on the door signage.

Some toxic gases have pungent smells at low concentrations, well below their exposure limits; these can be said to have good 'warning properties'. Gases with poor warning properties include carbon monoxide and carbon dioxide which cannot be sensed by human beings, while a gas with good warning properties is sulfur dioxide as its presence could be identified by smell at low concentrations.

If there are any toxic gases, appropriate monitoring is needed, particularly if the gas has poor 'warning properties' and it is important that the laboratories are labelled to indicate the risk, how people will recognise that there is a dangerous condition, and what they should do about it.

Guidance on special gases is available in British Compressed Gases Association code of practice CP18.[6]

The Fire and Rescue Service prefer gas bottles to be in the open, so that they can spray water on them from a safe distance. Gas cabinets make this difficult. Guidance is given in CP18[6] regarding which gases should be in cabinets. In some cases, insurance companies require gas cylinders to be located in ninety-minute flame proof cabinets in 'new builds' to ensure that the insurance is validated.

11.3.4 Chemical Hazards

Laboratories may contain a vast array of chemicals. Some of these will be in very small quantities, but some will be present in relatively large quantities. Note that the Fire and Rescue Services are not a substitute for your own local chemical spillage emergency response team!

Signage on laboratory doors is very important, but it can be overzealous. If the door of a room is decorated with the hazard warning sign for every one of the chemical hazards in the room it can be daunting. Emergency personnel are likely to refuse to cross the threshold until assured that it is safe to do so. A small bottle of correction fluid is labelled with a highly flammable sign, but the presence of this in a room is clearly a negligible risk. On the other hand, several Winchesters (2.5 Litres) of ether or hexane would present a real risk in the event of fire and, if spilt in very large quantities, might also present an explosion risk. This was the cause of a serious fire at Ohio State University[7]. The account of this fire tells us that the damage was minimised because there was a professor on hand who could give the fire fighters the information they needed to enter safely.

A puddle of liquid on the floor could be water, but it is helpful to know whether it could be something much more hazardous, such as concentrated sulfuric acid.

It is recommended that you consider how much of any particular chemical substance should trigger a warning sign on the door. If thresholds are set, and applied consistently across the site, this is of benefit to everyone. The author has found that estimates of what quantities of chemicals pose a danger to emergency services, and what are not, vary enormously from people in one discipline to another and even within the same discipline. They can range from the overcautious to the 'gung-ho' approach. It is worth trying to reach a common agreement.

It is debatable whether you should attempt to list all the chemicals in any laboratory. Most are likely to be present in quantities that pose insignificant risk, and the few that do pose a significant risk may well go un-noticed in the list due to its sheer length. You will probably come to the conclusion that listing everything is not a worthwhile exercise, and instead focus on any substances of high intrinsic hazard and those which represent a risk by virtue of their quantity.

The Institution's chemical store itself is likely to contain large quantities of hazardous substances. It may be appropriate to have a notice on the outside of the door to indicate the approximate current contents of the store. As an example, a store was observed which had a notice attached to the outside of the door to record the approximate contents of the store (Figure 11.2).

The notice was laminated to permit writing in the current figures in the appropriate places with a dry-wipe pen. In a fire situation, this information tells the emergency services the likely scale of any problem in the room. The chemical waste stores had a similar notice.

It is recommended that precise details of the chemicals within the store is not given on the sign for security reasons. For the same reason, although it is good practice to have a window in the door so that you can check for a casualty without entering the room, it is advisable that labels on cabinets naming substances such as sulfuric acid, acetone *etc.* cannot be seen from the door. These materials can be used for making improvised explosive devices by terrorists. If the labels can be seen it may render the institution vulnerable to break-ins and theft.

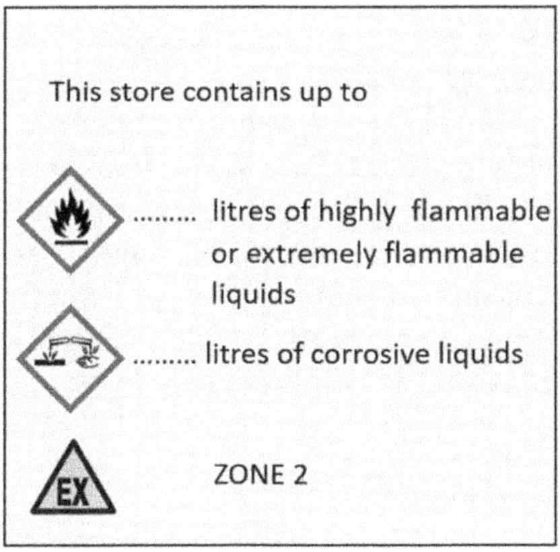

Figure 11.2 A possible sign for a bulk chemical store or waste store.

11.3.5 Explosive Atmospheres

Explosive atmospheres can be created by release of highly or extremely flammable gases such as propane and hydrogen, or by spillage of volatile highly flammable substances. In an ideal world, the people responsible for the laboratory should have addressed the possibility of a leak when compiling their risk assessment and safe systems of work. They should have limited the quantity of material in the room to minimise the risk of explosion and they should have planned what they should do in the event of a mishap. However, it is reasonably foreseeable that even if the bulk supply is outside the laboratory, there could be a substantial release inside the laboratory as a result of failed components.

High velocity oxyfuel spraying facilities, which are used to produce ceramic, metallic and composite coatings, use very large quantities of their fuel gases. They will often be supplied from palleted cylinders. A pallet of hydrogen consists of a number of hydrogen cylinders that are manifolded together to keep the spray gun supplied continuously for several minutes. The pallet of gas will be out of doors, and the storage area should be classified according to BS EN 60079 Part 10,[8] to indicate the extent of the possible explosive atmosphere. These areas should have prominent warning signs indicating that they could become explosive so that only the correct type of electrical apparatus is taken into the area ("EX" signs).

The Fire and Rescue Service are equipped to test atmospheres for flammable gases, when alerted to their possible presence.

11.3.6 Explosive Substances

A small number of institutions undertake research using explosives. This may include both primary and secondary explosives. They will have a secure dedicated magazine for the storage of the materials, with an alarm to alert the Police if the door is opened by an unauthorised person. Such installations are subject to licensing and are overseen by both the Health and Safety Executive and the Police force. An armed response is to be expected in the event of an alarm. Due to security considerations, the locations of these stores are generally known to a relatively small number of people.

In a fire situation, any primary explosives present an increased risk, but secondary explosives will burn quietly since they require a detonator to explode. In the case of the latter, continued security of the store is the primary concern. The Fire and Rescue Service will need to be briefed on the classes of substance that are present, in the event of a fire in the vicinity, particularly if the store contains primary explosives.

11.3.7 Ionising Radiation

Sources of ionising radiation that are electric, such as X-ray equipment, should have their power supplies interlocked to the doors so that it is not possible to open a door without de-energising the equipment. There should also be illuminated warning signs to indicate when the apparatus is energised.

Many research facilities work with radioactive sources. They should be stored as required by the Health and Safety Executive and Environment Agency. If a source is actively in use in a room it is good practice to have a label on the door to indicate its presence.

In the event of a fire it is possible that the shielding of radioactive sources is compromised. If sources are housed inside a lead castle, a fire might melt the lead. The Fire and Rescue Service will need to be warned of this possibility so that they do not expose themselves to radiation when they are conducting a survey of an area in the aftermath of a fire.

11.3.8 Lasers

Lasers are classified according to the potential to cause damage to vision and to cause fire. Beams from classes 3B and 4 should be enclosed, and the power supplies should be interlocked to the enclosure so that someone opening it will automatically switch off the laser. However, it is not unusual for the interlocks to be linked to the door to the room. In these circumstances there may be an over-ride code for authorised people to enter and leave the room, opening the door for a brief period. If a member of any of the emergency services is to be admitted the laser should be switched off in order to rule out any possibility of the person coming to harm.

11.3.9 Magnetic Fields

Many institutions have superconducting magnets. These will often be operating in persistent mode, so are effectively permanent magnets. Many of these have fields that extend to significant distances from the magnet. There may be markings on the floor to indicate the extent of the '5 Gauss' line. Before emergency personnel enter a room containing such a magnet they should be briefed about the extent of the significant magnetic fields and the risks that they face.

They should also be briefed about what happens if the magnet quenches. They need to recognise that it is happening, understand how the gas will be vented and what they should do in the event of a quench.

11.4 The Premises Information Box

The Premises Information Box (PIB), which will often be sited close to the control point in the department, can be a rich source of information for the emergency services. The amount of information you put in it may be restricted by security concerns. The following sections make some suggestions for its contents. In addition to details of the hazards and the utilities discussed below, in the author's department a list of the laboratories and the contact details of the people who should be called to each in an emergency was kept in the PIB.

It is recommended that an inventory of the documents is kept in the box, and that each document is dated. Some documents, such as those detailing the mains gas supplies, will stay valid for many years but others may need to be reviewed on a frequent basis. If they are dated and the inventory records the date of the most recent version, you can be sure that the correct document is being used.

Note that in an incident it may not be safe to access the PIB, so it is recommended that copies of all the relevant documents are also available online.

The following are discussed, in alphabetical order:

- biological hazards
- building plans
- chemical and gas stores
- electricity
- explosives
- mains gas supplies
- piped asphyxiants
- radioactive substances
- water supplies and drainage

11.4.1 Biological Hazards

The locations of Containment Level 3 and 4 laboratories should be clearly marked on the plans in the PIB.

11.4.2 Building Plans

In the event that the Fire and Rescue Service need to conduct a search, they will need floor plans. It is advised that two to five copies of the floor plans are kept in the PIB. They would use these plans to mark off each room that they have checked. If the building is very large, a floor may need to be searched by a team of several people, hence the need for several copies.

Site plans are also useful, as there are circumstances in which the Fire and Rescue Service will want to set up an exclusion zone of 200 m or more. A plan of the site, approximately to scale, with a marker showing 100 or 200 m, would prove useful. On this plan is it also helpful to indicate any adjacent residential areas or public roads, in case the incident puts them at risk.

11.4.3 Chemical and Gas Stores

The location of the bulk stores of chemicals, the gas stores and the waste stores should be marked on a plan.

As mentioned above, the doors to the chemical and waste stores should give some indication of the scale of the hazard.

The risks from gas stores with mesh walls or fronts are easy to assess during an incident as the bottles are readily visible. For gases that are in cabinets (*e.g.* ventilated cabinets commonly used for gases such as silane) then the plan should show what the gases are.

Any EX rated areas, which may arise from the use of palleted flammable gases, or the main bottled gas stores, should be marked on the plans.

11.4.4 Electricity

In an emergency situation the Fire and Rescue Service may need to know where the electrical substations are located on the site. It is useful to have a plan showing these, and a list of contact numbers for the people who are competent and authorised to access them.

11.4.5 Explosives

For reasons of security it is advised that the precise location of an explosives store is not marked on the plans. One way to alert the emergency services to the presence of danger is to have a label on the plan to indicate that for any emergency in that area of the building the named person in charge of the installation, or the Safety Officer should be contacted for advice.

11.4.6 Mains Gas Supplies

It is useful to have an outline of the buildings on site, with an indication of where the mains gas enters the site. This may be from more than one pipeline; in the author's department the gas came from several separate supplies, entering the site from three different directions. The destinations of the supplies could be included (*e.g.* supplies to the kitchens, supplies to the undergraduate teaching labs).

The positions of the main shut-off valves, both internal to the building and external to the building should be shown on this plan. The valves themselves should be clearly labelled to indicate what areas of the building are isolated by operating them.

11.4.7 Piped Asphyxiants

If asphyxiant gases are piped into the building from essentially unlimited sources (*e.g.* a nitrogen generator or the boil-off from a 30 000 Litre Tank) it is important to have shut-off valves in strategic positions (*e.g.* outside the rooms where they are used). The plans in the PIB should show the network of pipes, and the locations of the cut-off valves, to enable any leaks to be isolated. A simplified hypothetical plan is shown in Figure 11.3.

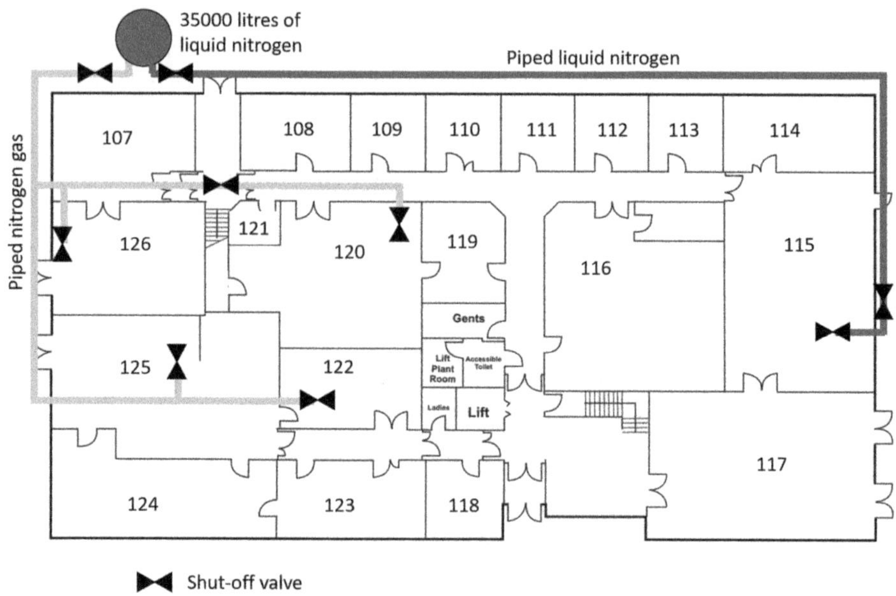

Figure 11.3 An example of a plan showing location of cut-off valves on the route of the piped asphyxiants.

11.4.8 Radioactive Substances and Radiation

The location of radioactive substances in storage should be available to the emergency services. It may be felt that, for security reasons, the information on the PIB plans should not be explicit. If this is the case, then the floor plan can be annotated in a similar way to the case for explosives stores.

11.4.9 Water and Drainage

The outline plans of the building(s) and nearby sites can be used to show the layout of the mains water supplies. Again, it is helpful to indicate the destinations of the supplies, and the locations of any valves where the supplies can be cut off.

The fire hydrants can be marked on this plan. It is advised that a check is made to ensure that they are all still available.

The site will have both surface water drains and foul water sewers. The destinations of the water from the surface water drains should be known. This may be to a balancing pond and then *via* a weir to a local river. In the event of an accidental spillage into the surface water drains information on the destination of this water could be crucial to preventing an environmental incident.

11.5 Door Signage

The signage on laboratory doors should reflect only what is in the room. It could be in the form of pictograms, with words to clarify the meaning. It might include the following, where appropriate:

- The location of gas bottles, and their contents.
- The presence of chemicals, above the agreed thresholds, *e.g.* highly/extremely flammable liquids to a combined total of 5 Litres, or toxic/very toxic/extremely toxic substances to a combined total of 1 kg.
- Lasers.
- Biological hazards and the classification of the room.
- Large magnets.
- Asphyxiants.
- Sources of ionising radiation.
- Electricity – the location of the master switches, if there is a distribution board in the room.
- Alarms, and any special instructions to go with them. If any alarms can be safely ignored this should be indicated.
- Contact details for people who can give specific information on the room, if needed.

11.6 Case Studies

The following are three real incidents. Each had the potential to cause serious harm or even a fatality. The aftermath of each resulted in a great deal of reflection on what aspects were well-handled and how procedures and performance could be improved. A no-blame culture can make this process much easier to achieve.

11.6.1 Nitrogen Escape

This incident took place in a department which had a piped supply of nitrogen to numerous laboratories, fed from a 35 000 Litre liquid nitrogen tank. A rough calculation showed that this quantity, if it boiled off and escaped into the building, could fill approximately half the building with nitrogen.

One evening some research students working with a glove box fed from the nitrogen line saw a glove blow off. They attempted to fit a blank plate over the opening, but found they were unable to do so due to the pressure from the gas flow. They alerted their supervisor who went to the nearest cut-off valve and stopped the flow to their lab. They then took an oxygen meter and went to another lab where they had further equipment attached to the nitrogen line. They heard hissing, and carefully putting the oxygen meter on outstretched arms through the door they noted that the oxygen level in the room had fallen significantly. They called the Department Emergency Team and the Fire and Rescue Service.

At that time, the leader of the Emergency Team was not aware of the presence of the nitrogen piping and was informed that the whole building had a network of nitrogen pipes, all connected to the bulk liquid nitrogen tank. Also, at that time, oxygen monitoring was not common in the building. The first priority was felt to be to get everyone out of the building and to check each room that had a nitrogen supply, to ensure there were no casualties. The fire alarm was sounded to signal to everyone to leave the building. A technician with knowledge of the layout pipework was found and he listed the rooms which needed to be searched. The Fire and Rescue Service were able to do this, wearing full breathing apparatus. They were reminded that an oxygen deficient atmosphere gives no warning and sudden death can take place. Therefore, they should not remove their breathing apparatus until after they had come out into the fresh air.

The building contained in the order of 300 rooms, but those with nitrogen supplies numbered only about thirty. Focussing on only these rooms enabled a quick reassuring search to be done by the Fire and Rescue Service. Fortunately, there were no casualties.

The next priority was to investigate what the problem was and restore the system to a safe condition. Inspection of the liquid nitrogen installation revealed that the main nitrogen tank had a pressure which was about four times the maximum stated on the gauge. This was clearly a concern! At that time neither the leader of the Emergency Team, nor the technician had information on the safe working pressure of the tank itself, or precisely what safety features it contained.

The gases technician opened valves to vent gas and reduce the overpressure. The main valves to the lines that fed the supply to the building were all closed, as a precaution, in case there were further leaks that had not been located.

Once the pressure had been brought down to safe levels, the supplies to the building had been terminated and it had been established that there were no casualties, the Fire and Rescue Service left the site.

For added reassurance a team of two people, carrying an oxygen meter, searched every room and space in the building, in order to check that it was empty. They then sealed off all entrances to the building, pending fuller investigations in the morning.

The following morning the gas supplies were restored to the building and a careful check of all the rooms and pipework was carried out to check for leaks.

In the investigation after the event it was believed that the fault lay in the system that maintains a small over-pressure in the liquid nitrogen tank. A small amount of liquid nitrogen is allowed to trickle out of the tank into a radiator, where it boils, and it is then fed into the top of the tank to maintain the pressure above the liquid. When the pressure reaches the normal working pressure the flow of nitrogen is stopped. It is believed that an ice crystal had become lodged in the valve causing it to be stuck in the open position. Consequently, too much nitrogen was turning to gas, raising the pressure in the tank, and raising the pressure in all the downstream pipework.

This incident was handled effectively, and there were no casualties. However, it highlighted the following important points, and our emergency procedures were reviewed and improved as a result:

- It is essential to brief the Fire and Rescue Service if they face unfamiliar risks. Smoke, while highly toxic, has very good warning properties but oxygen deficient atmospheres have none.
- Oxygen monitoring was essential in any laboratory fed from the piped gas. A programme was launched to ensure this happened.
- The nitrogen pipework to the building needed a pressure relief valve to prevent the pressure in the pipework in the building from rising above the normal level, in case the system ever developed a similar fault in the future. This was speedily fitted.
- The room where hissing had been heard was found to have a connection to the nitrogen line which consisted of a PVC tube pushed over a tapered plastic nozzle. This was easily blown off. A programme of education and awareness was begun to highlight the need for robust secure fittings to the nitrogen supply.
- A map of the lines, and the locations of all the valves was essential for the emergency team and was drawn up within a week.
- It was essential to know the maximum working pressure and the safety devices that the main tank possessed. This was so that the Fire and Rescue Service could make an informed decision about their response.

11.6.2 Uncontrolled Release of Extremely Flammable Gas

On one occasion there was a failure of a piece of gas control equipment in a laboratory. Enquiries on the university safety networks in the aftermath revealed that no one else had ever suffered such a failure, and it was therefore not possible to foresee. The consequence of the failure was that a very large quantity of an extremely flammable gas was released into the laboratory. If this gas cloud mixed with the air and reached a source of ignition, the consequence would have been a very large explosion, sufficient to demolish the building and send shrapnel over a very wide area.

The researcher workers in the lab attempted to stem the flow using the main valve on the cylinder but were not certain that they had been successful as they could still hear gas escaping. The valve was icing up and very stiff. They then evacuated the whole building by word of mouth (fearing that the fire alarm would be a potential source of ignition) and they called the Fire and Rescue Service. The senior researchers then reported to the reception desk of the adjacent large department. They recognised that there was significant danger, and their fire alarm systems were connected to the master panel in the large department. The large department's emergency team was summoned.

The risk of significant explosion was instantly recognised. The first consideration was whether to evacuate the adjacent buildings and, if so, how to do this. If there were to be an explosion anyone in front of the building would probably be killed, and the explosion would cause serious damage to the adjacent buildings. The roof of the adjacent building was rather thin and debris would almost certainly penetrate it. There was also a footpath and cycleway which would shortly be full of over a hundred undergraduate students on their way to their afternoon activities. One assembly area was dangerously close to the site of the incident and to get to it people would walk right past the most dangerous area. The general layout is shown in Figure 11.4.

This figure shows that there were six main entry points to the site, through which people on foot and on bicycles could approach the danger area. Several buildings were within approximately 200 metres of the incident. The site of the incident is a building that is cross-hatched and marked with a large star in Figure 11.4, and the buildings within an approximate 200 m radius are shaded in the darker grey. Buildings beyond this are shaded in the paler grey.

The decision was made to evacuate the buildings in the immediate vicinity of the incident. It was felt that this needed to be by word of mouth, so that we could direct people to exit the building, walk away from the incident and assemble at a named safe location. The department's core emergency team undertook this task. People were also stationed at the entrance to the site from the footpath (lower right of Figure 11.4), to prevent anyone from entering the site at that point. This prevented the large numbers of undergraduates from being put at risk.

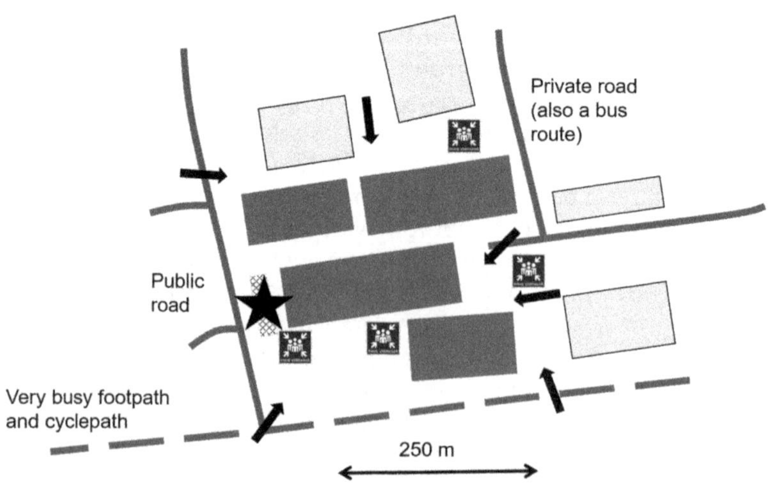

Figure 11.4 The site of the incident (starred), the surrounding buildings and the assembly areas.

Once all the people who might walk into the area which was close to the incident had been sent to a place of safety, the fire alarms in all the buildings belonging to the department were set off. People were sent from the assembly areas to locations far enough away for their safety.

By the time this was done the Fire and Rescue Service had arrived and were briefed. They were pleased that all personnel from the site had been safely evacuated and were now out of range. They asked for several other buildings to be evacuated, and this was done by liaising with the institution's security personnel, who had the authority to evacuate other buildings which belonged to other departments.

The Police attended and with their assistance it was possible to put a block on all the entrances to the site, including the public road at the rear of the affected building. Discussions with the Fire Service centred around whether the public road and houses to the rear of the site (on the right-hand side of Figure 11.4) should be evacuated. The decision was to leave them as the geography of the site would ensure that any explosion would be directed to the left of the figure, away from the public. The final exclusion zone is indicated in Figure 11.5.

The buildings cross-hatched in Figure 11.5 were all empty. An estimated 1000 people had been displaced.

There was also a discussion with the Fire and Rescue Service as to whether the electricity supply to the affected building should be shut off from the sub-station. While the electricity supply was a possible source of ignition, it was also needed in order to keep the ventilation system operating. The decision was made to leave the electricity on as clearly the ventilation system had,

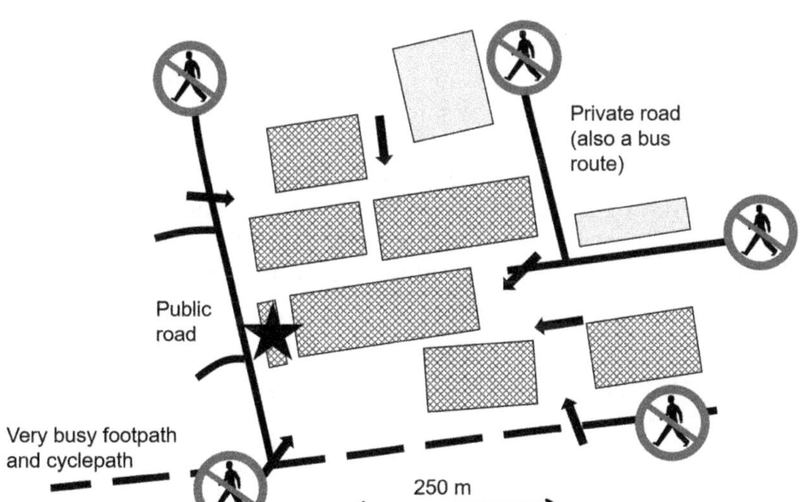

Figure 11.5 The final exclusion zone.

since the beginning of the emergency, demonstrably succeeded in keeping the concentration of the gas in the laboratory below the explosive limit. It was felt that the whole incident would come under control a lot faster if the electricity remained on.

After more than an hour elapsed, it was decided to find out whether the situation was now safe. The Fire and Rescue Service have flammable gas monitors and it was decided that they would enter the building and investigate whether any flammable gas remained. A complication was that the affected laboratory contained more than one superconducting magnet running in persistent mode. This building did not belong to the department which was handling the incident and their floor plans were not available. They were located in their own PIB, right next to the laboratory in which the gas release had taken place. It was felt too dangerous to attempt to retrieve these plans.

However, the senior researcher workers were still on hand and drew up floor plans for the Fire personnel, indicating where the room affected was located and what parts of the room they should avoid. The monitoring exercise was carried and, finding that the area was now safe, the Fire and Rescue Service and the Police were stood down and the staff and students were allowed to return to their rooms.

The key features that allowed this to be so efficiently handled were:

- The senior researchers attempted to stem the flow of gas before they left (later investigation showed they had been successful).
- They used their initiative in the initial evacuation and called the emergency services manually.
- An emergency team was available to relay verbal messages for the initial evacuation and to block the entrance to the site.
- Liaison with the Police allowed us to ensure that no one was at risk on the adjacent public road.
- The senior researchers remained on hand throughout and were able to inform the Fire service of the safe areas in their magnet laboratory.

It was, however, felt that it would be useful to have copies of the floor plans of the small building in the PIB of the large adjacent department, because all their fire alarms were routed to the large department's control panel. It would also be useful to have a copy of the major hazards within the building available from an online repository.

11.6.3 Flashover on a Three-phase Main

This incident concerned the three-phase mains to a large laboratory. Three-phase supplies are fairly common in scientific and engineering research laboratories, for running large pieces of equipment. The supply to this laboratory had fuses rated at several hundred amps, because one of the main activities in the laboratory was welding.

The start of the incident was a flashover which was inadvertently caused between two of the phases. This is essentially a short-circuit between them, allowing a current of several hundred amps to arc over for a short period. This comes to an end if the conductors move too far apart to sustain the arc between them, or the conductors vaporise, or the fuses blow. On this occasion we believe the flashover was terminated by the conductors moving apart.

An electrical flashover is capable of inflicting very serious burns to the skin and the eyes to anyone close to it (similar to a very serious weld flash) as it emits large quantities of visible and ultraviolet radiation. Direct contact with the three-phase main will also lead to electrocution. There was one casualty, who sustained serious skin burns to all his exposed skin, and all his hair was singed. Fortunately, he had not made direct contact with the electricity supply so was not electrocuted.

A first aider heard the event and noted that the lights dimmed momentarily, so guessed what might have happened. They immediately shouted for someone to call a further first aider to help them and went down into the laboratory.

On entering the laboratory, they asked the witnesses where the danger area was. It was noted that the wires were still smoking and the bystanders could not confirm whether the site was safe. It was therefore assumed that it was still live and lethal. The casualty was conscious, dazed, and still fairly close to the danger area. The first priority was to move the casualty well away from the live wires, give first aid treatment to his burns and summon an ambulance.

The first aiders asked for the electricity to be switched off, but no one could be found who was able to do this. There was also a fear, by some, that they did not have the authority to do this. The electricity supply remained live until after the casualty had been taken to hospital.

Once the ambulance crew arrived, they needed to be briefed about the location of the live wires and warned to keep well away from them. Since they needed to manoeuvre their metallic stretcher into position, the risk was substantially raised. The first aiders were aware exactly where the dangerous wiring was located and ensured that the ambulance crew remained safe. The casualty was placed on a stretcher and taken to hospital.

This incident was handled efficiently because:

- A first aider observed the dip in the electricity supply and heard the incident, so they were alerted to the possible danger.
- The witnesses were able to brief them on the probable danger area.
- It was assumed that the area was not safe, so a priority was to move the casualty further away.
- When the ambulance crew arrived they were also briefed. The risk to them was substantially greater as they had not experienced the events leading up to the injuries, and they needed to manoeuvre a metallic scoop stretcher into place.

It would have been very helpful to be able to isolate the electricity supply earlier, to remove the danger at source. The first aiders and ambulance crew were acutely aware of the danger beside them. In order to make the area safe there needed to be someone in the vicinity who knew how to isolate the supply and who felt that they had the authority to do so. This was rectified after the incident when the problem was brought to the attention of senior management. They stressed that safety for the casualty and for people handling emergencies was paramount. This was part of a cultural change that was already in progress in the organisation.

Note that fuses on three-phase mains do not necessarily blow during flashover events. Years later the author experienced the aftermath of a flashover event at another institution, and on that occasion only one of the three fuses was blown, leaving the installation live and in a very dangerous condition. On that occasion no one was injured but it did start a fire. Luckily a very levelheaded research worker was able to grab a special metal fire extinguisher and avert disaster.

11.7 Reflections on the Highlighted Incidents

Over the course of twenty years in a safety role in research institutions it has become evident that good targeted information available in the PIB or provided by witnesses can greatly reduce the time taken in handling an emergency, allowing a dynamic risk assessment to be carried out quickly and efficiently. It can reduce the controls imposed by the Fire and Rescue Service as, when in any doubt, they work to a 'worst case'. It therefore reduces the impact of the intervention. It reduces costs to both the emergency services and the institution.

References

1. BBC Scotland, Safety problems led to lab death [internet], 20 June 2000. Available from http://news.bbc.co.uk/2/hi/uk_news/scotland/798925.stm, last accessed on 3 March 2020.
2. J. Mills, Imperial College fined £70 000 after scientist suffocated in lab [internet] Metro News, 13 December 2017. Available from https://metro.co.uk/2017/12/13/imperial-college-fined-70000-scientist-suffocated-lab-7156620/, last accessed on 3 March 2020.
3. British Compressed Gases Association, Code of Practice CP27. Transportable Vacuum Insulated Containers of Not More Than 1000 Litres Volume, BCGA, Derby, Revision 1, 2004.
4. HSE, EH40/2005 Workplace exposure limits, TSO, 4th edn, 2020.
5. The Schedule 5 to the Anti-terrorism, Crime and Security Act 2001 (Modification) Order 2007. Available from https://www.legislation.gov.uk/uksi/2007/929/contents/made, last accessed 3 March 2020.

6. British Compressed Gases Association, Code of Practice CP18. The safe storage, handling and use of special gases, BCGA, Derby, Revision 3, 2014.
7. American Chemical Society, *Chemical and Engineering News*, Fighting lab fires [internet], Government and Policy, 23 May 2005, vol. 83, no. 21, pp. 34–35. Available from https://pubsapp.acs.org/cen/news/83/i21/8321ohio.html, accessed 3 March 2020.
8. British Standards Institution, BS EN 60079 Part 10-1 Explosive atmospheres. classification of areas, explosive gas atmospheres, 2015.

CHAPTER 12

When the Walls Come Down: Secrets of Open Space Safety

T. V. ASPINALL*[a] AND R. DERBYSHIRE[b]

[a]Manchester Institute of Biotechnology, University of Manchester, 131 Princess Street, Manchester, M1 7DN, UK; [b]University of Manchester, Safety Services, Room 4.81 Simon Building, Manchester, M13 9PL, UK
*E-mail: tanya.aspinall@manchester.ac.uk

12.1 Introduction to Open-plan Workplaces

The recent trend for more open, cross-specialism research, starting firstly with the introduction of open-plan offices and breakout spaces, and more recently, open-plan, shared scientific research laboratories, has mirrored the move away from traditional, single discipline research to a more collaborative and holistic approach.

The design and build of open space offices appears to be a natural development in today's architectural landscape. Less frequent are individual or smaller shared office spaces in new or refurbished buildings. Arguments for and against the effectiveness of open space offices are well documented. A recent study by Harvard University concluded that open space offices may not necessarily improve efficiencies or generate synergies.[1]

Purpose built, well designed, open-plan spaces can work effectively, promoting multifunctional working whilst saving on floor space, energy and generic equipment. The success of any open planned space is reliant on the

Challenges for Health and Safety in Higher Education and Research Organisations
Edited by Olga Kuzmina and Stefan Hoyle

Published by the Royal Society of Chemistry, www.rsc.org

quality of design, born from an understanding of the type of research work being undertaken, and how staff and students will function within the space. It is also imperative to allow the space to be adapted to meet the changing demands of research. Experience, however, has shown that adapting areas within existing buildings under the banner of "refurbishment", without a clear understanding of how the space will be utilised, or consideration given to the needs of those working within that area, tends to reduce efficiency savings and harmonised working.

Serious consideration should be given to how those carrying out research in a shared area can work safely, both in isolation (as individuals), but also collectively (as a group). Risk assessments tend to only focus on the work an individual plans to carry out, without any thought as to their neighbour's work as a combined risk. Cross-disciplinary working in shared spaces potentially increases the risk of incident compared to working in an open space carrying out similar work.

The Manchester Institute of Biotechnology (MIB) at the University of Manchester, built in 2003, was the first purpose-built University research Institute in the UK that focused on quantitative approaches to specific biotechnology challenges, at the interface between medicine and biology and the physical sciences, engineering, mathematics and computation and was specifically designed to remove the barriers between disciplines and to promote innovative science. The architecture of the MIB reflects the needs of interdisciplinary science, featuring open-plan, multifunctional laboratories and a wide range of high-tech facilities, as well as open-plan office and meeting areas to promote interaction, spontaneous discussion and shared research (see Figure 12.1). This cross-discipline approach has led to progress in the development of advanced materials, pharmaceuticals, value added chemicals and the next generation of biofuels, and has enabled the transfer of research breakthroughs into real world applications. The MIB was awarded the Queen's Anniversary Prize for Higher and Further Education in February 2020 for its pioneering expertise in industrial biotechnology. This move to cross-disciplinary research is mirrored in many other institutions worldwide, including Imperial College London, the University of Sydney, and Penn State, Philadelphia.

12.1.1 Multicultural Working

Within the MIB, multidisciplinary research is carried out within large, open-plan laboratories, enabling scientists and academics from a range of backgrounds to work together in the same shared space. Researchers from traditional "single science" disciplines, including physics, biology, engineering, chemistry, medicine and computing, bring with them different experiences, working cultures and standards, in addition to different perceptions of risk.

A common example of a difference in working cultures is often seen when comparing the working practices of chemists to those of biologists; a chemist will always wear safety glasses within the lab, but may not wear a lab coat, whereas a biologist will often do the opposite. Differences in the perception

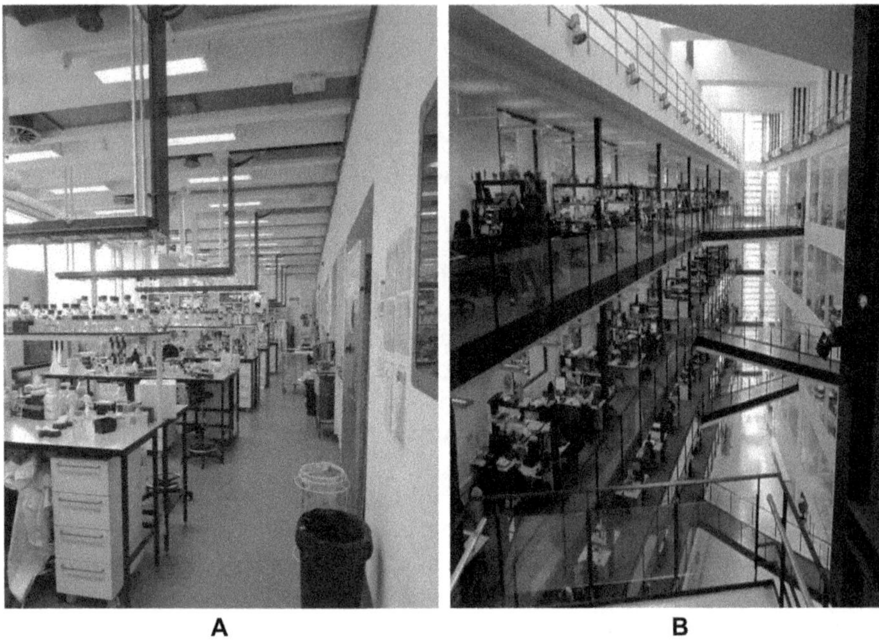

Figure 12.1 (A) – open plan, multidisciplinary laboratories within the MIB. (B) – shared office workspace within the MIB.

of risk are also regularly seen. A prime example was an engineering group within the Institute, who were developing a new technique to study liquids. When the group wanted to use human blood samples on their equipment to study fluidic properties, it did not occur to them to consider the hazards presented by the blood samples themselves (blood-borne viruses, *etc.*) as they simply considered the blood to be another "liquid".

In addition to the multidisciplinary nature of the work conducted within the MIB, it is also important to remember those who actually conduct the research. The MIB, like most universities, has an international workforce, with researchers often being employed on short-term contracts (usually 3 years) and students studying 3 year post-graduate programmes. This can present challenges for safety management, as different cultures have different "standard responses" to safety issues. A prime example was observed in response to fire alarm activations. While many workers respond to fire alarm activations by evacuating the building, others can perceive it to be an earthquake alarm and seek shelter under tables. Differences in risk perception are often compounded by difficulties in communicating safety messages clearly to those for whom English is not their first (or second) language. In addition, the transient nature of the majority of workers within the Institute brings further challenges, as it is difficult to develop and maintain a "corporate memory", meaning that lessons learned from safety incidents need to be repeated on a regular basis, and safety training needs to be repeated at regular intervals.

Understanding these differences is important when developing health and safety (H&S) management plans and identifying means of communicating them to the workforce. The following discussions represent the authors' personal observations and opinions, highlighting the H&S processes developed and implemented within the MIB.

12.2 Safety Management of Open-plan Workplaces

The successful management of safety in open-plan workplaces, and the development of a positive H&S culture, can be attributed to having clear and visual leadership, defined roles and responsibilities, and a sound safety management framework which sets out the standards required. While the principles of safety management remain the same for all workplaces, additional challenges are present when developing H&S management plans and a positive culture in open-plan, shared discipline workplaces. The safety framework implemented in the MIB was primarily developed using guidance from the Health and Safety Executive's document, "HSG65: Managing for Health and Safety",[2] focusing on the 4 Cs of safety management (see also Figure 12.2):

- control: allocating responsibilities, securing commitment, instruction and supervision
- competence: recruitment, training and advisory support
- co-operation: between individuals and groups
- communication: spoken, written and visible

12.2.1 Control

Within the MIB, most staff and students belong to a home School or Department, each of which has their own H&S policies and procedures, which can often differ. To address this, dedicated MIB-specific policies and procedures were developed and implemented. These override those of the individual

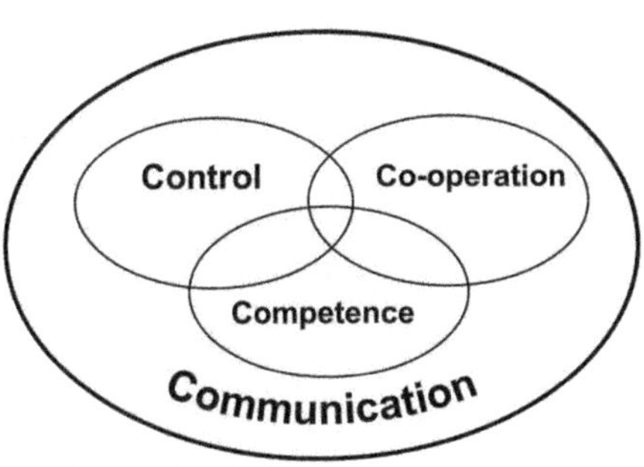

Figure 12.2 The 4C approach to safety management within open-plan, multidisciplinary workspaces.

Schools and Departments, ensuring that all work conducted within the Institute is governed by the same rules and to the same standards.

In addition to University research groups, the MIB also hosts external companies, who access facilities and rent research space within the Institute. This joined up working relationship is becoming more commonplace, with universities often renting out research space to industrial partners. Commercial partners are required to adapt their safety procedures to comply with the Institute rules, participate in H&S inspections of their areas, and have input into the safety reporting systems. It is important, however, that the commercial sensitivities of industrial partners are respected, meaning that specific information is shared with a limited audience.

Establishing effective governance arrangements is essential. Research staff have a tendency to bring with them their own equipment. The current procedure within the MIB requires each piece of equipment to be registered on a centralised asset system, where ownership is established along with any maintenance and statutory requirements. Statutory inspections are completed in defined timescales, with a strong emphasis that any defects identified with the equipment are rectified in a timely manner. Working in this way enables a unified approach.

12.2.1.1 H&S Policy

One of the most important steps in managing safety where staff and students work across organisational boundaries, in shared areas which promote multifunctional working practices, was through the development and implementation of an Institute-wide safety policy. The policy set out the organisational structure, clearly defining the roles and responsibilities of those in charge of safety, including the Institute Director, the Principal Investigators/research group leads and the safety team (which includes specialists in biological, chemical, UV, laser, radiation, and general safety). The policy sets out a transparent accountability system, where it is clear who to approach for competent advice and who is ultimately legally responsible for safety. Commitment was also secured from senior management, who demonstrate their visible commitment to safety by participating in safety inspections and audits, introducing safety inductions, and chairing the local H&S committees.

The H&S policy was developed in tandem with a compliance policy, which was supported by the Institute Director and Faculty Deans; this ensures that H&S policies are followed, and that there are serious consequences (including exclusion from the building) in cases where H&S procedures are ignored or compromised due to poor management.

12.2.1.2 Access Control

An important consideration during the initial planning stage of the Institute focused on the ability to maintain security of the building. The Institute installed a 2-stage swipe card access control system, allowing control of staff and students entering and leaving the building. The system allows different levels of access to be given, dependant on the individual (staff or student) and their competence levels. Access *via* the internal barriers is only given when an individual has completed a building induction, and this is only for a 3 month period. At the end of the 3 months, access is automatically cancelled unless the individual has registered with Occupational Health (OH) and provided proof of a fitness to work certificate. By linking the induction, access and OH registration processes, the management of individuals in the building can be assured. Access *via* the external barrier (which allows out-of-hours access to the building) is only permitted when an individual can demonstrate they are competent to work out-of-hours/lone work and where a risk assessment has been authorised at a senior level.

The MIB also has several specialist facilities (including a Nuclear Magnetic Resonance (NMR) hall, tissue culture suites, and mass spectroscopy laboratories), which are overseen by dedicated facility managers. Access to these areas is again controlled *via* swipe card access, which is only granted once an individual has received a facility induction.

12.2.1.3 Control of Shared Spaces

In addition to general safety rules governing the research within the MIB, other areas were identified where new control mechanisms were required. This is due to the physical layout of the open-plan labs, and shared office and kitchen areas. In a traditional research laboratory setting, individual labs "belong" to one or a limited number of groups, enabling a clear distinction to be made when identifying who is responsible for the safety of the area, with each group having its own equipment and facilities (such as fume cupboards, sinks, cold rooms, chemical storage areas and waste disposal). Within the MIB, the open-plan nature of the labs means that most equipment and facilities are shared, including open-access centrifuge parks, shared cold rooms and sink areas, and communal waste disposal routes. This cuts down on research costs and contributes to the development of a shared research community but introduces questions about how to control safety within a communal area where there is no clearly defined ownership of the space.

In order to address this, a "floor co-ordinator" system was developed, with a senior academic and senior lab-based experimental officer taking on responsibility for each floor within the Institute. These roles are appointed on a permanent basis *i.e.* not reliant on external grant funding. Their role is to support the safety team, and to respond to low level H&S concerns which the safety team cannot always respond to immediately. Common examples of this type of issue include poor housekeeping, such as items being abandoned in communal areas (*e.g.* by sink areas or in cold rooms, where there is often no clear ownership of items), and waste disposal *via* the communal waste routes. This system has proven to be extremely effective, as the "responsible" experimental officer is familiar with the people and research within their area and can therefore respond to problems as they arise. The senior academic provides the authority, supporting the experimental officer and ensuring that problems are dealt with in a timely fashion, and that the compliance policy is enforced when necessary. A system of this type is vital in ensuring high levels of effective H&S management within open, shared, multidisciplinary labs.

12.2.1.4 Experimental Control

The MIB uses a standardised approach to risk assessment. Before any experimental work can start, a risk assessment must be completed by the person(s) undertaking the work, and then authorised by the group leader, Principal Investigator or facility manager. Risk assessments are approved by the MIB Safety and Risk Manager and posted onto the MIB intranet. We found posting them for others to use as a starting point encourages researchers to complete their own assessments as they have something to work from. Group leaders, Principal Investigators or facility managers are deemed to be ultimately responsible for the safety of the work being conducted within their area(s) of responsibility, hence the requirement for them to authorise all risk assessments for the work in their area(s). All risk assessments for ongoing work are held within the lab area, where they are easily accessible. Where standard emergency procedures differ, discussions are held with individuals to remove any ambiguity. A review of the risk assessment is then undertaken by the individual to ensure any agreed changes are captured within the risk assessment.

In cases where the risk is deemed to be significantly higher, authorisation to enable the work to start requires a further level of approval. An example would be an experiment where there is the potential for cyanide gas to be released. In these instances, authorisation of the risk assessment is passed to the MIB Chemical Safety Advisor (a senior chemist) for approval. This type of work must also be registered with the MIB Safety and Risk Manager, who ensures that specialist training is completed, and that additional emergency procedures (such as provision of emergency oxygen and a trained first aider on standby) are in place during the course of the experimental work.

Shared waste disposal processes ensure that all waste is processed correctly by MIB technical staff. The MIB also has a centralised lab coat system, where all lab coats are for general use, rather than belonging to an individual. This system ensures that clean lab coats are always available, and that they are laundered effectively on a regular basis (rather than being taken home to be washed, as some students were found to be doing!).

In addition, the University of Manchester has introduced an online chemical inventory system, which is used throughout the University; this system records a chemical's storage location, expiry date and owner. This system provides assurance that all chemicals within the Institute have been identified, are adequately controlled, and are stored securely. In the case of controlled chemicals (*e.g.* drug precursors and controlled drugs), records are kept on the system, but access to this information is restricted to the MIB Safety and Risk Manager. This provides further assurance that certain chemicals are held securely, and also assists with statutory returns, as all information is held centrally. Different levels of access are built into the system, with group members only being able to view holdings and storage locations of chemicals belonging to their group. Higher levels of access are available to group administrators, who are able to view chemical holdings within the Institute. This feature has proved to be very useful, as it enables group administrators to identify chemicals which they may need, but which don't belong to their group. In the cases where small amounts of high-cost chemicals are needed, groups have started sharing chemicals (rather than ordering their own), thus reducing costs and fostering co-operation within the Institute. The Institute Safety and Risk Manager has the highest level of access and is able to view all holdings within the Institute, including any controlled chemicals. Checks of physical storage holdings compared to the online inventory are planned to ensure chemicals are stored correctly, and types and volumes of chemicals listed in the software are accurate.

All research groups and facilities have a safety file, which is prominently displayed in all labs. The safety files contain the MIB H&S policy, and a copy of all risk assessments for the work being conducted within the area. Audits of safety paperwork, and laboratory inspections, are completed on a regular basis, with a standardised template being used throughout the Institute in order to ensure consistency and transparency. The findings of the audits/inspections are presented at formal Institute H&S meetings, along with an action table and timeframes by which to rectify any identified actions.

The Institute also utilises a checklist contained within the Institute's group and individual exit policies. The compulsory group exit procedure generally works well, ensuring that research groups vacating the Institute remove all of their belongings, dispose of any waste in an appropriate manner, and don't leave any nasty surprises at the back of any cupboards! However, the individual exit policy is not compulsory, and has been used to varying degrees of success by individual group leaders/Principal Investigators when a staff member or student leaves the group. Consideration has been given for making the individual exit procedure mandatory but as yet, it remains voluntary.

12.2.2 Competence

12.2.2.1 Developing the Competence of New Starters

All new starters to the MIB receive a building induction before swipe card access to the building is authorised. As mentioned previously, this is also linked to registration with OH, thus ensuring that all individuals with access to the building have undertaken an induction and been deemed fit to work by the University's OH Service. In addition, all new students who intend to work within the MIB laboratories must complete a "lab passport" training course, which was developed in-house by the MIB Safety and Risk Manager.

Due to the multidisciplinary and open-plan nature of the MIB labs, it is not unusual for different types of work (*e.g.* synthetic chemistry and genetic manipulation of microorganisms) to be conducted within the same shared space. It is important that all new students have a basic understanding of hazards they may be unfamiliar with. The lab passport course covers general safe lab working practices, and an introduction to common chemical and biological hazards which are present within the lab areas of the Institute. New students are not permitted to start work within the MIB labs until they have completed this course.

12.2.2.2 Competence of Group Leaders and Principal Investigators

Following a serious incident in the MIB in 2008, involving an azide explosion in a chemistry lab, followed by a fire that caused significant damage, a key recommendation in the subsequent investigation report identified the need for mandatory H&S refresher training for all group leaders and Principal Investigators. The introduction of compulsory refresher training was supported by the Institute Director and the Faculty Deans and is repeated every 2 years. The training session is introduced by the Institute Director, to demonstrate leadership and commitment to H&S from the top. The training incorporates the legal and H&S responsibilities attached to each individual role, changes and updates to the Institute's H&S policies and procedures, and signposting to competent advice.

The investigation findings post incident acknowledged that while group leaders and/or Principal Investigators had in-depth knowledge of their individual research interests, there was a lack of understanding of the H&S and legal responsibilities of their role. Subsequent training has ensured that all group leaders and/or Principal Investigators now have a much better understanding of their H&S responsibilities and view H&S as an integral part of their work, from the planning stage, rather than as a last-minute "bolt on". It should be noted that group leaders and/or Principal Investigators are not always made aware of their legal and H&S responsibilities when taking up new positions within organisations. There appears to be an expectation that they will assume the role and understand their level of responsibility. Within the MIB this is addressed by ensuring that any new group leader and/or Principal Investigator receives training as soon as they take up their position within the Institute.

12.2.2.3 Competence of the Safety Team

Ensuring competent advice is easily available to staff and students is extremely important. At the Institute, this falls under the responsibility of the Safety and Risk Manager. It is recognised that individuals must have a high degree of knowledge in all aspects of safety, however, it is also recognised that softer but equally important skills, like influencing and mentoring, are as important for safety professionals to have or acquire. The University of Manchester developed an in-house competency framework, which identified which skills, knowledge and experience were expected of all staff, at all grades, who are responsible for providing H&S advice. This information is incorporated into the annual development reviews of all safety personnel and is a useful tool in identifying knowledge-gaps and training opportunities.

12.2.2.4 Documenting Competence

Equally important is to ensure that competencies are documented, and records are easily accessible. Individuals within the Institute who are leaving to take up positions in other institutions often request evidence of their training and competency levels. Training records also constitute an important factor in accident investigations, *etc.* – especially where external authorities are involved. Within the MIB, online training records are kept for all in-house training by the MIB Safety and Risk Manager.

A competency form is also used to document the level of competency an individual has attained. This is mainly used within the specialist research facilities (such as the tissue culture suites and NMR Facility). Following facility induction and initial training, an individual may be deemed competent to work on equipment under supervision. As competency develops, the individual may be authorised to work without supervision during normal working hours, and as competency develops further, they may then be permitted to work without supervision and out of hours. This development of competency is recorded and held within the facility for easy access. It is used by the facility managers to identify individuals who require a higher level of supervision or who may require additional training. These records are checked during regular lab inspections.

12.2.3 Co-operation

As previously mentioned, the MIB developed a system of "floor co-ordinators", with a senior academic and senior lab-based experimental officer taking on responsibility for each floor within the Institute. This system also allows members of the senior management team to demonstrate their visible commitment to safety. The floor co-ordinators meet regularly with the MIB Safety and Risk Manager, to ensure that any concerns are highlighted and that identified good practices are shared throughout the Institute; in addition, they hold monthly meetings (attended by a representative from each research group working on their floor), and conduct monthly lab hazard-spotting/housekeeping inspections. In parallel to this, shared processes (such as waste disposal and lab coat laundry) and shared systems (such as the chemical storage system) all help to develop a higher level of co-operation and community.

The Institute has a number of voluntary emergency response teams, including fire evacuation marshals, first aiders and a chemical spills team. All members of the Institute are invited to become a team member, with regular refresher training and meetings ensuring that the team members work well together as a cohesive unit.

The open-plan nature of the Institute, including the large atrium social space and lecture theatre, makes it an ideal location for conferences, poster sessions and social activities. Members of the MIB community regularly come together to discuss their research and forge bonds with other groups, sharing techniques and best practice.

On a wider level, the University encourages a high level of co-operation and collaboration, with a university-wide safety advisor network being convened and coordinated *via* the University's central safety unit. The aim of these network events is to bring together safety advisors from across the University, and acts as a forum for sharing best practice, disseminating lessons learned from incidents, and developing shared processes (such as the online chemical inventory system). The network aims to break down barriers and remove the traditional "silo mentality" which often exists in separate buildings and/or Faculties, ensuring wider learning and the development of consistent standards throughout the University.

12.2.4 Communication

Good methods of communication are central to the functioning of all effective H&S systems, without which control systems will weaken, competence will fail to develop, and co-operation will be lacking.

As discussed earlier, difficulties can exist when communicating safety messages clearly to those for whom English is not their first (or second) language. Experience has shown that providing information in a variety of formats tends to work best, with safety messages and training being provided face-to-face within a lecture theatre, as a hard copy, and online. This was confirmed by feedback provided by staff and students. Online testing is used to ensure that the information has been understood, with those failing to pass the test attending refresher training and being given more support. Regular safety training sessions are run by the MIB Safety and Risk Manager and are open to all MIB workers; this ensures that knowledge is frequently refreshed and reinforced.

In addition, formal inspections of all lab areas are conducted annually (as a minimum); the inspection team includes the MIB Safety and Risk Manager, members of the senior management team and floor co-ordinators. A standardised inspection tool (iAuditor) is used throughout the Institute, which examines not only the physical safety of the workplace (such as housekeeping and chemical storage) but identifies how well workers in the area understand safety instructions (*e.g.* how to respond to a fire alarm activation), thus enabling a measurement of how well safety messages are communicated and understood. All actions identified during the inspections are tracked through to completion by the safety team.

Within the open-plan, multidisciplinary lab areas, it is common for different types of work (*e.g.* synthetic chemistry and genetic manipulation of microorganisms) to be conducted within the same shared space. It is vital that workers in these shared spaces are informed of the work being undertaken near to them.

Formal lines of communication include regular floor meetings, distribution of safety information from the safety team by email, and the operation of the MIB H&S committee. The committee is chaired by the Institute Director and includes representation from the safety team, staff, students and Union representatives. The safety team within the MIB aim to act as a "critical friend", giving constructive advice and support to academics and researchers, rather than being perceived as adversaries (which, unfortunately, is how some H&S personnel are still viewed!). The development of soft-skills by the safety team, including good communication skills and a friendly persona, encourages people to engage with safety, demonstrates a visible commitment to H&S, and helps to foster a positive H&S culture.

12.3 Conclusion

Using the 4 Cs approach to create an effective safety management framework within the MIB has shown very positive results; it has led to a sense of community, which has broken down the traditional discipline barriers, leading to shared learning and the development of skills and competencies which may not have been available in traditional "single science" environments. The introduction of effective control systems, coupled with the development of competency, co-operation in using shared spaces and equipment, and the development of effective routes of communication, has resulted in tangible improvements to processes, procedures and behaviours within the Institute.

References

1. S. Stillman, *New Harvard Study: Your Open-Plan Office Is Making Your Team Less Collaborative*, Available from https://www.inc.com/jessica-stillman/new-harvard-study-you-open-plan-office-is-making-your-team-less-collaborative.html, last accessed on 24 March 2020.
2. *Managing for health and safety HSG65*, Available from https://www.hse.gov.uk/pUbns/priced/hsg65.pdf, last accessed on 24 March 2020.

CHAPTER 13

Health and Safety Considerations in the Design of Teaching and Research Laboratories

LOUIS DIBERARDINIS*

Director, Environment, Health and Safety Office, Massachusetts Institute of Technology, 77 Massachusetts Avenue, Cambridge, Massachusetts, USA
*E-mail: loudib@mit.edu

13.1 Introduction

The design and construction of new laboratories and the renovation of existing laboratories requires close communication and understanding between laboratory users, architects, design engineers, health and safety personnel, project engineers, and construction managers. One of the principle objectives of laboratory design is to provide a safe place in which scientists, engineers and their staffs and students can perform their work in a safe, efficient, and effective manner. Good laboratory design does not only provide a safe place to conduct research or education, it can facilitate productivity. It can reduce downtime and minimize laboratory mistakes. Conversely a poorly designed laboratory can be a barrier to good science. This chapter cannot address all

Challenges for Health and Safety in Higher Education and Research Organisations
Edited by Olga Kuzmina and Stefan Hoyle

Published by the Royal Society of Chemistry, www.rsc.org

the design features in detail but will provide an overview of key issues to consider and sources of information that can be consulted for advice.

13.2 How Laboratory Design Affects Operations

There are several aspects of laboratory design that can significantly affect how a laboratory operates and can either assist or deter compliance with health and safety regulations and best practices. It is important to consider these in the design phase and obtain feedback from those who will occupy the space. A few key examples follow.

13.2.1 Purpose of the Laboratory: Dry *vs.* Wet Laboratories

There is not a universally accepted definition of a dry laboratory. To many it is considered a laboratory that does not utilize hazardous chemicals. However, there are gray areas where the definition of "hazardous chemicals" is not always clear and there is sometimes a quantity limit below which the laboratory might be considered "dry". For example, a physics laboratory may only use solvents for cleaning equipment and not consider that a chemical use. Similarly a laser laboratory may use chemical dyes as well as cleaning solvents but if asked would say they do not use chemicals. Based on the criteria established for a dry laboratory, key safety features such as emergency eye washes and showers, flammable storage space and local exhaust ventilation may not be provided. This may restrict expanding of future "wet" activities in the laboratory or may require expensive renovation.

13.2.2 Type of Laboratory

Open laboratory *vs.* modular laboratory design: The concept of "open laboratories" has been around for decades. It has many advantages including facilitating collaboration, maximizing useful space, sharing equipment and possibly lower initial and operating costs. However, there are potential noise and privacy issues that may require different operational procedures than in a modular laboratory. A larger disruption of activities can occur in the event of a hazardous material spill as the whole space may need to be evacuated disrupting many occupants. The use of shared equipment can present many challenges related to maintenance and scheduling of use as well as potentially contaminating equipment of samples. Other potential safety issues are ergonomics and ventilation efficiency. See Guidelines for Laboratory Design, Chapter 21 for a more detailed discussion of open laboratories.[1]

13.2.3 Location of Desks

The location of desks inside *vs.* outside the laboratory has been debated for years. The preference for location varies between scientific disciplines and cultures. Recent experience had indicated that chemists and engineers generally prefer to have write-up spaces outside the laboratory while the biologists prefer to have their desks close to the experiment inside the laboratory. Renovation of existing laboratory buildings designed with the desks inside the laboratory may also be challenging as it may not be technically nor financially feasible to design with office or write -up spaces outside the laboratory. The location within the laboratory can reduce compliance with some regulations and good practices such as no food and drink in the laboratory and the wearing of required Personal Protective Equipment or PPE (eye protection and gloves/lab coat) when working with or near hazardous chemicals. Enforcing compliance with these operational procedures can become difficult and confrontational when the desks are in the laboratory. Locating desks outside but as close to the laboratory with a viewing window and easy access is desirable. A discussion with potential occupants can help to identify the best design for the particular use. Figure 13.1–13.3 (Harvard), (National University of Singapore) show laboratories recently constructed within the past 10 years where the desks are separated from the laboratory. The Massachusetts Institute of Technology (MIT) has also designed several laboratory buildings where the desks are separate from the laboratories but are located adjacent to and have a direct view into the laboratory. Additional benefits of this design is allowing more light into the laboratory and creating a greater sense of openness and collaboration. However, it should be noted that some types of laboratories such as laser labs require operations in the dark so careful consideration must be made in these cases.

Figure 13.1 Laboratory with desk space on right, separated but within easy access and viewing (Harvard University, Cambridge, MA, US).

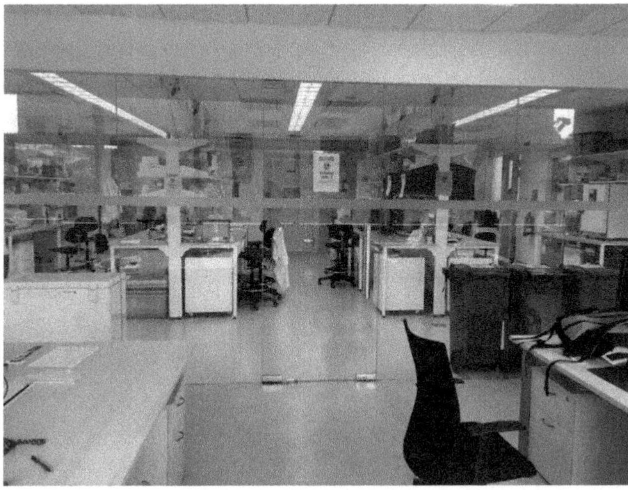

Figure 13.2 View of the laboratory from the office area separated by transparent wall (Singapore-MIT alliance for research and technology, Singapore).

Figure 13.3 View from the laboratory space to office/desk area separated by transparent wall (Singapore-MIT alliance for research and technology, Singapore).

13.2.4 Materials Management

A lack of adequate storage space can result in unsafe conditions as well as inefficient and costly materials management. This can result in purchasing excessive quantities of chemicals or materials because of the inability to locate the needed item. Storage of flammable materials is a major concern particularly in high rise buildings. Regulations and some building codes restrict the quantities of flammable materials allowed in the laboratory based on the class of the flammable materials and which floor above ground the laboratory is located. The allowable quantities can be increased somewhat by providing fire suppression systems in the laboratory as well as the use of flammable liquid fire-rated storage cabinets. Other hazardous materials such as radioactive or biological agents or highly toxic chemicals, may require specialized storage capabilities. Storage of compressed gases is also a concern. There have been several attempts to adapt LEAN techniques to laboratory operations and one recent successful effort at MIT is described in the paper by Dagdeviren *et al.*[2]

13.2.5 Electrical Outlets

The modern laboratory utilizes much equipment that uses electrical power. The lack of adequate power provision can result in potential fire and electrical hazards. The lack of required number of outlets can result in the use of extension cords and plug strips that can add to both electrical and fire hazards as well as tripping hazards. A complete inventory of current and potential future equipment and an understanding of the power requirements is needed in the design stage.

13.2.6 Laboratory Layout

The laboratory layout defines the laboratory area requirements and special organization of each type of laboratory with special regard to means of egress and access, equipment and furniture location, ergonomics of movement and work, and ventilation requirements. The location of exhaust hoods, biological safety cabinets, clean benches and safety equipment are given special attention. Many of the items listed in Table 13.1 will be located on the laboratory design drawings. Materials of construction are discussed with respect to sustainability as well as resistance. The goal is to have a laboratory layout that will allow the researcher to efficiently and effectively carry out their research in a safe manner. It should facilitate work flow and minimize the potential for accidents.

13.2.7 Heating Ventilation and Air Conditioning (HVAC)

The HVAC system is designed to control the comfort and cleanliness level in the laboratory. This includes temperature and relative humidity. Excessive temperature and humidity can not only affect comfort but may also adversely affect research samples or sensitive equipment. High humidity can also introduce mold issues. Cleanliness is controlled by selection of proper filtration, location of outside supply air and control of airflow within the building. Within the laboratory the location of the tempered supply air and the exhaust is critical to prevent or minimize interference

Table 13.1 Design topics to consider in lab design.

Distribution of mechanical, electrical and plumbing services	Directional air flow in laboratories
Supply air velocity and quality	Location of supply air intake to the building
Location of supply air in room	Temperature control
Humidity control	Emergency power supply
Fire detection system	Fire suppression system
Fire alarm system	Special lighting
Laboratory egress	Universal access for disabled persons
Access restrictions	Laboratory furniture location
Location of ventilated enclosures	Laboratory chemical fume hood
Specialized fume hoods	Radioisotope fume hood
Biological safety cabinet	Glovebox
Local exhaust ventilation	Filtration of exhaust air
Emergency gas shut-off	Ground fault circuit interrupters
Master electrical disconnect switch	Emergency showers
Emergency eye wash	Handwashing facilities
Hazardous spill control	Fire resistive construction methods and materials
Experiment alarm systems	Hazardous materials disposal
Hazardous waste treatment	Flammable liquid storage
Special hazard chemicals	Compressed gas cylinders
Emergency cabinet	Change or gowning room
Durable, cleanable work surfaces	Durable, cleanable floor and wall materials
Protection of laboratory vacuum system	

with contaminant control devices such as fume hoods and biological safety cabinets. It can also enhance the ventilation effectiveness by providing better air mixing in the space to dilute any fugitive odors and maintain temperature distribution.

Operations involving the use of hazardous materials (chemicals, biological agents or potentially airborne radiation sources) must be controlled by ventilated exposure control devices. These devices must be selected based on the source of the exposure. Typical devices include laboratory fume hoods, biological safety cabinets, glove boxes, canopy hoods, slot ventilation and specially designed enclosures. For more detail on these devices see Scientific Equipment and Furniture Association (SEFA)[15] and Guidelines for Laboratory Design.[1] Figure 13.4 shows a typical laboratory layout with a peninsula bench and laboratory fume hoods and a snorkel type exposure control device above the bench to control emissions from benchtop operations.

The HVAC system is also used to isolate spaces from an airflow point of view by controlling the direction of the airflow. It can be either into the space to contain any airborne hazard that may be generated or out of the space to protect the processes or materials in the space form outside contamination.

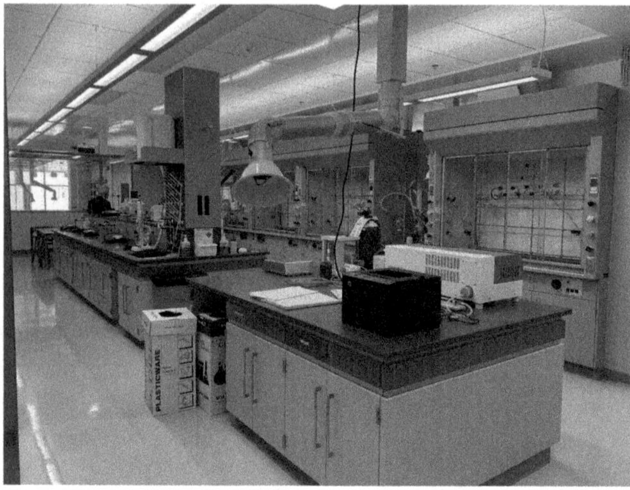

Figure 13.4 A teaching laboratory (MIT).

The use of Computational Fluid Design (CFD) in laboratory design has become more common. Memarzadah[4–6] used CFD as a method to optimize laboratory hood containment and design animal facilities. It has been used recently to select proper types and locations of supply air diffusers relative to exposure control devices.

Some key ventilation design concepts are provided in Table 13.2.

13.3 Common Mistakes Made in Laboratory Design

Unfortunately, there have been many examples of poor laboratory design decisions. Many of these occur because of poor communication between all the members of the design team. Others occur because fiscal constraints result in redesign or cutbacks without adequate consideration of the effect these will have on the ability of the laboratory group to function as planned. The most common mistakes are:

- Lack of adequate storage space as noted above resulting in poor housekeeping and inefficient use of materials.
- Failure to allow for growth or change in research activities.
- Lack of adequate exposure control devices such as fume hoods or gloveboxes.
- Inadequate flexibility designed into the space to allow for a repurpose of the type of use.
- Lack of adequate electrical power and outlets as laboratory equipment evolves over time and more automation is introduced into the laboratory.
- Location of emergency equipment (emergency eye wash, emergency showers, fire extinguishers) that are difficult to maintain and access in an emergency.

Table 13.2 Example of a laboratory ventilation summary: design considerations presented in June 2018 at the Harvard School of Public Health course "Guidelines for Laboratory Design".

Disclaimer: Generalizations have pitfalls as there are always exceptions to the rule. Operations and personnel in the lab can have critical impact on performance. Future changes in materials and research can have critical effects. *Ability to monitor and adapt to changes is necessary*. The main goal is always to protect people and the environment. With that said, here are some things to consider:	
Fume Hood Face Velocity	*Face velocity alone is not a good predictor of containment*. 80–120 feet per minute (fpm) average is typical. A well-designed hood in a well-designed lab with minimal cross drafts can contain hazards sufficiently at 80 fpm. Face velocity above 120 fpm does not help in a poorly designed hood and/or lab and otherwise is not necessary and a waste of energy. If acceptable velocity cannot be reached then consideration to limiting the hood opening or restricting use of hazardous materials or certain processes must be considered. Note: High face velocities may create turbulence and cause more leakage.
Air Changes per hour	Should not be used as a health and safety criteria. Most should be within 6 + 2 ACH. Higher values "generally" do not increase health and safety protection but may be needed for heating or cooling, odor control or cleanliness reasons. See Sept/Oct 2009 issue of the ACS Journal of Chemical Health and Safety for discussions on ACH. Note: 6ACH = 1 CFM/ft2 for a 10 ft. ceiling (approximately).
Minimum Hood Flow Rate	~10–25 CFM/ft2 of work surface (Use 10 with caution) See ANSI Z9.5–2012 for guidance on when to use 10 *vs.* 25. Could consider using 10 when no volatile flammables are used (or very small quantities).
High Performance Low Volume Hoods	Generally they are acceptable but only if performance tested. Determine what is meant by low flow *vs.* low velocity. Location of hood relative to air supply, doors, and egress routes is more critical. They are more susceptible to walk by and other cross drafts so location is critical. (See NIH Design Guidelines). The value of these hoods is to reduce the volume of airflow required thus reducing energy consumption. However, this could also limit the ability to increase exhaust airflow in future expansions as research needs change.
Performance Testing	Biocabinets: Use NSF standard 49 for Class II Fume Hoods: Modified ASHRAE 110 or equivalent (*i.e.* NIH requirements) Owner selects acceptance criteria. Ventilated cages: Owner decides. Look at leakage from cages and velocity animals will feel. Enthalpy wheels: Owner decides. Look at leakage. Need method for performance testing. Do not use with fume hoods. Ductless Hoods: Owner decides (see ANSI Z9.5–2012) Remember to distinguish between the equipment performance and the system performance. Think of my "car" analogy! Excellent car may not compensate for a poor driver!

(*continued*)

Table 13.2 (*continued*)

Biological Safety Cabinets	Follow NIH's BMBL and NSF Standard 49. Note new NSF requirement that Class II Type a cabinets cannot be hard ducted. Can be canopy style (thimble) But if so, must have an audible and visual alarm.
Glove Boxes	Follow ANSI Z9.5 and ACGIH Industrial Ventilation Manual. See American Glovebox Society guidelines. If this is a Class III Biosafety cabinet then it must meet the requirements set forth in NSF 49.
Specially designed local exhaust hoods	Enclosures, exterior or receiving hoods. Follow design guidelines in ANSI Std. Z9.2 and ACGIH Industrial Ventilation Manual. Contact manufacturer of equipment for advice as well.
Hot Exhaust (Autoclaves, ovens, *etc.*)	Follow design guide in ACGIH Ventilation Manual. See "hot processes" in index. See also NIH Design Guidelines, Appendix 1 "design of canopy hoods".
Ductless Hoods	Use care, applications must be carefully evaluated. Better when number of chemicals used are small and each one can be removed by appropriate filter material. Less application in research labs, more for process, QC/QA or clinical labs. See ANSI Z9.5–2012 for testing requirements. See standards: AFNOR NF X 15–211, ANSI/ASME N510, ANSI/ASHRAE 52.2 See also Article in JCHAS by DiBerardinis.
Recirculation of room air	See ANSI Z9.5 and Z9.7. Generally accepted *within the lab* (chilled beam, fan coil unit). These are OK to do and have many advantages but there are cases where they are not needed or not applicable. Must evaluate others.
Directional Airflow	Generally from low hazard areas (corridors, offices) to high hazard areas (labs, equipment rooms). Exceptions are clean room facilities which require more complicated flow patterns. Typical negative pressures are 0.03 to 0.05 'H_2O. Above 0.10 'H_2O can make it difficult to open doors.
Variable Air Volume Manifold or Ganged Systems Diversity	There are OK to do and have many advantages but there are cases where they are neither needed nor applicable. See ANZI Z9.5.
Re-entry study	For new buildings or major renovations, perform a study of airflow from exhausts or around the building. It could be actual wind tunnel tests or computer monitoring. Look at effects on the building in question and surrounding buildings. See Z9.5 and ASHRAE Fundamentals.
Exhaust Stacks	Minimum of 10 ft above roof line and 2000 fpm discharge velocity unless wind tunnel or computer modeling shows higher values are needed or lower values are needed.

Air Cleaners	Keep as close to the source as possible to minimize contaminating duct but need to balance with space availability and maintenance needs. Examples are HEPA and Charcoal filters. Less common but possible to use a bag house filter or liquid scrubber.
Supply Air to Lab	Quantity: Generally 5–10% less than exhaust quantity to maintain directional airflow unless it's a "clean" room. Delivery: Don't want to create a cross draft >50 fpm at face of exhaust hood. See NIH guidelines. Do mock-ups if you can.
Other Considerations	Use of Computational Fluid Dynamics (CFD) may have application for large or critical projects. *Person* who does CFD must be qualified – check experience and education. CFD program used must be verified. See NIH publications of CFD.
Heat Wheels/Enthalpy Wheel	Heat wheels can be used effectively to recover heat (energy) from exhaust air. However, it is recommended that contaminated exhaust from fume hoods and other containment hoods not be included. See interpretation IC 62.1–2010 of ASHRAE standard 62.1–2010. In part "... this requirement appears to specifically preclude the use of total energy recovery devices serving fume hoods or chemical storage type spaces."
Opportunities	Use the energy savings of ACH as an opportunity to assure and require good lab design, especially location of hoods, supply air design, operations requirements, resources for monitoring performance and potential exposures.
Management of Change	Research equipment and materials can change rapidly and with it may be new requirements for ventilation and safety controls. Must have a way to identify and manage significant change.
Hazard and Risk Assessment	*It all starts here! Cannot provide the correct ventilation needs until you know the potential hazards and perform a risk assessment.*

13.4 Teaching Laboratories

Teaching laboratories present a unique challenge for safe design. The major differences between a research laboratory and a teaching laboratory is the number of occupants that can be in the laboratory at the same time and the relative inexperience of the students. Special attention must be given to provide adequate egress during emergencies and adequate space to perform experiments safely. Ease of movement around benches and equipment is important to prevent accidents. Complete line of sight for teaching assistants or instructors to all students must be maintained. There may also be a need to provide instructional or lecture capabilities within the same space. Otherwise all the elements discussed below should be addressed and the same design process followed. See chapter 16 on teaching laboratories in Guidelines for Laboratory Design.[1]

13.5 What is the Process for Laboratory Design?

Before design can begin there must be a clear understanding and agreement of the purpose of the laboratory or facility, the materials to be used, the processes to be conducted and the operational procedures. Table 13.3 provides guidance on the information that is needed at the beginning stage.

The main design elements that effect safety and health as well as usability are (1) laboratory or facility layout, (2) heating ventilation and air conditioning, (3) building systems or laboratory emergency equipment that enhances loss prevention or workplace health and safety, and (4) and special requirements that are unique to the specific laboratory type such as high containment lab, clean room, *etc.* These items will be discussed in more detail.

Generic laboratory types such as chemistry, biology, and engineering are too broad and have different meanings to different professional disciplines. Architects and design engineers may have a very different vision of a "chemistry laboratory" than the chemist who will occupy the laboratory. This is particularly challenging in modern times where there is much multidisciplinary research occurring. Chemists are often working with biological agents while engineers may work with both biological and chemical agents. Thus a "chemistry" laboratory may also need a biological safety cabinet which may not be intuitive to an architect or engineer who has a preconceived notion of what a chemistry laboratory looks like. Other terms such as "dry and wet" labs and "open labs" as noted above can be misleading and lead to some design deficiencies. Table 13.4 provides a list of some commonly used "names" associated with laboratory research. In the design stage it is critical that everyone on the design team agrees with what

Table 13.3 Information needed before design can begin.

Type of research being conducted	Hours of operations
Materials and quantities to be used–	Processes to be performed
chemicals, microbiological agents, radiation producing materials	Central *vs.* local storage space
Number of people	Amount and type of hazardous waste generated
Equipment to be used	Power requirements

Table 13.4 Common names used for different types of labs.

General Analytical Chemistry	High Toxicity
Nanotechnology	Aeronautical Engineering – Wind Tunnel
Jet/Rocket Propulsion	Hydraulic Engineering
Material Analysis/Testing	Electrical Engineering
Foundry	Internal Combustion
Pilot Plant	Controlled Environment
High-pressure	Radiation
Biosafety	Clinical
Teaching	Gross Anatomy
Pathology	Autopsy
Morgue Facility	Open or Team
Animal Research	Microelectronics & Clean Room
Printmaking Studio	Photographic Darkroom/Imaging
Support Shops	Hazardous Chemical Waste Room
Radioactive Waste Room	Biological Waste Room
Laboratory Storeroom – General Chemical	Flammable Liquids/Bonded Alcohol
Compressed Gases	Biological Specimens
Archaeology, Geology Specimens	Laboratory Shop Storeroom
	MakerSpace

will be done in the space. This involves collecting and understanding all of the information in Table 13.3. Then an appropriate risk assessment can be conducted, and the correct design features needed for the specific activities planned can be selected.

13.6 Assembling the Design Team

The members of the design team must be engaged early and often and develop a good working relationship as a team. During the design and construction stages there will be compromises and changes that may be needed and it is important that everyone understand and trust the perspective of all the parties involved. One particular part of the design process that has the potential to severely impact the safety is sometimes called "value engineering". This is usually done at the later stages of design particularly when the budget needs adjustment. This is an appropriate process as long as everyone on the design team is involved and the consequences of any changes are well understood and accepted by all.

The key members of the design team are:

- laboratory scientists and workers
- design architect
- engineers of several disciplines may be needed (mechanical, electrical, plumbing, structural, fire protection)
- project managers
- administrators managing the budget
- environment, health and safety advisors
- commissioning agent
- regulatory experts

13.7 Sources of Information

There are many diverse sources of information on laboratory design, construction, and operation. These include country and local regulatory standards, international building codes, international as well a specific country standard that are considered best practices unless adopted as a regulation by the authority having jurisdiction. Several of these standards and codes are referenced in the following sections and appear in the references. There are many sources of good information on the safety and health aspects of good laboratory design as well as the regulatory requirements. Some of them follow.

- The National Institutes of Health (NIH),[3] US. Their website provides a comprehensive set of design guidelines that goes well beyond safety and health issues.
- Dr Farhad Memarzadeh,[4–6] Director of Technical Services at NIH has published several studies using computational fluid dynamics (CFD) that identify good design practices as it relates to exposure control ventilation in laboratories as well as animal facilities.
- American National Standards Institute (ANSI)[7] has adopted several standards that are directly related to safety issues in laboratories including laboratory ventilation, laboratory decommissioning, emergency eye wash and showers.
- American Society of Heating Refrigeration and Air Conditioning Engineers.[8] ASHRAE publishes guidance on designing sustainable laboratories.

- International Building Code (IBC).[9] This is a model code produced by the International Code Council. It focuses on a variety of building functions. Those that relate to laboratories include means of egress, fire safety, ventilation, lighting, and sanitation among others.
- International Mechanical Code (IMC).[10] This code establishes minimum requirements for ventilation systems in buildings. Section 4 specifically addresses the design and installation of laboratory ventilation systems.
- International Fire Code (IFC).[11] This code provides design requirements related to fire prevention, detection and suppression in buildings as well as life safety and safe storage.
- British Standards Institution (BSI).[12] This is the national standards body of the UK. It has produced several standards related to laboratory ventilation.
- International Institute for Sustainable Laboratories (I2SL).[13] This is an organization that focuses on designing and operating sustainable laboratories and was originally called Laboratories for the 21st Century. It provides many case studies on successful applications of design and operating principles.
- UCLA Center for Laboratory Safety.[14] The University of California at Los Angeles established this center after the death of a laboratory researcher. Its mission is to advance the knowledge about laboratory safety.

Table 13.1 provides a checklist of items that should be reviewed and discussed during the design stage.

13.8 Loss Prevention, Safety and Health

The specific laboratory operations will determine the safety and health design features required. The operational factors include what materials, equipment and processes will be used as well as the number and qualifications of the occupants. The health and safety design features are listed in Table 13.1 and all should be carefully considered before rejecting as not applicable to the particular design.

Key items are emergency equipment (eye wash, showers, Fire detection and suppression and emergency shut-off systems) and operational (storage, hand washing, facilities, access restrictions). Figure 13.5 shows a "safety station" in a teaching laboratory. All the key emergency equipment including emergency eye wash, shower, fire extinguisher, spill control material, safety data sheets and communication devices can be located in one location. In a teaching laboratory or open laboratory multiple stations may be needed.

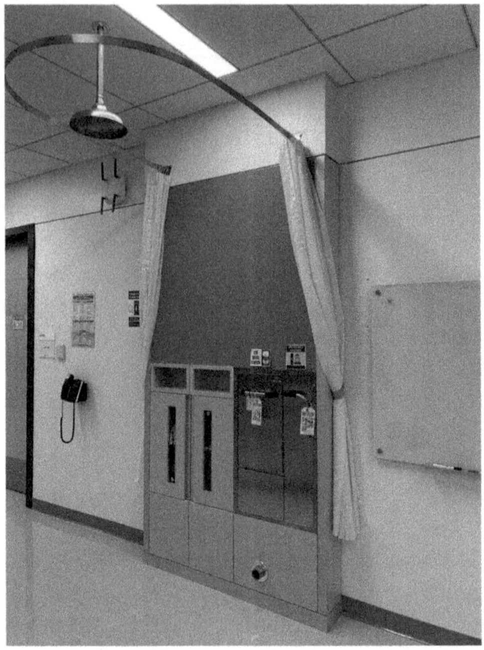

Figure 13.5 Safety station in a teaching laboratory (MIT).

References

1. L. J. DiBerardinis, J. S. Baum and M. W. First *et al.*, *Guidelines for Laboratory Design: Health, Safety and Environmental Considerations*, John Wiley & Sons, New York, 4th edn, 2013.
2. C. Dagdeviren, T. Durak and D. Sadat, Research resiliency through lean labs, *Adv. Intell. Syst.*, 2020, 2000074.
3. NIH, *NIH Design Guidelines*, 2012. Available at htttp://orf.od.nih.gov/PoliciesAndGuidelines/BiomedicalAnimalResearchFacilitiesDesignPoliciesandGuidelines/, last accessed 16 June 2020.
4. F. Memarzadeh, *Methodology for Optimization of Laboratory Hood Containment*, National Institutes of Health, Bethesda, 1996, vol. 1 and 11.
5. F. Memarzadeh, *Ventilation Design Handbook on Animal Research Facilities Using Static Microisolators*, National Institutes of Health, Bethesda, 1998, vol. 1 and 2.
6. F. Memarzadeh, Effect of reducing ventilation rate on indoor air quality and energy cost in laboratory, *J. Chem. Health Saf.*, 2009, **16**(5), 20–26.

7. American National Standards Institute (ANSI), 25 W 43rd St., NY, NY 10036 – Z9.5/AIHA 2012 Laboratory Ventilation; Z9.11/AIHA 2008 Laboratory Decommissioning.
8. American Society of Heating Refrigerating and Air-Conditioning Engineers (ASHRAE), 1791 Tullie Circle, N.E., Atlanta, GA 30329 – Standard 189 2011 Standard for the Design of High Performance Green Buildings.
9. IBC, *International Building Code*, International Code Council, Inc., 4051 West Flossmoor Road, Country Club Hills, IL 60478, 2012.
10. IMC, *International Mechanical Code*, International Code Council, Inc., 4051 West Flossmoor Road, Country Club Hills, IL 60478, 2009.
11. IFC, *International Fire Code*, International Code Council, Inc., 4051 West Flossmoor Road, Country Club Hills, IL 60478, 2012.
12. BSI – British Standards Institute British Standards Institute, Chiswick, High Road London, UK W4 4AL, UK – (2004) Fume Cupboards Part 4: Onsite Test Methods, BSI Document 4175-4:2004DC; (2007) Specifications for recirculating filtration fume cupboards, BSI 389.
13. International Institute for Sustainable Laboratories, https://www.i2sl.org, last accessed 16 June 2020.
14. Center for Laboratory Safety, *Fume Hood Summit*, http://cls.ucla.edu/cls-publications, last accessed 16 June 2020.
15. SEFA, *SEFA 2010 Desk Reference*, Scientific Equipment and Furniture Association, 1205 Franklin Avenue, Suite 320, Garden City, NY 11530, 3rd edn, 2010. Available at www.sefalabs.com.

CHAPTER 14

Innovation Spaces: the New Campus Risk Paradigm

TOLGA DURAK*

Managing Director, Environment, Health and Safety Programs, Massachusetts Institute of Technology, 77 Massachusetts Avenue, Cambridge, MA 02139, USA
*E-mail: tdurak@mit.edu

14.1 Designing Spaces for Safe Innovation

Facilitating safe innovation begins with the design of innovation spaces and encompasses a vast range of topics from design standards and considerations to training and usage procedures to supervision, oversight, and regulatory compliance. Innovation spaces do possess all characteristics of a traditional laboratory. They are, however, set up to operate with the main purpose of innovating rather than replicating or duplicating as in the case of traditional labs. While a full explication of the various laboratory types and their respective design and procedural requirements could fill several books, Table 14.1 provides an overview of common types of innovation spaces and the United States regulatory bodies that oversee each.

Within each of the innovation spaces listed in Table 14.1, users should meet all applicable standards and requirements of the regulatory agencies involved. For Example, Good Laboratory Practice (GLP) or Good Manufacturing Practice (GMP) facilities are strictly regulated by the FDA. GLP facilities are regulated closely for the processes and conditions under which clinical and non-clinical research is planned, performed, monitored, recorded, archived, reported and how these facilities are maintained.[2]

Challenges for Health and Safety in Higher Education and Research Organisations
Edited by Olga Kuzmina and Stefan Hoyle

Published by the Royal Society of Chemistry, www.rsc.org

Table 14.1 Innovation space overview

Innovation space type	
Makerspace	• **Overview definition** Term covering a variety of spaces, principally machine shops, project makerspaces, and community makerspaces, designed to offer access to tools and space to invent, prototype, and bring ideas to physical life;[1] recent scope and use expansion of these spaces includes biologicals, chemicals, and radioactive materials laboratories, as well as "hackspaces" or "hackerspaces", devoted to community-based technological innovation • **United States Regulatory Bodies Involved** OSHA NIH NRC
Incubator	• **Overview definition** A safe innovation space often designed and operated to enable the initial stages of a commercial start up, within a larger structure. These spaces are usually not purely academic research space even though they can be a part of institutes of higher education. • **United States Regulatory Bodies Involved** OSHA EPA DPH
High Containment Laboratories	
Biolaboratory	• **Overview definition** Space in which biological research, including cellular and microorganism, is completed with inherent risk of infection or environmental damage • **United States Regulatory Bodies Involved** OSHA EPA FDA NIH
GLP/GMP Practices	• **Overview definition** Good Laboratory Practices (GLP): Regulations for the processes and conditions under which clinical and non-clinical research is planned, performed, monitored, recorded, archived and reported; Also governs how research facilities should be maintained[2] Good Manufacturing Practices (GMP): Regulations enforced by the FDA requiring that manufacturers, processors, and packagers of drugs, medical devices, some food, and blood take proactive steps to ensure products are safe, pure, and effective[2] • **United States Regulatory Bodies Involved** FDA

(*continued*)

Table 14.1 (*continued*)

Innovation space type	
Cleanroom	• **Overview definition** A contained environment free from dust and other contaminants, classified according to the cleanliness level of the air within the space[3] • **United States Regulatory Bodies Involved** OSHA EPA FDA
Nanotechnology Laboratory	• **Overview definition** Space dealing with the manipulation of matter on an atomic, molecular, and supramolecular scale • **United States Regulatory Bodies Involved** OSHA EPA NRC DHS
Microelectronics Laboratory	• **Overview definition** A semiconductor research and fabrication facility that supports the design, fabrication, and packaging of novel devices[4] • **United States Regulatory Bodies Involved** OSHA EPA DHS

Similarly, regulations for GMP facilities require that manufacturers, processors, and packagers of drugs, medical devices, some food, and blood take proactive steps to ensure that products are safe, pure, and effective.[2]

Agencies and organizations with codes and standards dictating the design of laboratories in the United States include:[5]

- 29 CFR 1910.1450: OSHA Occupational Exposures to Hazardous Chemicals in Laboratories
- ANSI/ASSE/AIHA Z9.5 Laboratory Ventilation
- ANSI/ISEA Z358.1 Emergency Eyewash and Shower Equipment
- ASHRAE 110 Method of Testing Performance of Laboratory Fume Hoods
- ASHRAE Applications Handbook, Chapter 16 Laboratories
- ASHRAE Laboratory Design Guide, 2nd Edition
- Association for Assessment and Accreditation of Laboratory Animal Care (AAALAC) Standards
- Biosafety in Microbiological and Biomedical Laboratories (BMBL) 5th Edition, Department of Health and Human Services, Centers for Disease Control and Prevention and National Institutes of Health
- Guidelines for the Laboratory Use of Chemical Carcinogens, Pub. No. 81-2385. National Institutes of Health

- NIH Design Policy and Guidelines, National Institutes of Health
- NFPA 30 Flammable and Combustible Liquids Code
- NFPA 45 Fire Protection for Laboratories using Chemicals
- Unified Facilities Guide Specifications (UFGS)

14.1.1 Initial Design Steps

This chapter will not investigate the specifics of a particular innovation space design, but the overview below is intended to provide a general understanding of the recommended steps for preliminary design.[6]

(1) Analyze the existing facility or compare with a similar facility.
(2) Identify and interview stakeholders (including users) to determine the requirements for equipment, chemicals, infrastructure, and usage patterns.
(3) Establish standards for the space according to current best practices and established sector design standards.
(4) Develop a room type list.
(5) Create room type diagrams.
(6) Determine the quantity and area of each room type.
(7) Determine the room type performance specifications and data.
(8) Estimate building area.
(9) Determine an overview of mechanical, plumbing, and electrical systems.
(10) Estimate construction costs; depending on institutional procedures, costs may instead be fixed before any other steps are undertaken. In this case, the project budget significantly informs all design considerations and choices.

14.1.2 Key Considerations

In this initial design phase, among the many considerations that must be taken into account, a few stand out as particularly worthy of note, detailed below.

14.1.2.1 Chemicals, Bioagents, and Gases

The first of these noteworthy considerations is the presence, type, and quantity of laboratory chemicals and gases – each of which is a factor in determining the critical design elements of control areas, as defined under local fire codes, and chemical and gas storage facility. Designated biosafety levels should also be selected and incorporated into the building design process carefully in order to facilitate specific research and activities involving different bioagents. Current and emerging laboratory trends will be discussed later in this section but suffice it to say here that transparency, or "research on display" is increasingly favored, often in large, open-plan laboratory spaces with transparent walls and established sight lines. Creating a laboratory space that takes into account this aesthetic but that also contains adequate control areas and spaces with biosafety designation requires making determinations early in the design process. The need for control areas is largely dependent on the types of chemicals and gases to be present in the facility. Early in the design process, in consultation with space users, Environment, Health, & Safety (EHS) should determine the types and maximum quantities of chemicals and gases in each laboratory. This information should then inform the establishment of control area requirements that affect the building design as a whole. Control areas should also have provision for changes that may occur in future laboratory use or chemical and gas quantity.[7]

Determining laboratory chemical type and quantity early in the design process also allows for appropriate plans to be put into place for chemical storage. EHS should collaborate with designers as well as other stakeholders, particularly laboratory managers, to ascertain the types of chemicals requiring storage as well as patterns for their usage. Doing so mitigates the possibility of chemicals being stored incorrectly on bench tops, inside fume hoods, or other unsuitable locations, and instead facilitates a safe and optimally functioning laboratory environment.[7] Additionally, chemical components must be considered when determining safety procedures. Any areas where chemicals must be stored in bulk, such as large volumes of regularly used solvents, must be taken into account. The handling and use of hazardous chemicals should be restricted to authorized personnel who have been given dedicated training in chemical safety, as well as access to data sheets with information about safety. These practices ensure that safe design elements carry through in practice.[8] Provisions must also be considered early in the design process for chemical waste collection and disposal, with adequate ventilation for waste container storage.

14.1.2.2 Equipment

Another key consideration in the early design phase of an innovation space is the equipment likely to be present in the facility. Again, making these determinations must be the result of extensive collaboration between EHS, designers, and researchers. Additionally, an inventory lists from comparable facilities can offer a useful starting point.[8] Beginning the design process with as comprehensive an equipment list as possible facilitates the effective design of crucial fundamentals such as Heating, Ventilation and Air-conditioning (HVAC), power, and plumbing, as well as appropriate spatial design. The choice of equipment will be informed by the reagents present in the space, and specialized equipment, such as fume hoods, gas manifolds, air monitors, gas detectors, and interlock systems, must be chosen in consideration of hazardous substances type and quantity. In addition to the choice of equipment type, consideration must also be given to the materials used.

For example, certain types of spaces may require non-reflective surfaces, or anti-mold wall paints, depending on their planned usage.

Fume hoods and other local exhaust ventilation for operator protection are critical and complex pieces of laboratory equipment, and the necessity that they be coordinated with HVAC control systems makes early planning crucial. In planning ventilation structures, flow rate volumes must be taken into consideration. Gases and non-condensing vapors, for example, require a different minimum duct velocity from process dust. Such considerations significantly impact the design of air handling units, and, if ignored, can result in high costs to correct errors. Designers must begin with a determination of the types of fume hoods necessary, as well as the nature of the controls needed for both the quantity and type of hoods. Throughout the design process, EHS and/or facility design guidelines must be incorporated.[7]

14.1.2.3 *User Movement and Circulation in the Space*

Predicting user and portable equipment circulation patterns within any space is complex, but it presents a special challenge in laboratory design. However, much as with other design considerations, designing with circulation in mind begins with communication. Soliciting detailed input from laboratory managers and users provides the necessary information to model circulation in a given laboratory type. Additionally, when possible, observation of similar facilities is a valuable source of information. These approaches can enable for adequate spacing to be provided for key portable equipment along with access to the power source where specific tasks need to be completed. The implications for incorporating such approaches during the design phase can range from reduction of accidental loss of samples to personnel injuries when the lab is fully operational.

Design for people circulation should take into account direction of movement, type of use, frequency of use, and time of use for laboratory users and for equipment. Key avenues of circulation should be as direct as possible, following the principle of the shortest distance between two points, and as unobstructed as possible. Spatial layout and equipment locations must take into account normal workflows within the lab, as well as emergency eventualities. When these workflows are not given adequate consideration, as is too often the case, the design of the space can be a barrier to efficient and safe use, rather than enabling it. For equipment that must move through the space, pathways to building exits should be planned and examined. These routes should take into consideration all doors, turns, and other building feature from "point of delivery to point of use and back again".[7]

14.1.3 Emerging Trends in Laboratory Design

In recent years, laboratory design has evolved along with researcher needs, institutional priorities, and available technologies. Three major trends have emerged:

(1) a shift towards "open" laboratory spaces
(2) an emphasis on flexibility
(3) a prioritization of sustainability

The driving force behind the shift towards open, transparent laboratory spaces is a focus on collaboration. In open laboratories, researchers share space and equipment, with communication and collaboration encouraged and facilitated by the design. "Modern science is an intensely social activity", writes Daniel Watch for the *National Institute of Building Science*.[9] "Thus, science functions best when it is supported by architecture that facilitates both structured and informal interaction, flexible use of space, and sharing of resources." In addition to the design of the laboratory facility itself, the inclusion of auxiliary spaces such as break rooms and other gathering areas are opportunities to design for collaboration.[9]

Flexibility, in itself a major emerging trend, is an important aspect of designing spaces ideal for collaboration. A flexible space can accommodate interdisciplinary research, as well as evolving team-based research. The concept of flexibility in laboratory design carries a variety of meanings, including space for expansion, the capacity to support diverse uses, and the ability to be reconfigured or otherwise adapted.[9] It is not uncommon for many university lab spaces to be used for different purposes during different periods of time. Teaching laboratories that are used as an experiential learning environment during the school year are often utilized for workshops or short courses during the breaks or holidays. A key component of a flexible laboratory space is its engineering system. This means that connects and disconnects should be readily accessible at ceilings or walls, in order to allow for expeditious and inexpensive equipment connections. Additionally, flexibility is increased when the design allows for the addition or removal of fume hoods or other local exhaust ventilation. In this way, the space can shift seamlessly between serving as an office or laboratory.[9] The use of mobile casework also allows for flexibility in equipment inventory, placement, and storage. Similarly, making use of all available space, including space above head height, increases the available uses of a laboratory.

Sustainability, the third significant emerging trend in recent years, continues to gain prominence. Since laboratories are by nature far more energy-demanding than spaces such as classroom and office buildings, designing for sustainability presents unique challenges. Typical elements such as containment and exhaust devices, heat-producing components, and fail-safe redundancies such as power supply mean that laboratories continue to lag behind simpler workspaces in terms of sustainability, however research and educational institutions are working to meet these challenges. Increasing energy and resource efficiency, reducing harmful waste, and making use of recycled elements are key goals for increasing laboratory sustainability. Additionally, several federal mandates have recently gone into effect concerning sustainable laboratory design in the United States, including the Energy Policy Act of 2005 and Executive Order 13693, "Planning for Federal Sustainability in the Next Decade".[9] As one of the many responses to such policies, many institutions of higher learning have included sustainability performance expectations for laboratories as a section in their institutional climate action and resiliency plans.

Particularly in university settings, the trend towards sustainability has led to the emergence of green laboratory facilities and programs. Whether in the design of new sustainable facilities, in retrofitting existing facilities with upgraded equipment, or in implementing environmentally friendly practices and procedures, many universities are working to reduce emissions and work towards sustainable innovation. At MIT, the Green Labs project was founded in 2016, with the goal of a 32% overall reduction in greenhouse gas emissions by the year 2030. The initiative seeks to ensure that "every laboratory within the MIT community has the knowledge, resources, and support to enable a sustainable approach to world class scientific research".[10] In addition to resources for design and training, the Green Labs project, and other similar initiatives at peer institutions, offers the means for laboratories to work towards official Green Laboratory Certification by the Institute.

14.2 Assessing and Managing Risk in Innovation Spaces

The assessment and management of risk is arguably one of the core responsibilities of an EHS professional, particularly within an academic institution. In fact, risk management is so deeply interwoven in the role of an EHS professional that the terms "EHS professional" and "risk manager" will be used interchangeably throughout this chapter. Risk management is a vast discipline, and its principles are the subject of countless books, articles, courses, and expert discussions. However, given that most risk management texts are written through the lens of managing risk in a corporate, for-profit setting, it is very fitting to discuss the principles of risk management in the context of innovation spaces in academic institutions.

14.2.1 What is Risk?

In broader context, risk is generally viewed as a purely bad thing: "a chance or possibility of danger, loss, injury, or other adverse consequences" as defined by the *Oxford English Dictionary*. More broadly still, risk is understood to be "human interaction with uncertainty", as described by Preston B. Cline in his 2004 paper "The Etymology of Risk".[11] When we talk about institutional risk however, the term "risk" carries a slightly, though crucially different meaning. The Risk and Insurance Management Society (RIMS) defines a risk as "an uncertain future outcome that can either improve or worsen your position".[12] Similarly, the International Organization for Standardization (ISO) 31000:2018 Risk Management – Guidelines makes a key revision to the definition of risk: while the previous ISO risk management manual used the definition "chance or probability of loss", the 2018 revision calls risk the "effect of uncertainty on objectives".[13]

While this evolution of meaning might at first rankle some language purists, it is worth recalling the word's etymology. As Peter L. Bernstein points out in *Against the Gods: The Remarkable Story of Risk*, "the word 'risk' derives from the early Italian *risicare*, which means 'to dare'". He continues, "when we take a risk, we are betting on an outcome that will result from a decision we have made, though we do not know for certain what the outcome will be. Uncertainty about outcomes that can be either negative or positive".[14]

In returning to the roots of the idea of risk as an act of daring with uncertain, multivalent outcomes, we are recognizing a key component of risk, which is at the core of innovation: it has the potential to be positive. Throughout this chapter, this is the definition of risk from which we will operate.

14.2.2 Pure Risk

While the risk management field has devised extensive classifications of risk, each of which has practical applications for innovation spaces, the comparison between pure and speculative risk is most useful to examine in this context. "Pure risk" most closely resembles what people have traditionally thought of as risk in general. A pure risk carries the potential for "loss or no loss, but no chance of gain".[15] The Committee of Sponsoring Organizations of the Treadway Commission (COSO) calls pure risk "loss only risk", or "risks that produce only negative effects".[16] Examples of pure risk include the chance for injury or disease, fire, crime, natural disasters, and other occurrences that may or may not happen, but carry no potential positive outcome. Of course, pure risk is not desirable and is best to be avoided whenever possible.

Traditionally, EHS professionals have focused predominantly on categorizing every hazard as a pure risk. This traditional risk assessment is conducted using one or more of the techniques listed in the following table (see Table 14.2). When the assessment of risk is completed, risk is then minimized or mitigated using hierarchy of controls, a system used to ascertain how to put into place control solutions that are both effective and practicable.[17]

An additional new variation of traditional risk assessment tools is currently in development at MIT. Created by Nancy Leveson of MIT, Partnership for Systems Approaches to Safety and Security (PSASS) takes "a systems approach to safety and security by applying systems thinking to develop practical methods".[27] Essentially, PSASS is based on the traditional safety approaches described in Table 14.2, but with significant enhancements, including considerations for human factor challenges, business continuity, and opportunity cost. PSASS operates on the following eight principles:[27]

(1) collaboration
(2) systems thinking
(3) solving real problems

Table 14.2 Traditional analysis techniques.

Tool	

Technique for Human Error-Rate Prediction (THERP)[18]	• **Definition**
	Technique for evaluating the probability of a human error throughout the completion of a specific task
	• **Capabilities**
	Allows measures to be taken to reduce the likelihood of errors occurring within a system, leading to an improvement in overall safety levels
	• **Applicable Circumstances**
	Introduced for three primary reasons: error identification, error quantification, and error reduction
	• **Advantages**
	  • Can be used at all stages of design; compatible with Probabilistic Risk Assessments such as those used to evaluate risks inherent in a composite technological unit
	  • Process is transparent, structured, and provides a logical review of the human factors considered
	  • Can be used within a wide range of differing human reliability domains
	• **Disadvantages**
	  • Highly resource-intensive
	  • Does not lend itself to system improvement
	  • Large discrepancies in practice with regard to risk assessment of specific tasks
	  • Does not provide guidance in how to model the impact of performance shaping factors

Failure (Hazard) Modes and Effects Analysis (FMEA)[19,20]	• **Definition** Systematic, proactive method for evaluating a process to identify where and how it might fail and to assess the relative impact of different failures, in order to identify the parts of the process that are most in need of change • **Capabilities** Provides a documented method for selecting a design with a high probability of successful operation and safety, and for assessing potential failure mechanisms, failure modes, and their impact on system operation • **Applicable Circumstances** Can be performed at system, subsystem, assembly, subassembly or part level

(*continued*)

Table 14.2 (*continued*)

Tool	

	• **Advantages**
	  • Collects information to reduce future failures, capture engineering knowledge
	  • Early identification and elimination of potential failure modes
	  • Emphasizes problem prevention
	• **Disadvantages**
	  • Has limited validity when used in isolation
	  • May only identify major failure modes in a system
	  • FMEA worksheet is difficult to produce, understand, read, and maintain

Hazard and Operability (HAZOP) Review[15]	• **Definition**
	A controlled, systematic investigation designed to identify and assess risks to equipment or personnel
	• **Capabilities**
	Reviews the design to identify issues that may otherwise not have been found; Technique is based on breaking the overall complex design of the process into a number of simpler sections called 'nodes', which are then individually reviewed
	• **Applicable Circumstances**
	Appropriate for projects and systems in which virtually all risks must be eliminated
	• **Advantages**
	  • Identifies causes and potential outcomes for each risk
	  • Subdivides the project or system design into small components
	  • Develops a solution for each risk
	• **Disadvantages**
	  • Requires significant time and expertise

Job Hazard Analysis[21]	• **Definition** Technique that focuses on job tasks as a way to identify hazards before they occur; Focuses on the relationship between the worker, the task, the tools, and the work environment. Ideally, after identifying uncontrolled hazards, users can take steps to eliminate or reduce them to an acceptable risk level • **Capabilities** After identifying uncontrolled hazards, allows for steps to eliminate or reduce them to an acceptable risk level

	• **Applicable Circumstances**
	  • Jobs with high injury or illness rates
	  • Jobs in which one simple human error could lead to a severe accident or injury
	  • Jobs new to an operation or that have undergone changes in processes and procedures
	  • Jobs complex enough to require written instructions
	• **Advantages**
	  • Findings can be used to eliminate and prevent hazards in workplaces
	  • Fewer worker injuries and illnesses
	  • Safer, more effective work methods
	  • Increased productivity
	• **Disadvantages**
	  • Confined to a specific job function

Human Error Analysis[18,22]	• **Definition**
	Technique used in the field of human reliability assessment for the purpose of evaluating the probability of a human error occurring throughout the completion of a specific task
	• **Capabilities**
	Allows measures to be taken to reduce the likelihood of errors occurring within a system, leading to an improvement in the overall levels of safety
	• **Applicable Circumstances**
	Conducted for three primary reasons: error identification, error quantification, and error reduction
	• **Advantages**
	  • Quick and straightforward to use and has small demand for resource usage
	  • Provides useful suggestions to reduce the occurrence of errors
	  • Allows cost benefit analyses to be conducted
	  • Highly flexible and applicable in a wide range of areas
	• **Disadvantages**
	  • Relies to a large extent on expert opinion and is therefore sensitive to both optimistic and pessimistic assessors

(*continued*)

Table 14.2 *(continued)*

Tool	

Technique of Operations Review (TOR)[23]	• **Definition**
	Theory of accident causation and control that states that management and supervision weaknesses are the root cause of workplace injuries and illnesses
	• **Capabilities**
	Designed to uncover management oversights and omissions instead of hardware or operator problems
	• **Applicable Circumstances**
	When management shortcomings cause accidents or hazards
	• **Advantages**
	  • Improved job procedures with a safety focus
	  • Elimination of workplace disorder
	  • Enhanced accountability for all employees
	  • Better management and supervisor training
	• **Disadvantages**
	  • Takes into account only management failings

Risk Analysis[24]	• **Definition**
	The process of assessing the likelihood of an adverse event occurring within the corporate, government, or environmental sector
	• **Capabilities**
	Designed to predict uncertainties in processes and minimize their occurrence or impact
	• **Applicable Circumstances**
	Situations in which uncertainty is undesirable
	• **Advantages**
	  • Reduces uncertainty
	  • Allows procedures to be put in place to minimize impact of adverse events
	• **Disadvantages**
	  • Does not apply when uncertainty is a desirable component of a project

Total Safety Management[25,26]	• **Definition**
	Performance-oriented approach to safety management that involves the total organization in establishing and maintaining a safe and healthful work environment
	• **Capabilities**
	Gives organizations a sustainable competitive advantage in global marketplace
	• **Applicable Circumstances**
	Most useful in a quality-management setting
	• **Advantages**
	  • Establishes a healthy and safe work environment conducive to consistent peak performance
	• **Disadvantages**
	  • Requires full executive buy-in, which can be difficult to obtain at the outset

(4) guidance from stakeholder needs
(5) enhancing communication and cooperation
(6) technology transfer
(7) consideration of all aspects of system safety
(8) socio-technical view of safety

Currently, Leveson and her team are focusing on evaluation of existing practices, focused research, mentoring and internships, information dissemination, sponsored research, and educational activities, as PSASS is tested and refined.[27] When testing of the tool is complete and PSASS is ready for publication, it will be widely acceptable for use on large and highly complex systems. As compared to the traditional risk assessment tools described above, PSASS will use systems thinking to "provide more comprehensive, more efficient, and more effective results",[27] improving the safety of existing spaces and systems, and informing the design of new innovation spaces.

14.2.3 Speculative Risk

As opposed to pure risk, "speculative risk" carries "a chance for loss, no loss, *or gain*".[15] COSO calls speculative risk "loss or gain risk", with "possibility of negative effect (loss) and positive effect (gain)". In other words, the three possible types of outcome from a speculative risk are negative, neutral (no change), or positive. In a traditional business environment, speculative risks can sometimes be advisable – for instance, investing in a new property, piece of equipment, *etc.* that has the potential for downsides, but that also carries possible gain. However, a traditional business can conceivably operate with some success while undertaking minimal speculative risk.

Here is where conventional risk management wisdom does not necessarily apply in an innovation or research setting: speculative risk is at the heart of innovation. By its very nature, the outcome of scientific endeavor is uncertain. Arguably, every research or innovation project is a speculative risk, with the possibility of a negative outcome, a neutral outcome, or a positive outcome in the form of a significant discovery, breakthrough, or invention.

Risk managers and EHS professionals working with innovation spaces must recognize and embrace speculative risks, while attempting to mitigate as much as possible the potential negative outcomes. All the while, they must be clearing the way for the vast positive potential inherent in these speculative risks. Yet there is another wrench: returning for a moment to the previously outlined examples of pure risk, a closer inspection reveals that the distinction between these two types of risk is hardly simple. Consider the well-known case of the University of California (UC), Davis' olive tree problem. The campus contained a large population of olive trees, which dropped their fruit, creating trip and fall hazards as well as issues of sanitation and property damage. In a sense, these olive trees represented a "pure risk". However, when a facilities management employee approached the university's risk managers with the idea of harvesting the olives for olive oil, what was previously understood as a pure risk situation was reframed as a speculative risk. Or, as the case is described in Jean-Paul Louisot and Christopher H. Ketcham's *Enterprise Risk Management: Issues and Cases*, "this endeavor represents a situation where what was considered to be a net-loss risk scenario was managed into something that not only mitigated the risk of loss but turned the risk into opportunity that could be exploited for its value-add to the organization and its stakeholders".[28]

This olive tree example illustrates the complexity of the risk manager's task. The approach can and should be applied to innovation spaces, with an appropriate balance when evaluating operations for hazards as compared to potential gain situations. It is worth pointing out that the speculative risk considerations of the olive trees at UC Davis do not negate their pure risk aspects. Nor is this example meant to suggest a cavalier approach to risk management, based on the potential for a positive outcome in nearly any risk, given the inherent uncertainty in innovation. It is meant to illustrate the complexity of determining the nature of risk in innovation spaces, and the crucial importance of speculative risk as a driver of innovation.

14.2.4 Enterprise Risk Management

Following the spectacular collapse of Enron, WorldCom, and financial institutions during the 2008 crisis, the concept of enterprise risk management (ERM) emerged as a more holistic way of tackling an organization's risk management, including strategic, operational, financial, and other risks.[15] For the colleges and universities in the U.S., the University Risk Management and Insurance Association (URMIA) appointed its own taskforce of risk managers and authors to prepare resources on ERM in university settings. URMIA's authors point to factors that are "increasing pressure to transform higher education", and the resultant need for "a 'risk-conscious' tone at the top" of universities.[29] They also point out the numerous financial benefits of ERM for educational institutions, as well as identifying other ways in which ERM can benefit an educational institution, including:[29]

- sustaining its competitive advantage
- solidifying its integrity and reputation
- responding effectively when a significant event occurs
- avoiding financial surprises
- effectively managing all of its resources

While traditional risk management practices focus largely on pure risk, ERM stipulates a holistic approach, and seeks to dismantle traditional siloed approaches. Much could be said about ERM, and indeed tremendous amounts of useful material have been written on this concept. In the United Kingdom, the British Standards Institutes (BSI) has published a panoply of documents containing risk management guidelines, in essence setting up ERM as standard practice for countries in the EU and round the world.[30] While a full discussion does not fall within the bounds of this chapter, suffice it to say that ERM is far more effective than its traditional predecessor and should be implemented by research and educational institutions.

14.2.5 Assessing Risk in Innovation Spaces

Measuring or assessing risk is both difficult and crucial in innovation spaces. There are six measures traditionally applied when assessing risk, along with a basic operating definition of each:[15]

- exposure: a condition that contains a possibility of gain or loss, regardless of whether an actual loss takes place
- volatility: frequent fluctuations
- likelihood: as opposed to "probability", likelihood refers to the degree to which an outcome is possible or probable
- consequences: "the measure of the degree to which an occurrence could positively or negatively affect an organization"[15]
- time horizon: the estimated duration of an event or series of events
- correlation: the relationship between any variables

The ISO 31010:2019, entitled *Risk Management – Risk Assessment Techniques* recommends the following steps for implementing risk assessment:

(1) plan the assessment
   a. define the purpose and scope of the assessment
   b. understand the context
   c. engage with stakeholders
   d. define objectives
   e. consider human, organizational, and social factors
   f. review criteria for decisions
(2) manage information and develop models
(3) analyze data
(4) apply risk assessment techniques
   a. identify risk
   b. determine sources, causes, and drivers of risk
   c. investigate the effectiveness of existing controls
   d. understand consequences and likelihood
   e. analyze interactions and dependencies
   f. determine measures of risk
(5) review the analysis
(6) apply results to support decisions
(7) record and report risk assessment process and outcomes

In research settings, exposure to risks, or any circumstance that offers the possibility of either gain or loss, is often very high. However, the consequences are often less severe in academic settings than in industries where frequency is low, but potential severity is high. For instance, while an individual researcher may be severely impacted by an adverse event in academic research, this incident will have much less severe consequences than an adverse event in an industrial environment, which, due to the scale of the operation, could cause massive injury or equipment loss. Whether in makerspaces, high containment labs, or other research spaces, exposure to harm is a risk that must be controlled if innovation is to be successful. However, volatility, or recurrent fluctuations representing uncertainty or risk related to the size of changes, can generally be understood to be low in innovation settings. While volatility certainly does exist, the research sector is not subject to the sorts of market pressures and fluctuations faced in the for-profit sector. Nevertheless, over the past several years, the decline of public funding provided to research institutions, the increasing reliance on private and alumni donations, the need to diversify portfolios and endowment funds into accommodation and other services for private hire and the social and political pressures to reduce student fees and overhead expenses add some unique risks to the research sector as well.

The next measure of risk, likelihood, is inherently difficult to measure in innovation spaces. "Likelihood can refer to the likelihood of an event or to the likelihood of a specified consequence".[31] Innovation trades on uncertainty, making it challenging for risk managers to determine the likelihood of a given occurrence, or even to determine a list of possible occurrences. Likelihood must be balanced with the consequences, or "the measure of the degree to which an occurrence could positively or negatively affect an organization".[15] Once again, consequences are inherently difficult to predict in innovation settings. However, a key rule for corporate risk management holds true in research settings: the higher the degree of likelihood and consequences of a risk, the greater the degree to which it merits substantial and continuous management.[15] These types of risks are frequently encountered in the context of innovation and must be treated with special care.

Time horizon is another measurement tool for risk assessment. It is generally understood that the longer the exposure, the more likely the risk will be realized. In a research context, risks often carry a significant time horizon, as in the case of long-term clinical trials and studies. An additional risk with regards to time horizon is the loss of competitive advantage – in other words, another researcher or institution can "beat you to the punch". In this globalized setting in which Intellectual Property (IP) theft concerns are heightened, it is crucial for research to develop expeditiously. Institutions must find the appropriate balance between being deliberate in risk assessment and management while being timely and efficient or in essence "cut corners" in order to avoid being outpaced.

The final measure of risk is correlation, or "a relationship between variables",[15] which should be used to evaluate the entire risk portfolio of a research organization. When risks are similar, it is likely that they are highly correlated. As correlation increases, so too does risk.[15] Correlation is often a significant factor in research organizations, since they engage in so many similar activities in close proximity. Traditionally, the remedy for correlation is to mitigate it through diversification. However, risks encountered in innovation spaces are often not diversifiable. Makerspaces are a particularly useful example here, as the uses of the spaces are by nature disparate, but there are similar risks. At MIT, Project Manus, an administrative oversight body, has taken up the challenge of makerspace safety through a set of programs, courses, and diverse resources designed to "see to the continued function and evolution" of the campus' some 45 different makerspaces. By situating the lion's share of the campus' makerspaces under one umbrella, Project Manus allows for consistent and uniform management of the same risks.[32]

14.2.6 Risk Appetite

When determining the appropriate risk management approach for a specific organization, that organization's risk appetite must be taken into consideration. Risk appetite is the "amount and type of risk that an organization is prepared to pursue, retain, or take".[31] According to the ISO 31010, this determination falls within the consideration of the significance of certain risks. "Once risks have been evaluated and treatments decided, the risk assessment process can be repeated to check that proposed treatments have not created additional adverse risks and that the risk remaining after treatment is within the organization's risk appetite".[31] It is safe to say that the risk appetite changes from institution to institution are varied, as senior leadership and boards of governance change.

14.2.7 Goals and Objectives for Risk Management in Innovation Spaces

Traditionally, it is the task of a risk manager and EHS professional to ensure that risk management objectives are in accordance with the overall objectives of an organization,[15] and this principle holds true for risk managers in the research and academic disciplines. To enable innovation, safety professionals must undertake the delicate task of managing risk and reward in individual innovation spaces. They should support effective decision-making in design, resource allocation, research direction, and personnel training.

In the for-profit sector, goals and objectives for risk management are oriented in the greater service towards minimizing or managing organizational risk to profits and shareholders. While the goals for risk management for higher education are not oriented towards shareholders or profits, there is commonality. These shared goals include:

- tolerable uncertainty
- legal and regulatory compliance
- earnings stability
- profitability and growth
- preservation of health and safety

An effective risk management strategy raises an organization's threshold for tolerable uncertainty, which allows for speculative risk taking, which in turn carries the potential for innovation. In addition, legal and regulatory compliance is an important goal for risk managers in the innovation sector.

While the previous two goals are relatively straightforward in their applications for risk managers in innovation spaces, the traditional goals of earnings stability and profitability and growth are no less important, though their applications for higher education differ from those in a for-profit organization. "Earnings" as such are not a factor in non-profit educational and research institutions, but fiscal prudency is. Risk managers should always consider avoiding excessive expenditure and limiting risk management and compliance costs to what is reasonable. Similarly, profitability and growth do not apply to higher education as they do to for-profit businesses, but even in a not-for-profit research setting, a certain type of Return on Investment (ROI) comes into play. Effective risk management allows for efficient deployment of resources and capital, which in turn allows research organizations to grow in reputation, income or endowment potential, and research funding and scope.

Of course, the primary objective of risk management in innovation spaces is the preservation of health and safety. This final and overarching objective is necessary for the achievement of any other objectives and is likewise an outgrowth of these objectives. The job of a risk manager and EHS professional involves countless and diverse goals, but the grounding principle is the preservation of safety.

14.2.8 The Benefits of Managing Risk

While the merits of risk management for ensuring and preserving health and safety are hardly in contention, it is useful to explore the benefits traditionally associated with risk management, and how they apply to the innovation and research in higher education.

One of the primary benefits of risk management that is traditionally recognized is its ability to reduce the costs associated with "risks from accidental loss, including the possibility of loss or no loss".[15] Risk management can also mitigate the deterring effects of hazard risks, another term for pure risks. Doing so carries important benefits in terms of enabling innovation. When hazard risks are unmanaged, their deterring effects can place a damper on research, as decision-makers lose their appetite for the speculative risk necessary for innovation. An effectively implemented ERM structure reduces uncertainty, creates tolerable uncertainty, and reduces the deterrent effects of risks.

Downside risks, or the potential negative outcomes of speculative risk taking, are a necessary byproduct of research endeavors. Not every speculative risk pays off, and the management of downside risk is another benefit of a robust risk management structure for educational and research institutions. Additionally, risk management allows for intelligent risk taking. Stakeholders are empowered to decide if downside risks outweigh potential positive outcomes, or *vice versa*. When downside risks are effectively managed, and protections against them are in place, stakeholders are more willing to take on the risks inherent in innovation, and to allocate appropriate resources for doing so.

Risk management also carries significant benefits for research institutions when it comes to meeting legal and regulatory requirements. The more effective the risk management strategies, the more easily institutions can comply with these requirements. These risk management strategies should include avoidance, mitigation, transfer, and acceptance coupled with a robust EHS program. While compliance and risk management go hand in hand, it is important to point out that the highest goal of a risk manager or EHS professional seeking to enable innovation should not simply be compliance. Instead, regulatory compliance is one of numerous positive outgrowths of a robust risk management strategy. This distinction is important and will be further discussed in the final section of this chapter.

The key takeaway for risk managers seeking to enable innovation is that speculative risk is to be embraced, as it is an essential driver of innovation.

14.3 Enabling Innovation

Having discussed some of the broader themes in designing innovation spaces and in managing risk sections, this section will in a sense "zoom out" and introduce some useful considerations for safety professionals. These rules of engagement with innovators, researchers, and other stakeholders should offer some tricks of the trade for safety professionals, geared toward enabling and facilitating safe innovation.

14.3.1 The Dangers of a Solely Compliance-based Philosophy

By its very nature, the direction of innovation cannot always be predicted. Indeed, it is when innovation outpaces established models that important breakthroughs and discoveries take place. However, this reality is not without its challenges for EHS professionals or the field of compliance in general.

Particularly with makerspaces, the issue of innovation outpacing existing regulatory frameworks and standards is a significant one considering all of the rapid prototyping efforts involving the use of different equipment such as 3D printers, laser cutters, water jets as well as different materials. In fact, if makerspaces as defined briefly in the first section of this chapter are functioning optimally, these catalyst spaces for discovery are inherently enabling innovation to redefine existing frameworks and create new ones. Thus, a significant argument can be made that inviting regulators to participate in some of these activities can allow them to better understand the compliance implications allowing for regulatory innovation to occur concurrently as well.

As such, a closer examination of the types, purposes, and unique challenges of makerspaces serves as a useful model for the perils of designing spaces and procedures with compliance foremost in mind. At MIT and other similar higher education institutions, makerspaces generally fall into one of three categories. MIT's Project Manus has created a useful set of definitions, guidelines, and goals for each of these categories, a summary of which follows.[32]

The three categories of makerspaces are defined by differences in the primary purpose and management models for the space, despite the fact that all three share predominantly similar sets of tools and aim to solve real world problems by innovative means. The types of makerspaces are:

- traditional machine shops
- project makerspaces
- community makerspaces

Machine shops are makerspaces in which the main purpose is the creation of complex systems and/or components with fine details, with the help of skilled staff members present in these spaces. Although machine shops specialize in training and mentoring makers in the use of particular equipment and systems, they are supervised environments with specific guidelines and rules that follow traditional safety compliance models.

Project makerspaces are quite similar to machine shops, with the main distinguishing factor that project makerspaces primarily support projects for certain classes and attempt to facilitate collaboration between classmates. They are operated similarly to teaching labs that follow academic programing models with integrated curriculum-based training.

Lastly, community makerspaces are unique maker-ecosystems governed by the community of members. The members of these spaces can range from current students to alumni, or even members of the extended community such as friends, family, residents of the nearby towns, cities, *etc.* These maker environments prioritize collaboration, community effort, and unrestricted making. The members are stewards of the space who govern, train, and self-proctor consequences and corrective action for their community. Faculty members and skilled staff are usually more involved during the initial start of these spaces, establishment of the management framework, selection, and training of mentors. The community approach creates more member ownership in the overall safety and compliance expectations, while concurrently allowing the community the flexibility to practice non-traditional approaches. These spaces are designed to be agile in their capacity to innovate. A strong ecosystem of dedicated members allows for a very strong safety culture.

Makerspaces make a useful test-case at the frontline of innovation spaces for examining problems with a solely compliance-based philosophy. If EHS professionals design, equip, maintain, and train personnel for these spaces, particularly community makerspaces, with a philosophy only rooted in compliance, rather than integrating safety culture through user ownership and design that embraces safety features, several dangers emerge. The first of these dangers, perhaps the most common scenario, is the high barriers for entry or use of the space resulting in reduction in users of the space, the potential for innovation being tamped down, and the very purpose of the space eroded. This danger is particularly present when it comes to user training especially in community makerspaces. When students, faculty, and other users are subjected to training that prioritizes regulatory compliance over safe use of equipment or operation of the space, a culture of fear and restriction builds, and innovation is unable to flourish. However, the most significant concern is the opportunity cost that comes at the expense of lack of use of the space and the related lack of innovation. This cost can ultimately lead to an institution losing its competitive edge to its peers, along with its ability to recruit and retain the best and the brightest faculty, staff, and students.

The second, and more insidious danger of a compliance-only philosophy is that spaces and procedures will be ill-equipped to handle unexpected eventualities, which are the inevitable byproducts of innovation. One particularly useful example to illustrate this concept is related to the use of a utility knife (pen or olfa knife) in makerspaces. At an initial glance, this tool is not necessarily associated with very specific compliance standards or guidelines. Thus, a compliance-only philosophy can fail to capture clear expectations for its use. However, in eventuality the majority of significant injuries in makerspaces are often associated to wrong or careless use of this tool. Thus, careful consideration should be given and specific expectations should be set for safe use expectations for this and any tool.

Compliance is an effective guiding principal only as long as the regulated spaces function exactly as foreseen by their creators, management processes, and their regulatory bodies. Of course, in an environment of discovery, this may not always be the case, and it certainly should not be assumed. Nevertheless, lab managers and technical staff who oversee activities, manage the space and play an active role in the implementation of safety culture should be provided with all the resources and support they need. This will allow them to be a part of the robust risk assessment and informed decision-making processes effectively in real-time.

The great danger of a solely compliance-based philosophical approach to safety is that without clear understanding of the purposes of these expectations in the context of safe operations, it can be seen as an obstruction or impediment to innovation. As a result, users usually find ways to bypass safety requirements and practices such as removing safeguards, retrofitting equipment – endangering the very personnel and spaces that regulations are put in place to protect. Thus, it is crucial for innovation spaces to have an integrated safety culture that embraces safe operation as one of the core missions of the space. This safety culture must be practiced by every user of the space individually and encouraged through community-based practices as much as possible. The community should also enforce personal accountability, provide robust training, and reinforce safe practices through sharing of best practices and lessons learned.

It is important to point out here that it is not compliance itself which poses these dangers, but rather the types of design, policy, and procedural decisions that are the natural outgrowth of a philosophy rooted solely in compliance. Such a philosophy is by its very nature limiting, both in its capacity to enable innovation and in its ability to mitigate the risks associated with discovery and maintain a safe working environment.

14.3.2 Safety *Versus* Compliance

The successful EHS professional will espouse a philosophy of safety beyond compliance. Readers may find this distinction trivial, or simply common sense, but it is of the utmost importance to the future of the EHS field. This distinction is about an ethos and a philosophy, but it is also about the specific practical manifestations of that philosophy. It may also be summed up in the idea of the letter of the law *versus* the spirit of the law: while a compliance-based philosophy adheres to the letter of the law or regulations, a safety-based approach honors the spirit of safety underlying approach of regulations and regulatory bodies.

When EHS professionals embrace safety as their guiding principal, one of the first things that commonly happens is an increased workload. As discussed in the previous section, a philosophy of compliance is inherently limited. A safety-based philosophy means working shoulder-to-shoulder with researchers, actively finding solutions to challenges, creatively working to foresee possible pitfalls and guard against them. It means creating spaces, equipment, procedures, and training that ensure the best chances of safety, even in unprecedented and unforeseeable circumstances.

It is useful here to examine some practical manifestations of an approach to EHS grounded in a philosophy of safety. Such manifestations can be seen on the granular level of increased demands on individual EHS professionals, for example. While a compliance-based, checklist driven approach generally places EHS departments and professionals at odds with researchers, becoming a part of and partner in the research community is a key component of a safety-based approach. Because researchers keep long and unpredictable hours, the successful EHS professional must embrace this reality, while making an attempt to mitigate an unhealthy work-life balance, both for themselves and for researchers. Researchers must be supported in their often unconventional working hours, which includes not only being available to them and planning for increased workload, but also discouraging practices that may put themselves and others in danger, with clear communication and respect. Overwork and the resultant fatigue can create its own risks, such as increased stress and improper procedures. The successful EHS professional must find a balance between a necessarily heavy workload for themselves and researchers and a balance that allows for safe working conditions with limiting unnecessary bureaucracy. In return, the safety professional will be seen as a colleague and a partner rather than a part of an administrative bureaucracy – and, to put it bluntly, a part of overhead expenditure that comes at the expense of available research funding.

In addition to such granular manifestations of a "spirit of the law" approach to EHS, there are larger structural and procedural outgrowths, often related to user training and inspection, maintenance, and repair procedures. While the idea of fostering a safety culture may seem amorphous, it is not only crucial for successful EHS management, but it is also the result of concrete steps. For example, putting into place procedures for sharing experiential information generated from the use of facilities such as makerspaces, and making sure those procedures are clearly communicated to all facility users allows for these spaces to be dynamic and evolving in regard to their safety equipment and procedures. Reporting structures, facilitating symposiums, forums, workshops, or other avenues for sharing knowledge can be an effective way to encourage and inculcate such a culture.

A simple but often overlooked step that EHS professionals can take towards ensuring the safety of innovation spaces is clearly defining and communicating the role and purview of the EHS office or department. Clearly defining the role and purview of EHS through institutional governance processes, in applicable program manuals and standard operating procedures and reaffirming during all training and outreach efforts are effective strategies to achieve synergy. This step not only ensures that the parameters of EHS involvement are set and understood, it also creates and defines channels of communication, resources, and training. When innovation space users are fully aware of how EHS professionals can assist them, a culture of safety and true partnership between academy and administration can flourish.

14.3.3 Meeting the Challenges of an "Invent or Perish" Mentality

One of the most difficult challenges for EHS professionals in higher education or research settings is the mentality of "invent or perish". The high-stakes cousin of the "publish or perish" axiom for which the humanities are notorious, the idea of "invent or perish" has only intensified in the past decade in higher education and research institutions.

Challenges arising from this mentality and its attendant pressures are perhaps easy to predict, but not so easy to fight. When innovation space users are under unrelenting pressure to achieve breakthroughs, laboratory safety can quickly become collateral damage. Regulations, requirements, and procedures put in place to ensure health and safety for researchers can swiftly stop being seen as crucial protections and gain a reputation as onerous impediments to innovation. With this attitude comes the tendency to treat safety requirements as obstacles to be skirted, avoided, or thwarted, to potentially catastrophic results. When EHS is seen as the enemy of innovation, laboratory safety is eroded, and in turn, EHS departments can be forced to crack down in the very ways that perpetuate this vicious cycle.

14.3.4 Strategies for Achieving Buy-in

Combating the challenges inherent in the "invent or perish" mentality must be an ongoing process for any EHS department. Communication and messaging are keys to achieving the buy-in from researchers that allows for effective EHS management. Researchers need to know that campus EHS professionals are on their side, and their primary goal is to enable innovation, and allow for research to take place safely albeit in a framework of safety and regulatory compliance. Essentially, this is about building true partnerships and achieving "buy-in" from the academic community.

For EHS departments to get buy-in from the research community, the first and crucial grounding step must be engendering mutual trust. As described in the previous section, one of the most unfortunate byproducts of the "invent or perish" mentality is the breakdown of trust in EHS professionals, as researchers increasingly view them as impediments to rather than highly skilled enablers of innovation. In essence, this truly crucial partnership is instrumental in the success of any academic endeavor. To improve this adversarial relationship, here are three important strategies for achieving buy-in from the research community:

14.3.4.1 Fostering Knowledge Sharing

When it comes to establishing credibility and achieving buy-in, one of the most effective tools in a safety or risk management professional's arsenal is knowledge and applicable experience. Disseminating this knowledge effectively throughout an institution of higher learning is key to being viewed not as doomsayers and dampeners of innovation, but as valuable sources of information. Sharing the latest knowledge in research disciplinary fields can be as simple and overarching as publishing a periodic newsletter. Not only does this sort of regular dissemination of information help to keep the community and stakeholders up to speed, it also helps to build confidence in EHS departments by assuring the academic community that they are well-versed in the latest developments.

In addition to sharing outside knowledge within institutions, EHS departments should ensure that effective mechanisms are in place for the reporting and sharing of internal, experiential knowledge, including lessons learnt and case studies from the specific campus or peer institutes. Where reporting mechanisms, such as easily accessible, confidential contact forms and web portals, are not already in place, they should be implemented. Where they are already in place, these mechanisms need to be publicized throughout the community, with assurances that knowledge reported is swiftly processed, acted upon, shared as appropriate. Often, implementing and sharing these types of voluntary reporting mechanisms is effective, but only on a limited basis. MIT is one of the many universities that utilizes dedicated email addresses, hotlines, web portals managed by third-party vendor to ensure anonymity when applicable. However, these mechanisms are still not a substitute for the trust built up by person-to-person contact. For these types of reporting mechanisms to achieve their maximum effectiveness, EHS professionals need to spend time engaging in person with members of the academic community, letting them know that the knowledge gained from their experiences is invaluable to EHS efforts and the institute overall.

Aside from voluntary knowledge sharing structures, the idea of shared knowledge should be built into research institutions from the ground up. This means that when laboratories or their operational procedures designed, a crucial component of these efforts should be the mechanisms for engaging with the appropriate subject matter experts and collecting all applicable knowledge. This type of approach should also be expanded to other facets of EHS roles and responsibilities. As appropriate, information from lessons learnt, case studies, and specific projects from the individual campus and peer institutions should be incorporated in all modes of information sharing.

14.3.4.2 Inclusion and Meaningful Engagement

When researchers find themselves alienated from EHS departments and decisions, achieving their buy-in is nearly impossible. To combat the pressures of the "invent or perish" mentality, the research community must find their perspectives, ideas, and representatives meaningfully included in the work of EHS departments. While the idea of including researchers in EHS decisions may feel somewhat like leaving the fox in charge of the hen coop, it is an important step in creating and maintaining a productive relationship between research and EHS departments, and lead to safer laboratory design and greater respect and buy-in for procedures.

To be most effective at achieving buy-in, including the research community in EHS operations needs to begin from the ground up. While a representative researcher does not necessarily need to be present in every internal EHS meeting, inviting members of the research community to sit in and offer their perspectives in meetings to determine crucial procedures or laboratory design decisions can be extremely valuable. Indeed, inviting the perspectives of researchers to inform the work of EHS departments not only helps with buy-in, it also contributes to EHS work in meaningful ways, informing decisions about design, procedures, training, and implementation.

Many institutions have governance models that utilize faculty and researcher participation in decision-making and even day-to-day activities, through committees in certain areas of specific expertise. EHS programs are inherently integrated with these committees, thanks to research activities involving biological materials, radiative materials, chemicals, and human subject and animal research. However, more successful and integrated programs expand the engagement with faculty and researchers beyond what is required by research compliance.

Regardless of whether faculty and researchers are invited to be physically present in EHS operations, it is crucial that their perspectives be solicited and incorporated. The successful EHS professional will personally seek out the opinions and experiences of researchers and allow those conversations to shape operations. Inclusion is a two-way street, and when researchers find themselves included by EHS departments, they are far more likely to include those departments and their personnel in their own internal decision-making. At MIT, we strive to engage with faculty members and researchers as often as possible, including them in campus planning activities, lab design, and new pilot projects as well as decisions and activities related to the EHS Management System. This level of engagement not only allows for strong relationships in the academy and informed decision-making, it also enables synergy with academic partners, particularly in times of crisis.

MIT EHS is among peer institutional safety departments that use their own campus as a living lab. In this context, campus challenges such as climate resiliency, carbon neutrality and safe innovation are often studied through formal research methodology, which usually includes the involvement of a research partner from the academy. These efforts commonly include student employees hired or post-doctoral fellowships created by EHS in order to support the scholarly activities. The end product typically includes development of a novel methodology or solution in order to address or overcome a challenge faced by peers. The formal research methodology ensures the solution is supported by empirical data, contributes to the body of domain knowledge, and can be published in a peer-reviewed journal. Needless to say, the most significant benefit of this approach is the peer status EHS professionals achieve in the eyes of the research community.

14.3.4.3 Enhancing Training

The widely held perception of institutional training, particularly in the realm of compliance, is notoriously dull, redundant and at times, unnecessarily (or necessarily) overwhelming. In institutions of higher learning, it carries the added stigma of being adversarial to innovation. To ensure the future of safe innovation spaces, it is crucial that EHS departments combat this reputation and utilize the opportunity to give the research community the best first impression. Every training session should be carefully designed to ensure that:

- the learning objectives are carefully crafted to meet the specific needs
- the content is current, timely and appropriate in length
- the presenter is knowledgeable, engaging and possesses good presentation skills
- the mode of delivery meets the needs of the trainees by striking the appropriate balance between in-person, online, and blended learning as needed

Additionally, until risk mitigation and EHS training integrated into subject training for degrees, dedicated EHS training runs the risk of being seen as a dreaded and onerous "add on". To the extent it is possible, EHS professionals should provide materials and guidance to subject instructors on safety procedures and best practices, encouraging and supporting this type of integrated training. The training for laboratory personnel must emphasize the role of EHS: namely the enabling of safe and productive research and innovation. It must stress that innovation is achieved through the symbiotic relationship between researchers and EHS professionals. As such, the training itself should also be innovative when possible. User engagement can be enhanced by involving subject matter expert researchers, interested and engaged student groups through partnerships on content development. Non-traditional approaches to training, including the use of gaming or game theory, augmented reality, or virtual reality, can be used to enhance training effectivity as well as user engagement.

In addition to the proper emphasis for regular training, EHS departments should also consider having a broader impact on education by offering courses, seminars, and other presentations for laboratory professionals and additional stakeholders. For example, MIT EHS offers professional education courses in the areas of crisis management and design of research facilities. The interactive course in design of research facilities uses case studies, discussions, and group activities to examines the roles, alignments, expectations, and required skills of key stakeholders involved in innovation processes. In addition to offering a resource for directors and other senior leadership involved in the design and operation of innovation spaces, courses like this one help to educate the research community about the role and operations of EHS departments.

14.3.5 Into the Future: Trends and Implications in Innovation Spaces

Globalization is yesterday's news, but the landscape of research compliance continues to undergo drastic shifts. As Intellectual Property (IP) rights, export control, and relationships to foreign entities become increasingly at issue, EHS professionals must keep abreast of these changes, and look to designing innovation spaces and procedures that are viable into the future.

The landscape of scientific research is simultaneously opening up as physical distances are becoming less and less meaningful and globalization marches on; and becoming more restrictive, as national and other granting agencies crack down and tighten regulations. In addition, new technologies are enabling stricter enforcement, and are able to root out IP theft in new ways.

In 2019, a neuroscientist formerly working at Emory University in Atlanta, GA was charged with defrauding the U.S. government by being compensated by a research institution in China while simultaneously receiving funds from research grants from the National Institutes of Health (NIH). The researcher and his wife, both employed by Emory University, were implicated as part of a wider NIH probe of close to 200 researchers suspected of violating "policies requiring disclosure of all sources of research support".[33]

Perhaps even more dramatically, a Harvard University department chair and a Boston University researcher were separately charged in late 2019 each for concealing ties with China. In the case of the former, Harvard chemistry department chair Dr Charles Lieber has been charged with receiving a salary through China's Thousand Talents Plan and the Wuhan University of Technology, where he was allegedly set to establish a nanotechnology research laboratory. Like the Emory University researcher, Lieber is charged with failure to make proper disclosures about his income sources while receiving NIH funding, and with making false statements concerning his connections with China. The Boston University (BU) researcher, Yanging Ye, is accused of "failing to disclose her status as a lieutenant in the Chinese People's Liberation Army," from which she was allegedly secretly taking orders while carrying out her research at BU.[34]

While the two above examples are more concerned with financial fraud than with safety and compliance, they are meant to illustrate the diversity and complexity of the role of EHS professionals into the future.

Yet another similar scandal, and perhaps the most dramatic of all, occurred in late 2019 when a Chinese graduate student "who had been doing research at Beth Israel Deaconess Medical Center was arrested at Logan Airport on December 9 [2019] after customs agents found 21 vials of an unknown brown liquid wrapped in a plastic bag in a sock in one of his checked bags as he prepared to board a flight back to China".[35] Remarkably, his arrest was one of 18 similar seizures made in 2019 at Logan Airport alone. Peer institutions hastened to warn their researchers against similar behavior.

These incidents point to the ever-expanding future of research compliance and the associated purview of safety professionals. Currently, particularly the U.S. focus is largely on these illicit ties with China. While government regulations are in place that prohibit this sort of financial double-dealing and IP theft, EHS professionals could potentially be asked to ensure that innovation spaces are as impervious as possible to such emerging threats in addition to the existing compliance responsibilities.

Innovation spaces are without question vectors for risk. Risks increase and morph, often at breakneck pace, as technologies progress and as global interconnectivity flourishes. To ensure the ongoing integrity, productivity, and safety of innovation spaces, safety professionals must strengthen and consistently learn from their relationships with researchers, continually evaluate shifting risks, and retain the agility to adapt to the swift and ongoing changes that impact the research sector.

Abbreviations

AIHA	American Industrial Hygiene Association
ANSI	American National Standards Institute
ASHRAE	American Society of Heating, Refrigeration, and Air-conditioning Engineers
ASSE	American Society of Safety Engineers
DHS	Department of Homeland Security
EPA	Environmental Protection Agency
FDA	Food and Drug Administration
ISEA	International Safety Equipment Association
NFPA	National Fire Protection Association
NIH	National Institute of Health
NRC	Nuclear Regulatory Commission
OSHA	Occupational Safety and Health Administration

References

1. 2020 MIT Facts, https://web.mit.edu/facts/makerspaces.html, last accessed March 2020.
2. U.S. Food and Drug Administration, Facts About the Current Good Manufacturing Practices, https://www.fda.gov/drugs/pharmaceutical-quality-resources/facts-about-current-good-manufacturing-practices-cgmps, last accessed March 2020.
3. Mecart Cleanrooms Cleanroom Classifications, https://www.mecart-cleanrooms.com/learning-center/cleanroom-classifications-iso-8-iso-7-iso-6-iso-5/, last accessed March 2020.
4. Lincoln Laboratory Microelectronics Laboratory, https://www.ll.mit.edu/about/facilities/microelectronics-laboratory, last accessed March 2020.
5. Whole Building Design Guide Academic Laboratory, https://www.wbdg.org/building-types/research-facilities/academic-laboratory, last accessed February 2020.
6. L. DiBernardinis, J. Baum, M. First, G. Gatwood, E. Groden and A. Seth, *Guidelines for Laboratory Design Health and Safety Considerations*, John Wiley & Sons, Inc, New York, NY, 2nd edn, 1993.

7. Clark Nexsen Successful Lab Design, https://www.clarknexsen.com/blog-top-10-tips-successful-lab-design/, last accessed March 2020.
8. MIT Shop and Makerspace Safety Manual, MIT Environment, Health & Safety, 2017.
9. Whole Building Design Guide, https://www.wbdg.org/resources/trends-lab-design, last accessed March 2020.
10. MIT Green Labs, https://greenlab.mit.edu/, last accessed March 2020.
11. P. Cline, *Wilderness Risk Management*, 2015, vol. 5, p. 43.
12. Risk and Insurance Management Society Risk Knowledge, https://www.rims.org/resources/risk-knowledge, accessed February 2020.
13. ISO 31000 Risk Management, https://www.iso.org/iso-31000-risk-management.html, last accessed March 2020.
14. P. Bernstein, *Against the Gods: The Remarkable Story of Risk*, John Wiley & Sons, Inc, New York, NY, 1998.
15. M. Elliot, *Risk Management Principles and Practices*, The Institutes Risk & Insurance Knowledge Group, Malvern, PA, 2016.
16. COSO Enterprise risk management – Integrating with strategy and performance, https://www.coso.org/Documents/2017-COSO-ERM-Integrating-with-Strategy-and-Performance-Executive-Summary.pdf, last accessed March 2020.
17. Center for Disease Control and Prevention Hierarchy of Controls, https://www.cdc.gov/niosh/topics/hierarchy/, last accessed March 2020.
18. P. Humphreys, *Human Reliability Assessor's Guide: A Report by the Human Factors in Reliability Group*, United Kingdom Atomic Energy Authority Safety and Reliability Directorate, 1995.
19. Institute for Healthcare Improvement, http://www.ihi.org/resources/Pages/Tools/FailureModesandEffectsAnalysisTool.aspx, last accessed March 2020.
20. H. W. W. Potts, J. Anderson, L. Colligan, P. Leach, S. Davis and J. Berman, Assessing the validity of prospective hazard analysis methods: A comparison of two techniques, *BMC Health Serv. Res.*, 2014, **14**, 41.
21. OSHA Job Hazard Analysis, https://www.osha.gov/Publications/osha3071.pdf, last accessed March 2020.
22. B. Kirwan, *A Guide to Practical Human Reliability Assessment*, Taylor & Francis, Ltd, London, 1994.
23. IRMI Technique of Operations Review (TOR) System, https://www.irmi.com/term/insurance-definitions/technique-of-operations-review-system, last accessed March 2020.
24. Project Management Institute, *Risk Analysis and Management*, https://www.pmi.org/learning/library/risk-analysis-project-management-7070, last accessed March 2020.
25. X. Zhi, L. Chunming and C. Zhang, Model of total safety management (TSM) and its application, *Procedia Eng.*, 2012, **45**, 204–207.
26. CU Pathways Total Safety Management, http://pathways.cu.edu.eg/subpages/training_courses/Healthl7/Chapter5_s.htm, last accessed March 2020.

27. Partnership for Systems Approaches to Safety and Security (PSASS), https://psas.scripts.mit.edu/home/, last accessed March 2020.
28. J. P. Louisot and C. H. Ketcham, *Enterprise Risk Management: Issues and Cases*, John Wiley & Sons, New York, NY, 2014.
29. URMIA White Paper – ERM in Higher Education, https://www.odu.edu/content/dam/odu/offices/risk-management/DOCS/erm-urmia-white-paper.pdf, last accessed March 2020.
30. R. R. Moeller, *COSO Enterprise Risk Management: Understanding the New Integrated ERM Framework*, John Wiley & Sons, New York, NY, 2007.
31. International Standard 31010, *Risk Management – Risk Assessment Techniques*, Geneva, 2019.
32. MIT Project Manus, https://project-manus.mit.edu/about-us, last accessed March 2020.
33. Science Magazine – Ex-Emory scientist with ties to China charged with fraud, https://www.sciencemag.org/news/2020/02/ex-emory-scientist-ties-china-charged-fraud, last accessed March 2020.
34. WCVB – Harvard department chair, BU researcher charged with concealing ties to China, https://www.wcvb.com/article/federal-charges-in-investigation-into-china-biomedical-research-infiltration/30689217#, last accessed March 2020.
35. Universal Hub – Chinese cancer researcher about to fly home from Logan not allowed to leave after customs agents find vials of brown liquid in his checked bag, https://www.universalhub.com/2019/chinese-cancer-researcher-not-allowed-leave, last accessed March 2020.

CHAPTER 15

Translation of Safety from University Research to Spin-out

SURRINDER JOHAL*[a] AND ANA PEDRERO-LLAMAS[a]

[a]Imperial College London, South Kensington Campus,
London SW7 2AZ, UK
*E-mail: s.johal@imperial.ac.uk

15.1 Introduction

Launching a company is becoming increasingly attractive for students, academics and researchers around the world. Considering this, universities often provide guidance on strategies and tactics for launching a start-up with the backing of university programmes, public grants, incubators and private partnerships. They provide details on how to test ideas for their marketability, develop products out of research projects, and engage partnership companies and investors.

What is often less well acknowledged is that once a spin-out is registered in Companies House, it is no longer covered by the university's Health and Safety Management System (HSMS). As most of the initial staff working in the spin-out will have worked for the university, and probably still do, they mistakenly think that the University's Health and Safety (H&S) policies, programmes and procedures will continue to apply and cover their activities.

The current advice on H&S regulations and responsibilities offered by universities to spin-outs appears highly variable and often sparse; it ranges from glossy and well-structured leaflets providing brief information on the management of H&S to verbal advice about the need for policies and risk assessments.

Challenges for Health and Safety in Higher Education and Research Organisations
Edited by Olga Kuzmina and Stefan Hoyle

Published by the Royal Society of Chemistry, www.rsc.org

If you are a member of a university encouraging the creation of start-ups, if you are thinking about forming a start-up company or if you are already managing one, this chapter will provide you with basic steps and useful tips to help you develop a safety management system, such that entrepreneurs can establish a safe foundation for their company's growth and development.

15.2 Health and Safety: The Sometimes Forgotten Compliance Element

Within most higher education institutes, H&S tends to be managed by a dedicated team of professionals who provide active support, obtain permits and licences and complete other relevant paperwork. This team often includes locally based safety professionals as well as a central provision. These professionals have a broad and detailed understanding of the legal requirements applicable to the wide range of activities delivered across universities, and their role is to facilitate these activities while ensuring compliance with legal requirements. This can involve completing forms and documenting processes so that there is evidence of compliance with the law leading to 'health and safety' being commonly associated with bureaucracy and perceived as 'holding people back'. It is therefore unsurprising that the very mention of H&S can be a turn-off for the dynamic and enterprising academics and researchers. The management experience and approach of individuals involved in setting up spin-outs varies greatly; from those who are highly committed and are thoughtful about H&S to those who perceive it as an administrative burden. In the latter case it is no surprise that when mentoring and providing support to budding entrepreneurs, the focus will be on market research, obtaining funding and the quality and marketing of the product or the service to be delivered. However, it is critical that entrepreneurs understand that having a H&S management system is as essential as ensuring the quality of the services and the products they are planning to provide and hence H&S must be integral amongst other compliance elements.

There is no business success without a good product or service, this is one of those facts of life everyone is fully aware of. What is not always obvious is that, in order to maintain sustainable growth, keep staff healthy, engaged and motivated, and to ensure the smooth delivery of activities, a well embedded H&S management system is desirable. Where instructions are clear, people will generally adhere to them and where individuals are empowered to report short falls as they arise, such matters can be resolved before they become a real issue. There is a line of thought that insurance policies will cover a company's H&S shortfalls; while the Employer Liability Compulsory Insurance will cover the cost of injuries and ill health suffered by employees at work, there are other costs that may not be covered, such as repairs to plants and equipment, overtime working and temporary labour, production delays and legal costs. Most of those costs are likely to come straight off the company's bottom-line profit. There are also other costs that are less tangible but just as crucial: the loss of reputation and the physical and mental impact that an accident can have on the injured person, their family and colleagues. Recent costs of H&S fines have shown that they far outweigh the cost of compliance; in 2019 businesses paid on average £150 000 for H&S breaches[1]. In the US the cost per worker in 2018 was $1 100[2]. This includes the value of goods or services each worker must produce to offset the cost of work injuries.

During the start-up period individuals establishing a company often focus on exploiting their academic ideas to develop high quality products or services and afford less attention to H&S. When, as often happens, there is a sale this can prove problematic as the potential buyers will be very conscious of not increasing their overall risk profile by acquiring undesired liabilities. Corporations interested in acquiring start-up companies tend to have stringent well-defined due diligence processes to test and challenge compliance with ALL regulations, including health and safety. The due diligence exercise is formulated to understand the weaknesses of the company they are buying, and they will use each of these weaknesses to bring down the purchasing price of the start-up company.

Years ago, one of the authors was involved in the acquisition of a small niche company producing heavy metal reference materials. The Directors' motto was "Health and Safety is everybody's responsibility" and they literally meant it! They did not understand that they were accountable as well as responsible for the health and safety of their employees. The two Directors had bought a safety manual from the bank that provided them with financial and insurance advice, and they firmly believed that having a file with safety documents made them compliant. Their focus was on running the business; they had not read the file containing the safety policy, training plan and risk assessments. If they had read the documentation, they would have noticed that the dust produced during their manufacturing process was highly toxic. During multiple discussions regarding the large number of H&S and environmental breaches identified during the due diligence process, they mentioned in passing that their only female employee had suffered several miscarriages. It was then pointed out to them that this could have been linked to her exposure to heavy metals; they kept repeating that "masks" and ear protection were available, but nobody wanted to use them so "it wasn't really their fault". They did not acknowledge the use of local exhaust ventilation for systems to prevent exposure to respirable particles containing heavy metals, and the use of dumpers for noise reduction. They were also unaware of work exposure limits, the health surveillance and the environmental monitoring legal requirements surrounding the work they were delivering. They did not understand *Ignorantia juris non excusat*: that 'ignorance of the law is not a valid defence'.

The corporation ended up acquiring the start-up company, only because their plan all along was to move the manufacturing process to a different country, and because the sale price negotiated was significantly lower than the original asking price. The two Directors spent years developing and manufacturing a high quality unique product, and it was only towards the end of the sale process, that they realised the moral and financial implications of not managing H&S.

15.3 Health and Safety Made Easy

H&S is often dismissed as "it's common sense". Unfortunately, common sense is often lacking, particularly at times when the pressure is on delivery, when new, inexperienced or young people are employed and during emergency situations. The purpose of a H&S management system is to define and instil "safety common sense" into the business from the very beginning, when the tone is being set, when it's easier to control the hazards and create the right culture; one that will positively grow with the company.

There are no two ways about it: in most countries the law requires all legal entities to have a H&S policy and to conduct risk assessments evaluating hazards present in the work environment and the associated work activities. The operational agility with which spin-outs evolve their products and services presents a particular challenge in keeping H&S documentation relevant and up to date. The key for developing and implementing an effective H&S management system is to keep it simple and to the point.

There are some aspects of H&S that apply to all employers, no matter the type of company:

1. H&S policy: The policy should include who is responsible for what, when and how. If the company has fewer than five employees, the policy does not need to be written down.
2. Emergency arrangements: It is difficult to imagine a company that does not require the use of some sort of technology; computers as a minimum. Where there is electrical equipment, there is always a risk of fire. Emergency arrangements are required to ensure that everyone knows how to act and stay safe in foreseeable emergency situations.
3. Compliance with fire regulations: If the company is leasing space within a building, the landlord may be responsible for ensuring that relevant fire-fighting equipment is available within communal areas. In most instances, the company leasing the space is responsible for the provision and maintenance of fire extinguishers; ensuring that fire routes and fire doors are free from obstruction, *etc.*
4. First aid or first response: Every company must have trained first aid personnel, in some countries referred to as 'first responders', to support medical emergencies. There are companies that rely on their landlord to provide this aid. It is always advisable to check the lease contract in case this provision is included.
5. Risk assessments or job hazard analysis: A risk assessment is an evaluation of the aspects of the work environment and activities that could cause harm. It details the controls in place to ensure that the risk is

reduced to an acceptable level: this is commonly referred to as being "as low as is reasonably practicable" (ALARP). The evaluation of the hazards will identify those who could be at risk, including visitors, contractors and other relevant parties. Each country defines the minimum set of safety-related documentation applicable to employers; These typically include:

- Process and Workplace risk assessments, which provide an overview of the hazards present within any given activity/area as a result of the range of simultaneous processes/activities being undertaken.
- Display screen or visual display unit risk assessments are required for each employee who spends a considerable amount of time in front of a computer, tablet, mobile phone, or other digital display equipment.
- Young person's risk assessment is required if you employ and/or allow work experience for those under 18 years old.
- Risk assessments for new and expectant mothers; this is an essential requirement in most European countries, where expectant mothers are obliged to inform their employer about their pregnancy at least 15 weeks before the baby is due.
- Travelling for work policy, including risk assessments for overseas travel and when driving a vehicle for work purposes. The spin-out company needs to develop policies and procedures to evaluate and mitigate the risks associated to these activities. It is important to mention that employers owe the same duty of care under health and safety law to staff who drive their own vehicle for work as they do to employees who drive company owned, leased or hired vehicles.
- Personal Emergency Evacuation Plan (PEEP) for those who need assistance to evacuate premises during a fire or other emergency.
- Specialisms. Additional risk assessments or job hazard analysis that are specific to the type of work being conducted, for example: biological, electrical, chemical, radiological (including lasers), pressurised systems or vessels, *etc.*

6. Consultation and training: In most countries, the law requires employers to consult with their employees on safety matters, including risk assessments/job hazard analysis, and safety training. The approach taken to fulfil these requirements should be proportionate to the size and nature of the business; start up is the time to define the consultation process and training needs of those who will form part of the company.

Training is one of the best ways to guard against accidents in the workplace; when individuals are trained to high standards, they are less likely to make mistakes. Unless workers understand the hazards and risks and know how to protect themselves and others from danger, directors could be held accountable and prosecuted for putting them at risk.

7. Equipment service records: Servicing and calibration of equipment is essential to demonstrate the well-functioning of any instrument. Some of the testing may need to be conducted to specific schemes of examination, as required by the law, such us: lifts and other types of lifting platforms and hoists, pressure vessel systems, *etc.*
8. A system for reporting accidents and near-miss incidents, so they can be investigated fully, corrective measures identified and implemented to prevent recurrence.
9. Audits and inspections: Conducting short inspections of the working area(s) and full audits of the H&S management system supports the early identification of issues, so they are resolved before someone gets hurt. Audits and inspections are a key tool to demonstrate the proactive management of safety within any work-setting.
10. Management of contractors: Employing a competent contractor to work on your premises is, in most instances, not enough. There is a need to ensure that their work will be conducted in a safe manner, so that your employees are not exposed to avoidable risk.

Spin-out companies that are a direct translation from a higher education set-up, will have access to all the information relating to the requirements discussed above and other additional information relevant to their activities. The documents would form part of the university's Health and Safety Management System; anything specific to the activities of the spin-out company would have been developed while the company was a research group, or similar, within the university. Where individuals work for both the university and the spin-out, the organisational boundaries become blurred as many connections and overlaps remain. This leads to confusion over accountabilities and responsibilities, and individuals do not always realise when they are crossing the legal boundaries and become personally responsible and accountable, and the university advisers can no longer support them.

A key aspect worth remembering is that once established as a separate company, the university processes and infrastructure will no longer be valid; the director of the spin-out will be accountable for the H&S performance. The day-to-day running of H&S can be delegated, but not the accountability. The directors must be the driving force behind safety, in the same way they won't hesitate to drive the company's quality and financial performance.

The question they might ask is 'How do I develop and embed a H&S management system effectively and efficiently?'. Like with most certified management systems, the 'Plan, Do, Check, Act' approach can be applied.

The Health and Safety Executive's Managing for Health and Safety (HSG 65) can be downloaded for free from the HSE website.[3] This guidance document provides a framework for setting up H&S standards within a company. The principles of Plan, Do, Check, Act steps are common to most ISO Standards, as represented in Figure 15.1.

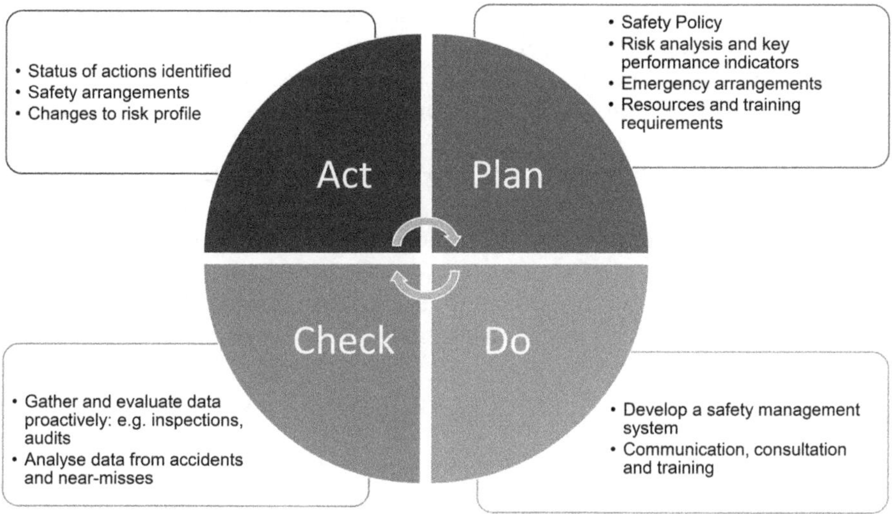

Figure 15.1 The principles of the PDCA cycle.

15.3.1 But What Does This Mean and What Really Needs to be Done?

Here is a check list that we have specifically developed to aid each stage of the Plan, Do, Check, Act cycle and enable the development and implementation of a safety management system.

15.3.1.1 Plan

- A safety policy which defines responsibilities and contains a statement of intention has been developed.
- Roles and responsibilities have been defined and documented and there are enough resources dedicated to the management of H&S.
- Hazards within the work environment and external hazards linked to the activities undertaken have been identified.
- A mechanism for consulting with employees on H&S has been defined.
- Safety legal requirements specific to the activities undertaken have been identified, are understood and there is a mechanism for monitoring compliance.
- There is a mechanism for evaluating safety implications linked to changes in activities *e.g.* safety features in every change control and budget review process.

15.3.1.2 Do

- The person responsible for the day-to-day management of H&S has received suitable and sufficient training and they are competent.

- Safety documentation has been developed, including risk assessments, and employees have been involved.
- All employees have been trained on safety matters, including emergency arrangements, use of equipment, how to report accidents and incidents, *etc.*
- There is a mechanism for reporting accidents and incidents; staff are trained on how to conduct investigations, so root causes are identified, controls implemented, and reoccurrence of accidents prevented.
- There are processes embedded for undertaking inspections and audits, and staff have been trained to conduct these.

15.3.1.3 Check

- Accidents and incidents are being reported and investigated. Lessons learned are shared.
- H&S is part of team meeting agendas, improvements are suggested and acted upon as relevant.
- Equipment and working space *e.g.* lighting, temperature, is fit for purpose, staff are motivated and have a sense of belonging.
- Actions identified through audits, inspections and accident and near-miss investigations are dealt with in a timely manner and checked for effectiveness.
- Checks are done for trends of staff sickness or absence that could indicate the environment is not conducive to well-being, health and productivity.

15.3.1.4 Act

- The data obtained in the 'Check' phase above are used to identify improvements. Safety management systems evolve over time, in the same way the company evolves; it's a dynamic process and it needs to be kept up to date.
- Legal and regulatory changes applicable to the spin-out are kept up to date and the processes and the safety documentation are amended accordingly. Changes in requirements must be communicated to all relevant staff and incorporated into training programmes.

Under H&S legislation, business owners are obliged to designate a 'competent' person to be responsible for H&S on site. Given the size and infancy of spin-outs, this resource is unlikely to be available in-house; if there is no appetite for training those with technical knowledge on H&S regulations and requirements, an external resource (consultants) with relevant knowledge of H&S regulations could be appointed.

15.4 Conclusion

Despite entrepreneurial programmes within universities, many spin-outs do not understand the H&S accountabilities they automatically inherit. The law states that every business must have a policy for managing H&S, which sets out its general approach and organisational structure. Being a start-up means being at the edge of business and entrepreneurship, and this must also include H&S. By making sure H&S requirements are addressed, a safe and conducive work environment is created for all employees, visitors, and contractors. Nevertheless, there is no need to overestimate the risk nor overspend on the rectifying solutions but aim to make safety an asset and not a burden.

References

1. The Health and Safety Executive, *Enforcement Statistics in Great Britain*, 2019. Available from www.hse.gov.uk/Statistics/enforcement.pdf, last accessed 15 June 2020.
2. United States Department of Labor, *National Safety Council Injury Facts*. Available from https://injuryfacts.nsc.org/work/costs/work-injury-costs/, last accessed 15 June 2020.
3. Managing for Health and Safety, *Health and Safety Executive Guide 65 (HSG 65)*. Available from https://www.hse.gov.uk/pubns/priced/hsg65.pdf, last accessed 15 June 2020.

CHAPTER 16

Please Throw it Away: Dealing with Legacy Chemicals, Equipment, and Infrastructure Safely

C. HARRISON*

University of Exeter, Exeter, EX4 4PY, UK
*E-mail: c.harrison@exeter.ac.uk

16.1 Introduction

Universities often trade on their great age, prestige and success. Whilst walking their halls you can reflect back on all those great minds of the past who have stood and contemplated where you now do the same.

The problem is that it seems that those great minds of the past have left most of their stuff behind! It is one thing working in the office used to discover a Nobel Prize winning theory and writing on the desk of the founder of your field but it is another one entirely when you are still using their chemical stocks! The issue is not just limited to chemicals, universities also tend to have older buildings, infrastructure and equipment.

This chapter will highlight some of the issues which arise from such legacy items and hopefully offer practical advice in managing the risk you have inherited from the past.

Challenges for Health and Safety in Higher Education and Research Organisations
Edited by Olga Kuzmina and Stefan Hoyle

Published by the Royal Society of Chemistry, www.rsc.org

16.2 Legacy Chemicals

"Death sits in the chemistry laboratory and a million people sit with him and don't mind. They forget he's there. Look at him there, sitting in half a hundred reagent bottles of brown glass, crowding the shelves. Each bottle of death was clearly labelled, each filled in varying amounts with its particular kind of fine, pure crystals. Most of them looked like salt. In every way death was there and no one minded."[1]

Perhaps a touch dramatic but in the decades since those words were written it sometimes seems like not much has changed, there are more analytical techniques, computational chemistry and entire new fields of research. The wet chemistry lab still feels fundamentally the same. The glass bottles have been partly replaced by plastic and less chemicals are left in the open but fundamentally the same hazards can still be found.

My introduction to chemistry safety was as Assistant to the Safety Officer. In theory this was to support them in a wide variety of duties, in practice this was dominated by the disposal of legacy chemicals. The largest job I ever dealt with was the laboratory of a retired academic who had been in the building since it was built. Every cupboard, shelf and fume cupboard was full of chemicals and old equipment. Over the next couple of years I got more than 3000 chemicals out of that lab either by finding new owners or disposal. As an introduction to chemistry safety this was something of a baptism of fire (not actual fire although it came close once!) but gave me an appreciation of a problem common to almost all research heavy institutions and a wide knowledge of chemical safety.

The Control of Substances Hazardous to Health (COSHH) regulations[2] extend the protection offered in law in the United Kingdom from the Health and Safety at Work Act (HSWA).[3] As well as reiterating requirements such as risk assessment, training and proactive safety management to make several provisions about the control of exposure to chemicals, this is primarily the requirement to 'ensure that the exposure of the employees to substances hazardous to health is either prevented or, where this is not reasonably practicable, adequately controlled'. This is normally achieved by the application of a risk assessment.

A common tool for achieving this is the hierarchy of controls of which many iterations exist but I think the one shown below in Table 16.1 is most useful in a laboratory context, it is often abbreviated to ESREAPS or similar. The uniting principle is that the higher up the hierarchy the more effective the control, it is not a cast iron rule but a useful aide memoire.

Other legislation such as the Dangerous Substances and Explosive Atmosphere Regulations (DSEAR)[4] operates on the same principle. DSEAR is primarily concerned with the creation of explosive atmospheres and when accompanied by written advice from the HSE offers a sound principle for operating a safe lab. One item I would particularly like to bring to your attention is the requirement to try and keep highly flammable solvent stocks within a work space (not already stored in high specification fireproof cabinets) to

Table 16.1 The hierarchy of control (ESREAPS).

Hierarchy (high to low)	Key actions
Elimination	Get rid of the risk entirely, do not perform the experiment, dispose of the chemical.
Substitution	Swap something more hazardous for something less hazardous. This could be a chemical, machine or entire process.
Reduction	Reduce exposure by limiting time exposed, frequency or quantity. Universities generally operate at reduced laboratory scale by default.
Engineering controls	Engineering controls such as fume cupboard or laser enclosures provide mechanical means to protect the user.
Administrative controls	Rules, safe systems of work, training and experience are all administrative controls.
Personal protective equipment (PPE)	Gloves, safety glasses and lab coats are all essential PPE for laboratory work as well as more specialist items.
Signage	Signage helps indicate the presence of a hazard or that you are entering a working area.

less than 50 L.[5] In my experience it is rare that a working chemistry laboratory is not pushing up against this limit if not exceeding it, often because of old legacy stocks.

It is on these principles that this chapter is based. Even where the law differs around the world the basic principle of protecting people from chemical hazards is a fundamental issue particularly in the complicated world of research. This brings with it many challenges for the safety professional but also the moral obligation to help those working in research to be able to go about their jobs in safety and go home in good health.

16.2.1 Sources of Chemical Stocks

When a new chemical laboratory is first opened it may be hard to believe but the shelves and cupboards are bare of chemicals. Within a few short years those laboratories will be full of containers of substances, some innocuous some hazardous.

Chemicals do not start as legacy chemicals, at some point in their lives they were valuable research assets which were purchased with a project or experiment in mind. A new lab may buy basic stocks of standard chemicals when setting up in the same way you might buy bread and milk when moving into a new house. Even though I have sometimes speculated that mysterious bottles spontaneously arrive in the middle of the night, they must all have been purchased by someone at some point.

Some chemicals are purchased and synthesised with the intention of long term storage. Sample libraries for research purposes whether for spectroscopic reference or for running assays were once common but have to a certain extent gone out of fashion being replaced with digital databases of properties. That said, the same underlying thinking often leads researchers to keep hold of chemicals whether they are unusual or not, 'just in case' they are needed, much to the despair of those overseeing safety.

Sometimes chemicals bought with the best of intentions simply become redundant due to the relentless march of science. A once essential reagent becomes obsolete due to a new synthetic route or an analytical technique is no longer required after the acquisition of a new spectrometer. How many laborious processes using many reagents have been replaced by simple, single use kit? Many laboratories have a cupboard full of old indicators which were once essential but now represent a slow and inaccurate alternative to modern methods. I was once charged with disposing of no less than twelve bottles of calcium metal from the same laboratory where it had once been essential but had become redundant decades before and left on a shelf to slowly absorb water from the atmosphere.

The laws of economics can make this harder to manage. It is often considerably cheaper to buy a larger quantity of chemicals than it is to buy more than one smaller container. If an experiment requires 10 g of a salt and may need repeating several times it may be cheaper to buy a kilo rather than multiple containers of 25 or 50 g.

Whilst this makes financial sense, it increases the amount of chemicals stored and therefore the inherent risk of storing them. Whilst blame can to a certain extent be laid at the feet of the increasingly small number of chemical suppliers, other factors are definitely contributory.

The economies of scale issues are often compounded by how annual budgets are calculated. It is common practice for institutions to not roll over savings into the next financial year instead keeping any underspend as a saving. This incentivises last minute buying of stock chemicals to carry over as an asset to the following year. When inspecting a lab I happened to open the cupboard under a sink to find it crammed full with bottles of acetone. When I asked the researchers how quickly they got through a bottle it transpired that they had enough to last them over a year all because of a budget deadline.

Modern grants from national and international bodies are more aware of the need for efficiency than ever. In order to prevent frivolous spending outside of the scope of the award only chemicals needed for the specified experiments are budgeted for.[6] This makes sense for the organisation giving out the grants but encourages retention of chemicals as researchers are aware there will not be excess in the future and any unfunded research must make do with leftovers. It is increasingly difficult because of this to build up cash reserves to fund unexpected expenses or undertake a novel line of research with no attached funding, and this further discourages the disposal of anything which could be considered useful in the future.

When a prestigious researcher leaves an institution their chemicals are often left behind due to transport difficulties or because of the time it would take to dispose of them. It may be that a deal is struck that they can leave behind chemicals in return for longer term collaboration or sharing of research impact. It is highly unlikely that serious contractual penalties will exist for leaving chemicals behind. This puts tension in place between those seeking to maximise research output and those charged with safety who may not find out until after it has occurred. There are no metrics for safety in this league table driven field.

The way jobs are structured within academic institutions means that it is common for those at the start of their career to move frequently, rarely staying for more than four years as a postgraduate student or postdoctoral researcher in the same institution. This along with the tendency for all researchers to seek opportunities all over the world leads to a high staff turnover and the accompanying high accumulation of chemicals with no one being explicitly responsible for them. These 'orphan chemicals' often remain within a laboratory or research group for far longer than their original owner. The primary problem with the lack of ownership is the diminishing likelihood of use along with no one having sufficient ownership or authority to dispose of them even if they have the resources. I encourage researchers to write their names on bottles as it gives accountability when they still work in the lab and also allows me to ask the lab manager how long ago they left!

16.2.2 Hazards Associated with Legacy Chemicals

Before I delve into specific hazards associated with legacy chemicals it is worth looking at the one which is, in my own experience, most common. Plastic containers will eventually go brittle and may crack/shatter depending on their contents, whether they are stored in direct sunlight or affected by other environmental conditions. This can obviously lead to container failure and exposure, or at the very least a fiddly clear up in a storage cupboard. Any plastic bottles displaying tell-tale yellow colourisation should be either replaced with a newer one or got rid of entirely. On a similar note this can also happen with plastic storage trays commonly found in labs, I know of at least one occurrence where the bottom fell out of an acid spill tray being moved across a lab leading to a nasty spill where aqua regia was formed.

It goes without saying that the very existence of chemicals within a lab represents a risk to the institution. This is however a risk which is readily managed by a competent scientist. The risk is enhanced by both the severity of individual compounds and overall quantities. It is rare for individual chemicals to be banned by a university as the nature of what we do is to use those properties as tools to further research. It is also rare for anyone to look at the risk presented by all the chemicals in a holistic manner. When you consider the gulf between these two factors it is understandable how legacy chemicals can become a problem over time.

The most obvious hazard with legacy chemicals is those which form peroxides upon contact with air over time. The most rapid and subsequently common of these is diethyl ether which can form explosive peroxides in a matter of months. Eventually this can be seen as a crystalline solid either suspended in the remaining liquid or dried out entirely. As a shock sensitive explosive it is very dangerous at this point, although the hazard starts long before crystals are visible.[7] The high turnover of many common solvents protects the end users from this much of the time particularly with those which decompose slower. This however compounds the issue as few practicing chemists have ever seen the phenomenon or are even aware of the process. When a less commonly used solvent is purchased it is more likely to be retained in the longer term just in case it is required later. I am more concerned with the single cupboard containing half a dozen small bottles of unusual solvents in terms of peroxide formation than I am with multiple cupboards containing many containers of the same chemical.

Peroxide testing strips can be purchased from chemical suppliers and are an inexpensive way of testing a solvent's viability before any visible crystals have appeared. It is not unknown for seemingly new chemicals to arrive in the lab having had time to decompose with the supplier. It is possible of course that peroxides never form in a particularly well stored unopened bottle however this becomes increasingly unlikely over time.

The other type of high hazard legacy chemical commonly found in laboratories is explosives which are desensitised with water. The most common is picric acid which despite the innocuous name is a highly shock sensitive explosive when dry. As it is sold wet or in solution the MSDS may not give sufficient information which may lead to improper storage, as the explosive properties are not true of the product 'as sold', *i.e.* wet.[8] Picric acid is now most commonly sold in solution but large containers of solid can still be found in store cupboards and on one memorable occasion the fume cupboard of a teaching lab I was about to teach in.

Once an explosive chemical has been found it must be disposed of, but this presents a problem. The knee jerk reaction of many unfamiliar with such chemicals is often panic and ordinance (bomb) disposal is called in. This was a particular issue in 2017 when 2,4-Dinitrophenylhydrazine (DNPH), more commonly known as Brady's reagent, was reintroduced to the school curriculum. On checking old stocks they were found to be dry and ordinance disposal was called. This resulted in abundant press attention and small holes in school playing fields due to controlled explosions of old stocks nationwide.[9] I have heard some argue this was an overreaction but the school technicians involved would not be expected to be competent to make such a decision and erred on the side of caution.

Even when a knowledgeable chemist is on hand the handling of explosives is still uncommon. The ability to render safe may require specialist facilities or at least a large amount of space! The cost of hiring specialist contractors often runs to the thousands of pounds even for limited quantities. This was however apparently preferable to my proposed solution which involved digging a hole in a football pitch!

Potentially explosive situations are also possible if proper segregation of hazards is not instituted. Many common chemicals are incompatible and could react in a dangerous way if mixed. Pressure on storage space can lead to mixed cupboards of chemicals where there is a much higher risk of injury in the event of a spill or leak due to incompatible chemicals.

16.2.3 Unknown Chemicals

Finding unlabelled white powders in a laboratory is a staple scenario for undergraduate chemistry teaching laboratory sessions. In the real world it is often a far more serious concern. Sometimes it is possible to narrow down the type of chemical but it must be identified before it can be disposed of. Identifying an unknown is a labour intensive or highly expensive process. You must treat it as if has a full set of hazards until you are able to eliminate properties or fully identify.

Hazardous waste disposal companies are often unwilling or unable to take unknowns without knowing how they will ultimately dispose of the substance according to legal regulations. They do not however always need to know the exact composition of a substance, instead they need to know their disposal route whether it be incineration, neutralisation or long term isolation. It is worth consulting with your waste contractor to find out their exact requirements as this can save considerable resources.

The arrival of the Classification Labelling Packaging (CLP)[10] system of chemical labelling was a huge boost to those working with chemicals. Not only did it refine and improve the system it also allows you to check the relative age of chemicals at a glance. Any chemicals which have the old style Chemicals (Hazard Information and Packaging for Supply) Regulations (CHIP)[11] labels (orange squares) are instantly recognisable as being at least five years old at the time of writing. Over time this will become an increasingly small minority but the relative age will also increase. This along with spotting defunct chemical suppliers or outdated terminology can provide a quick assessment of how well managed laboratories' chemical stocks are.

Of particular concern is the practice of using codes for research samples. This is in itself fine but puts a requirement that the code system is maintained for as long as the samples are kept. For research in high hazard substances such as toxic heavy metals it is essential that appropriate safety labels with hazard information are also added. This can be directly or by having small sample vials stored together inside a larger container. A code system without its key is no better than no label at all.

Most unknown legacy chemicals become unknowns by the label being removed and lost or rendered illegible. This can happen through glue being removed by age or chemical action. Handwritten labels can easily be removed through the use of acetone or UV light. Poor storage of iodine can also render labels irreversibly stained in a distinctive shade of brown familiar to many organic chemists. The modern standard for chemical labelling[12] removes most of these risks however it does not protect older stocks or research samples.

Chemical waste is often overlooked even if it has not yet been handed over to a waste contractor. It must be stored properly until it is taken off site and this can cause problems if there is insufficient space for the waste produced, which can end up being left in the open lab or other unsuitable area. As well as the residual risk of storing chemical waste poor labelling practices can also cause problems. It is important that no ambiguity remains over the contents of a chemical bottle pressed into service as a waste container. By placing a label over the top of the old one or by insufficiently erasing existing labels ambiguity is created which could cause a dangerous mistake. The adoption of British Standard for pressure-sensitive adhesive labels used for marine applications (BS5609)[12] has had the unintended consequence of making labels harder to remove when using an empty bottle for waste.

16.2.4 Non-hazardous Issues with Legacy Chemicals

Given the issues presented above you would think that those responsible for a laboratory area would be keen to get rid of old chemicals and minimise their stock levels. As I have often experienced one of the biggest challenges of dealing with legacy chemicals is persuading their owners to dispose of them. A helpful technique for doing this is to emphasize their limited use in research.

Over time many organic chemicals will decompose. Leaving aside those which form explosives, many will change over time leaving them useless for research. Most people will be familiar with alcohol forming vinegar over time and this type of oxidation process is a constant reality with laboratory chemicals. For some simple processes, such as making bacterial media, using slightly decomposed sugars will make little difference, but using them for any quantitative analysis will render doubt upon the result.

Metal salts may seem to be immune to decomposition as their stable crystalline structures may endure for millennia. They can however absorb water over time which may make them less reliable for use in standard solutions or highly sensitive work. You can often tell the quality of the sample with a quick shake of the bottle – fine grains means it has likely remained dry, large clumps or completely solid means it is now useless and may be very difficult to get out of the bottle.

It is of course possible to test the integrity of old chemicals using the full range of experiments in a chemist's toolbox; this is, however, time consuming and laborious. If an experiment using old chemicals fails it often takes less time to order new stock and re-run the experiment than it would to definitively test the quality of existing stocks. This practice is common in some wealthier lines of research and it is essential that the older stock is disposed of as this is often forgotten. Of course, the chance that an experiment succeeds because of a decomposed organic compound is also a reality and is one of many excellent scientific reasons that studies should be confirmed by reproduction.

16.2.5 Disposal

Before I discuss disposal of chemicals it is worth exploring the cheapest method of getting chemicals out of the lab – give them to someone in a different lab. I myself have used this technique when faced with bulk disposal. An emailed list of chemicals to the department resulted in about a third of the chemicals finding new owners. I did receive a pointed email from the head of school questioning whether I should be offering 500 g of potassium cyanide around the department but after a quick edit this was largely seen as a success. This also made many of my colleagues happy particularly those swift of foot who claimed the more expensive catalysts, but in hindsight it was perhaps not completely thought through. My focus was very much on dealing with the problem in the lab in front of me however as time has gone on my view has broadened and I think that perhaps I just shuffled the risk around a bit but left it mostly intact from a departmental point of view. This is not to say that it is not a good technique for saving money on both disposal and new chemicals, but in retrospect it may have been prudent to remove some of the less useful or more hazardous items before offering them to the rest of the building.

Disposing of legacy chemicals is often the primary method of control particularly in cases where stocks have been left to build up over many years due to lax management. Whilst it would be nice to envision a workplace where only what is necessary is purchased this is somewhat unrealistic and the reality of inheriting stock chemicals must be acknowledged.

Depending on the chemical there are several routes available for disposal. I once went to an academic who I knew had worked on phosphorus compounds during their PhD and asked how they disposed of them. Their answer was 'you don't want to know!' perhaps hinting at previous poor practice when once upon a time it was acceptable to put everything down the sink or perhaps bury them in a little-used corner of campus.

Disposing of chemicals down the sink and into the wastewater stream is still a valid route but cannot be done as indiscriminately as in the past. It is essential to check what the local rules and regulations are but generally neutralised (or diluted) acids and bases can go down the sink with plenty of water, as can salt solutions which don't contain heavy metals.

In some cases this can, however, be poor advice as bad habits may form or this could be perceived as irresponsible to those who do not understand the nuances. It is important that putting items down the sink is never considered the default option. Gases and volatile solvents can also be disposed of through the local exhaust ventilation system or into open air but again this can be dangerous and should only be done if you are absolutely certain that no harm can come of it.

For most hazardous chemicals the normal disposal route is through some kind of waste contractor. This may be directly or *via* a local hazardous waste team. If you are in a position to obtain the services of such a contractor it is important to perform sufficient duty of care checks to ensure that they are upholding the standards expected of them and those written into a contract. This may involve site visits which I enjoyed immensely much to the bemusement of my non-scientist friends, or even following a waste vehicle to its destination (much less exciting than it sounds). Waste contracts may be split up in several ways such as toxic chemicals, biological waste for incineration and general waste to cost effectively cover the full spectrum of waste produced by a research building.

Some contractors may not take higher hazard or reactivity chemicals and this can be frustrating particularly if it is beyond your control as this is usually the preferred option for safety and cost reasons. I was put into exactly that situation early in my career and many days were spent in the laboratory working reactive chemicals up into a form they would be taken by the contractor. Disposing of acid chlorides onto soda ash one spatula full at a time is exciting at first but quickly becomes tedious when you have almost a kilo to get through. Whilst I am grateful for the experience which taught me a lot of practical and pragmatic techniques, I do wonder if it was cost effective for the department when my salary was considered as opposed to the cost of bringing in a different contractor.

It is important to understand that other than specialist toxic waste contractors it is unlikely that those taking away your waste will understand exactly what it is. Be clear with your general waste contractor exactly what is acceptable to go into large bins as the presence of chemical containers may cause alarm even if they have been cleaned. Even non-hazardous chemicals such as sugars and calcium carbonate can cause distress and alarm if a cleaner or waste operative comes into contact with them.

As an aside, I have had a recurring conversation with several large waste contractors who have promised total recycling solutions. I always ask them about ‘blue roll’ and nitrile gloves. These are both theoretically possible to recycle but cannot be if they are contaminated. In every case once I have explained that we simply cannot guarantee that they are free of chemical contamination the desire to separate them for recycling wains and they generally end up being sent to energy recovery incineration facilities. Do not mistake this for scepticism – I am forever optimistic about future recycling options but I feel it is important that the reality of waste coming out of a lab is understood.

16.2.6 Cost

The disposal of chemicals has an associated cost where it is not appropriate for them to be released into the environment. This can be significant, particularly for large scale clear outs which may be necessary for a facility which has not been closely managed for some time. This can act as a disincentive of its own for a cash strapped department where the temptation to leave it to the next financial year or even the next set of staff is an understandable temptation.

The cost of disposing of legacy chemicals often defaults to the institution as associated research grants are long gone. This can be challenging within the current culture of money saving as many departments have very limited reserves. The technical staff who will inevitably be brought in to assist are often overworked and struggle to fit in additional duties around their normal work.

The small physical size of chemical containers and often high associated cost means they are often looked upon as an asset. The cost of disposal which is inevitable once the delivery has been accepted is rarely considered at point of purchase. This cost remains throughout the life time and is more likely to increase than decrease. Due to the reasons described above chemicals will often depreciate, rapidly becoming more of a liability than an asset.

There is a second harder to spot cost associated with storing legacy chemicals. The space they occupy. This may not be an issue for smaller stocks which fit within the same cupboard as working stocks, but when this space is exceeded the institution is no longer getting value for money from its expensive laboratory spaces. The same argument can be made for the increasing cost in time for working around or dealing with legacy chemicals. In both cases I have often stated to researchers that storage costs money.

16.2.7 Hints and Tips

16.2.7.1 Controls on Acquisition

The most effective control where chemicals are concerned is to eliminate the hazard entirely and not buy them in the first place. This is obviously unrealistic so a compromise is found and depending on the size and complexity of the organisation there may be different levels of controls for what users can buy. In smaller organisations this is easy to manage with a smaller number of people to manage and less to buy. Most universities and research institutes are large and often have a high turnover of staff. This means effective controls must be instigated. This may be simple financial controls where the budget holder who is in control of a space approves purchases or a team of safety professionals approves hazardous purchases.

The key is ensuring individuals cannot buy hazardous items through ignorance or malice, whilst providing a smooth and efficient ordering process that permits oversight of the institutional risk which comes with the storage of chemicals. Historically this trust in the professionalism of their researchers has been rewarded, however it does not take many mistakes or malicious purchases to make life difficult for those who manage risk.

16.2.7.2 Leaving Procedures

In order to minimise the accumulation of orphan chemicals it is essential to have proper procedures for people to follow when a researcher leaves. As well as making sure there is not a significant unknown hazard being left behind any remaining useful chemicals can be absorbed back into a general stock for use or disposal. This can be a completed PhD or an established academic moving to another department or retiring. The main difference is when a more junior member of a group leaves there is a good chance that the chemicals will be used in the future; this is much less likely when an entire group moves on in addition to the much larger scale.

Group ownership of chemicals tends to be the responsibility of a lab manager or member of technical staff. This can be a good way of achieving economies without removing controls so long as the lab manager has sufficient authority to dispose of chemicals without being overruled.

Ensuring effective leaving procedures is in my experience a big challenge. The size and complexity of even a single department makes this hard without considerable cooperation and goodwill. It is however time and effort worth investing in the long run.

16.2.7.3 Inventories

You cannot start to deal with a problem with any hope of success without sufficient information about it. At its most basic level this is the purpose of a chemical stock inventory. A simple list of chemicals is invaluable to anyone assessing chemical safety but other features can improve its usefulness. The figure below (Table 16.2) contains additional fields which are either essential for a chemical inventory or helpful additions. The more of this information is available, the more readily an assessment of safety can be made, particularly by someone working outside of the area. A detailed inventory is of course useful to the researcher as well.

An inventory is only as good as its accuracy, the longer it has been in place the more likely that errors have crept in. The worst case scenario for many people is when an inspector (usually me) fishes around in the back of a cupboard to produce a yellowing bottle and asks 'can you show me this on your inventory?' I have uncanny luck with finding the one anomaly on the spreadsheet but I temper this with sympathy for the lab manager who has been put on the spot. When checking your own inventory it is helpful to check both

Table 16.2 Chemical stock inventory fields.

Essential fields	Optional fields
Chemical name	Company of origin
Building and room	Product or CAS number
Date purchased/expiry date	Hazard symbols
Container size/quantity	Price
Owner/person responsible	
Storage requirements	
H (hazard) phrases	

that selected chemicals are on the inventory and to check that items on the inventory are where it says they are in real life. An outside pair of eyes can also be invaluable as it is very possible that you develop blind spots for certain chemicals when you pass them on the shelf every day.

16.2.7.4 Laboratory Management Software

The inventory has been enhanced in recent years with the slow spread of the laboratory management system. The features and capabilities vary between the various pieces of software available commercially or built in house but they all fundamentally open up capabilities not possible with conventional inventories. I will not recommend any in particular here but I will say that they have improved immeasurably since I first looked at them in the early 2010s and I would recommend getting one to anyone managing more than a couple of labs. They can seem expensive but once time savings for those managing chemical stocks and reporting on risk are considered they are often excellent value.

With a fully implemented laboratory management system it is possible to share information in a more nuanced way than with simple documents. There are often concerns about sharing chemicals particularly those which are valuable such as platinum group salts used in organic chemistry or those with licencing restrictions such as controlled drugs. This is completely understandable; laboratory management software allows visibility to be changed for different items without compromising the inventory or hiding things from those who require oversight. It is particularly time saving when transferring chemicals from one area to another as instead of typing information into emails (or worse keeping no record at all) a few simple steps moves the record at the same time the chemical is moved. These features mitigate against the potential for chemicals to become orphan chemicals and issues relating to labels being lost.

16.2.7.5 Oversight and Auditing

It is sometimes helpful to have someone from outside the normal work of a research group to have oversight. This is often the role of a safety advisor or officer who is charged with auditing or reporting on the risk presented by the activities of a college or institute. This outside role has the benefit of a fresh pair of eyes not veiled by the habits or vested interests mentioned above. This objectivity should allow them to provide a second opinion about the relative safety of chemical stocks and, if necessary, report to senior managers.

The more information available the better informed their decisions will be. However, it is worth considering that they cannot be everywhere and someone intent on concealing something will likely be successful. That said, I have always preferred to trust people as being suspicious is rather exhausting!

Sometimes there is no substitute for an outside person visiting a lab in person and rummaging through cupboards. This is a favourite hobby of mine. The response you get to this varies as I try to keep an open mind, with some users the guilty conscience emerges as soon as I move towards the big yellow cupboard (especially if they know I will find something I had firm words about some years earlier) but many are completely blind to the scale of their legacy chemicals. It is a lot more persuasive to have a discussion about the value of older chemicals when standing next to shelf upon shelf of dusty bottles than it is to send an email reminding people to have a clear out.

16.2.7.6 Key Questions for Disposing of Chemicals

I have often found myself tasked with persuading reluctant scientists to part with their old chemicals. This can be a battle of wills unless an order comes down from on high so I have found it helpful to ask them to consider the following questions:

- Is it likely to be used again?
- How long will it last?
- How hazardous is it?
- Will it generate any new hazards?
- How much did it cost?

I find this helps people understand that this is not an indiscriminate process but a logical systematic one. This also helps people prioritise when faced with a task large enough to be insurmountable.

16.2.7.7 Positive Clear Outs and Amnesties

I have found that a simple method for increasing the effectiveness of chemical clear outs is to switch the question of what needs disposing on its head. Instead of asking what can be disposed of ask what needs keeping. Simply by asking this the easier option becomes to allow chemicals to be disposed of rather than working out which ones need keeping. If this is paired with the requirement to justify why chemicals need to be kept, with an obligation to take ownership, the likelihood increases of an effective clear out. A practical way of doing this within a large multi-user laboratory is to place items within a tray and ask users to claim items before a specific date and to dispose of anything remaining at the end of the time limit.

Although laboratories tend to be busy all year round there is often a lull in the summer due to the lack of undergraduates and annual leave being taken. This is a good time to have an annual clear out and amnesty of excess chemicals. This sets good expectations for the rest of the year as well and gives an excellent opportunity to check the inventory. The ethics of throwing away chemicals whilst the principle investigator is on holiday I leave to your own conscience!

The alternative to this when faced with a large number of chemicals accumulated over many years is to have a rolling programme of clear outs one tray at a time. If this is the approach taken it is important to prioritise on a risk basis. Depending on the laboratory this may mean age, hazard (such as poisons) or likelihood of being used.

Whilst it can be tempting to assign blame to the issue I have rarely found this helpful. I would recommend taking an amnesty approach to forgotten chemicals which has the added bonus of ingratiating you to the researcher coming forward as you relieve them of their chemicals and feelings of guilt.

The complexity of the modern chemical laboratory means that there will always be legacy chemicals and there will also always be work for those of us who have chosen a career in chemical safety. It has been postulated that the only way to be certain of having dealt with all of your legacy chemicals is to have the building demolished!

16.3 Equipment

Chemicals are by no means the only challenge which comes with an ageing laboratory. With the exception of the newest facilities, high shelves and dusty corners will have old equipment long past its useful life. Even older equipment still in use brings with it potential issues and expense.

Research grants often transfer ownership of equipment to the host institution when the research is complete.[13] This may seem generous, and often is, but it comes with a hidden cost: the liability of ongoing maintenance and eventual disposal.

Many improvements in the safety of machinery are obvious to the untrained observer; moving parts are now covered, blades and other working edges are protected by guards and safety labels sit on prominent parts. Not all such improvements are as obvious as many fundamental safety standards cover build quality, electrical safety or complex matters such as dust ingress. This is not to say that older equipment is inherently unsafe but more that you cannot make any assumptions about the construction or safety features.

The key piece of UK legislation covering machinery is the Provision and Use of Work Equipment Regulations (PUWER).[14] This legislation covers any and all equipment used within the work environment. It has been said that in theory even the humble pencil requires some sort of assessment under these regulations if you take the more literal interpretation! It is more common however to apply some common sense and only assess items which represent a significant risk of harm.

The principle of the hierarchy of control (Table 16.1) can once again be applied, however there is more of a focus on engineering controls with work equipment as well as the more fundamental processes such as substituting better equipment or eliminating tasks entirely.

16.3.1 Legacy IT

Universities have always been at the forefront of technological development particularly those with a heavy research focus. This has inevitably led to ageing IT equipment being a common sight. The retention of such high capital items seems to make sense from a financial point of view but their long term viability is much more variable. I know of one spectrometer sitting within a chemistry department which requires a chain of four computers to get the data from the original format 5¼ inch floppy disk on to a modern machine. This particular spectrometer is possibly unique in both the part of the spectrum it analyses and its research longevity. Most machines are lucky to have decade-long working lives and seem to go from invaluable to waste of space in not very long at all.

There is a certain charm to older equipment which can make it hard to dispose of, particularly those items featuring a lot of brass which come in bespoke wooden boxes. Unfortunately, museum-quality pieces are rare with most old equipment having little or no value. I could always justify keeping a few choice items for display but at some point a decision needs to be made.

Sources of legacy hardware are more varied than chemicals mainly due to their availability. Anyone can buy and bring in domestic equipment which can be a fruitful source of examples of unsafe case studies. This is particularly true to equipment coming in from abroad directly or *via* the internet. The differing standards, particularly in electronics in different countries, as well as counterfeiting, can render even brand-new items little more than landfill.

One peculiarity of large complex organisations is how some equipment gets used practically every day but no one seems to know who it belongs to. This almost always means that no one is maintaining it either. This is particularly important to resolve in cases of safety critical equipment such as lifting equipment or high hazard tools such as drills and electric saws. I have never worked somewhere where I did not know of at least one orphan ladder of dubious provenance sitting under a bench which got produced periodically to put up signs or flip circuit breakers in ceiling voids without having to call out estates. As useful as this can be there are safety rules which must be followed and items must be claimed or disposed of. One effective way of finding out who owns equipment is to lock it up and see who turns up to complain.

16.3.2 Hints and Tips

As with chemicals, no informed assessment of institutional risk can be made without first understanding what equipment is present within an area or building. An asset register is an essential tool and may already be required for the purposes of insurance or maintenance. Ensure that even obvious details are recorded, as serial numbers and product codes are of limited use if you only know you are looking for a drill.

Resale of equipment is an area fraught with difficulties. It is important to have clear procedures in place to prevent theft and to protect staff from the accusation of theft. When selling old equipment it is also important to consider the reason for sale. If something has been taken out of use because it does not conform to modern safety standards can it realistically be sold without incurring liability on the institution? The same thought process should also be considered if donating equipment to a school, charity or research organisation abroad, are you really donating something useful or are you pushing the burden of disposal on those less fortunate whilst stroking your own ego?

16.3.2.1 Disposal and Decontamination

Before equipment which has been used in a laboratory is disposed of it is important to ensure it is no longer contaminated with chemical and biological agents. Perception of chemical contamination is in many ways as bad as actual contamination particularly when an item is being handled by those without lab experience. That said, I am often nervous as well when I find lab equipment being used as decoration in trendy restaurants!

A decontamination procedure should be appropriate to the type of work involved in the life of the item. When I have been called upon to clean up old fridges I have erred on the side of caution and used common solvents, soap and water and household bleach for my own piece of mind as much as anything. This should be documented and communicated to the new owner so that they can appreciate the effort but also be aware of limitations. As a note of caution, it is important not to sign off 'certificates of cleanliness' when it is not possible for this to be verified. These are sometimes requested by contractors assigned to repair ducting which only they can access. It is better to inform them of positive steps to ensure their safety such as ceasing solvent work for 12 hours than to sign a piece of paper which will give them a false sense of security.

As responsible scientists, I think it is important that we lead by example in responsible disposal. Care should be taken to ensure that those taking equipment containing potential hazards know how to handle them safely. This may be refrigerant, asbestos in old ovens or mercury contained within old gauges. By performing your due diligence on a company you will also be assured that an item bearing the name of your department will not end up being found somewhere it shouldn't be by the local media.

16.4 Buildings

Over the years I have been to many university campuses in the UK, many have beautiful architectural features although not all. What they do all seem to have in common is at least one concrete and asbestos monstrosity built in the 1960s! This is normally the chemistry building. This sticks out between classical buildings in red brick and stone and the very latest facilities in glass and wood. The lack of uniformity is testament to not only the rapid expansion of universities but also how the older buildings persist.

Even the outwardly older buildings have often been heavily refurbished inside with use changing over time to meet the need of the wider institution. This is one of the greatest challenges for those overseeing safety. You will see both buildings built before modern safety standards and those converted significantly away from their original purpose and design. I have sat in many planning meetings for laboratory refurbishments where our hands were tied by unusual room sizes and utility configurations which forced us to compromise away from an ideal solution. I have speculated that in the world of industry they would never attempt to refurbish such difficult buildings but would instead start again from scratch, however this may well be wishful naivety.

Every generation of building seems to have its own set of problems. The oldest are cold and rarely have provision for disabled access. Those designed post war were often not built to last and are absolutely riddled with asbestos. Modern buildings are no exceptions, the recent fashion for atriums and cathedral-like wells running the height of the building create issues for noise and are often perceived as wasted space. In buildings of all ages it is rare to find provision of space to allow for future expansion of utilities such as power, water or ventilation all important to scientific spaces.

Managing buildings is no simple task and represents several different discrete careers which take many years of training and experience. It is important to work with those who have developed that expertise, but as someone who works in safety, there are key principles to ensure the best possible end results.

The larger and older the site of a university the more likely it is to have forgotten spaces. These often occur in basements, lofts and isolated outbuildings. Often no one is sure who they belong to and the estates department is too busy to go poking around as a matter of course. If you are fortunate these spaces only contain dust and the occasional rodent but I have stumbled into more than a few "Aladdin's caves of wonder". When I say "wonder" I normally mean "horror". I have found several dubious experimental set ups in service corridors and plant rooms, and a notable low point was several dozen poorly preserved animals in formaldehyde! It is important that when these spaces are found they are assigned an owner. This allows them to be maintained and for someone to be held accountable for them being used safely.

16.4.1 Hints and Tips

Without rigorous record keeping it is very challenging to keep track of the sheer variety of buildings and spaces at a large research university. Too often we are over reliant on the memories of long serving members of staff within a building who know all its quirks. This however will always present an institutional risk as staff move on or retire and do not always pass on all the knowledge they have picked up over the years.

Consider how the age of the building will affect any new project. Will you be able to control the temperature? Are you close enough to a fire escape route? Can you fit large pieces of equipment down the corridors, around the corners and through the doors? Can you control security? Are there enough toilets? Do you know where all of the utilities are (particularly important for excavation work)? What about disabled access? I have asked variants of all of these questions during refurbishment projects which, although challenging at the time, was a lot better than discovering a problem in retrospect after the money has all been spent and the builders sent home.

You should not have to compromise on safety simply because the situation is challenging. However, this is not to say the exact way of doing something is not up for discussion but that fundamental standards should not be ignored. A good example of this is how many desks you can fit in a room (or workstations in a lab). Ignore the advice to not fit in extra seating, and you end up with a space that is unpleasant to work in where people are constantly in the way and jostling each other. Over time this may develop into ergonomic issues, ill health and stress. As a safety professional I have always felt that it is my duty to be motivated to make the best possible space for those who have to use it, as often this is not the priority of those instigating the project or the contractors realising it.

16.5 Staff Attitudes to Legacy Issues

Universities are full of brilliant people and I have felt privileged to work with many of them having spent my entire career in the higher education sector. There is an amazing sense of ownership particularly for those working in the sciences, but on the subject of legacy chemicals and equipment this can cause problems particularly when coupled with some of our more eccentric colleagues!

Legacy issues can arise at any time however, due to the autonomy given to elite researchers, problems with legacy items often don't come to the attention of other services until a staff member is on the cusp of retirement. This can lead to a large amount of work needing to be done at one time if it has not been managed proactively over time.

This autonomy and the camaraderie which unites those working academia can lead to an 'us and them' attitude towards health and safety. Given how dependent their career is on research it should come as no surprise that many researchers are protective against anything they might see as a threat to this. Safety advisors often find themselves in a hostile environment as they are seen as an interloper who is there to stop research happening particularly when confronting legacy issues. In order to be effective it is essential that this hostility is resolved as quickly as possible as progress in making work safer is rarely made until everyone is on the same page.

Another common scenario is that whilst an academic is technically (or legally) in charge of an area the day to day management will fall to a lab manager or technician where they are available. An academic resistant to disposal of their legacy chemicals can seem to be an insurmountable obstacle to a much less senior member of technical staff.

One of the most effective techniques for defusing such tense situations is to unite in the common ground of science. We are all on the same team and are looking to advance research from different perspectives. I have often found simply by asking a researcher about their work that an understanding can be reached. Over the years I have also learned a great deal and find there are few things more enjoyable than watching a previously curmudgeonly academic's face light up when they take you through their latest breakthroughs. Common ground can also sometimes be found in mutual rivalry with different disciplines, this is a high risk technique if you use it in several different departments but is often an easy win. Of course, I would never stoop to such a level as I value all areas of science equally but if in doubt pick on the social sciences as they are unlikely to buy this book.

The perceived value of items, whether it be chemicals, computers or equipment, is often much higher than their actual values. The attitude of 'it might come in handy' is rarely realised. This is particularly true in the fast moving world of research. No one knows what direction the future of research will take although many would hazard a guess. Previously fruitful avenues can be rendered obsolete by alternative methods or theories or simple frustrations with inefficiencies. Old equipment and chemicals may not be transferable to new fields.

With people also comes data, depending on the old-fashioned habits of the data-owners this may just be digitally kept on physical media or it may be in the form of boxes and boxes of paper. In this modern age of data protection, backed up by increasingly rigorous data laws, stored data has become increasingly like the stores of aging chemicals described at the start of this chapter. There might be something valuable in there but it is more likely to be junk which might just blow up in your face. The sheer amount of time which is required to go through old paper records is too high a cost for any profit to be obtained from it.

Some safety and health data you are required to keep for a long time for legal reasons, and lab books and similar can become important evidence in patent disputes, but most of the time even the most dogged historian would struggle to find anything of interest in the aging papers and notebooks found in dusty corners of research buildings. In the event of long term illness due to asbestos or other chemical exposure you can be asked to produce safety information up to 40 years after the event, in my experience this is often almost impossible.

16.6 Conclusions

It can sometimes feel overwhelming for those tasked with managing safety at universities. Many of the issues faced are a result of the unique systems, infrastructure and staff. In some respects this means it is a system we are stuck with, however it can be improved.

In order to begin managing legacy items it is first important to have sufficient data. Knowledgeable staff may have good oversight but there is no substitute for a comprehensive inventory whether it be of chemicals, equipment or building features. Obtaining this information can be labour intensive but once it is established it is much easier to maintain and begin to effectively manage.

Once the overall picture of the challenge is available it is then possible to begin proactive management even if it is intimidating. By being systematic and proceeding on a logical basis whether it be by risk, convenience or ease, the results can quickly be obvious making the rest of the process easier. Whichever approach you take I can guarantee that it is more effective than doing nothing at all.

This does sometimes require a firm hand even with those who are not actively hostile. I have noticed the same legacy chemicals on multiple visits to a lab over several years with otherwise cooperative employees. The demands on university employees are many whether they be an academic or member of technical staff so I tend to view this with sympathy but look for solutions to overcome the roadblocks.

It is far more preferable to take a positive view and make it clear that we are all working together to the same ends. A positive attitude and the occasional bribe in the form of pizza or beer is far more conducive to a safe department in my experience than one dominated by blame and recriminations. If you are friendly and approachable you are far more likely to be alerted to small problems before they become big ones. Taking the role of breaking bad news to senior people has also served me well in the past, not only in currying favour with those who were reluctant to do it for fear of the reaction but also in ensuring those at the top understand the nature of our work and the issues arising.

The job of removing legacy chemicals and equipment can be a thankless task but it has its upsides. In the process of digging through cupboards you may find a few horrors but you will also find gems from the long scientific history of your institution. My personal low point was a fridge completely full of phosgene but this has since been compensated for by some simply beautiful chemical testing kits used for fieldwork identification of chemicals I found on top of a cupboard still complete in their wooden boxes! Old equipment can evoke feelings of nostalgia or bring appreciation of techniques which preceded modern analysis. That said don't fall into the trap of keeping something just because it is cool too often, or you may find yourself becoming part of the problem.

Acknowledgements

I would like to thank Dr David Kinnison for giving me such an interesting and varied introduction to chemistry safety as well as being an amazing source of advice and knowledge since. I would also like to thank the members of the Chemistry Safety Network mailing list who have provided authoritative and timely advice when faced with some of the trickier legacy chemicals.

References

1. I. Asimov, *A Whiff of Death*, Victor Gollancz Ltd, 1968.
2. The Control of Substances Hazardous to Health Regulations 2002 Provision 7, (1), UK, 2002.
3. The Health and Safety at Work etc. Act 1974, UK, 1974.
4. The Dangerous Substances and Explosive Atmospheres Regulations 2002.
5. Health and Safety Executive, Storage of flammable liquids in process areas, workrooms, laboratories and similar working areas, https://www.hse.gov.uk/fireandexplosion/storageflammliquids.htm, last accessed 15 May 2020.
6. European Commission, Horizon 2020 Online Manual, Reports and payment requests, https://ec.europa.eu/research/participants/docs/h2020-funding-guide/grants/grant-management/reports_en.htm, last accessed 15 May 2020.
7. Environmental Health and Safety, Berkeley University of California, *Lesson Learned – Peroxide Explosion Injures Campus Researcher*, 2018 https://ehs.berkeley.edu/lessons-learned/lesson-learned-peroxide-explosion-injures-campus-researcher, last accessed 15 May 2020.
8. Fisher Scientific, Safety Data Sheet: Picric acid, 2015, https://www.fishersci.com/store/msds?partNumber=SP9200100&productDescription=PICRIC+ACID%2C+SAT%2C+100ML&vendorId=VN00006705&countryCode=US&language=en, last accessed 15 May 2020.
9. L. Wright, Bomb teams sent to schools 600 times in chemical alert, *BBC News*, 2017, https://www.bbc.co.uk/news/uk-england-birmingham-38596153, last accessed 15 May 2020.
10. Classification Labelling Packaging (CLP) Regulation (EC No 1272/2008), European Chemicals Agency, 2008.
11. The Chemicals (Hazard Information and Packaging for Supply) Regulations 2009, UK, 2009.
12. British Standards Institute, BS5609 Specification for printed pressure-sensitive, adhesive-coated labels for marine use, including requirements for label base material, 1986.
13. UK Research and Innovation, Standard Terms and Conditions of Grant, RGC 9.2, https://www.ukri.org/files/funding/ukri-fec-grant-terms-and-conditions-pdf/, last accessed 15 May 2020.
14. The Provision and Use of Work Equipment Regulations Part II, UK, 1998.

CHAPTER 17

Green and Sustainable Whilst Avoiding Risks

H. J. SHEELEY

Public Health England, PHE Porton, Salisbury, Wiltshire SP4 0JG, UK

17.1 Introduction

In the period up to the 1980s laboratories, like other workplaces, saw significant changes to reduce workplace accidents.[1] In addition, the duty to protect those affected by the conduct of the work extends to include students, temporary and contract staff and visitors.

Changes have included the reduction of glassware use, mechanisation, electronic and automated safeguards and controls to prevent physical injury. The increase in standards and the standardisation of manufacture has made equipment and instruments safer with far fewer home-made devices. Implementation of electrical and pressure systems standards, regulations and quality control has significantly reduced many energy related injuries. Increased understanding of chemical toxicology has resulted in improved labelling standards, quality and environmental controls.

The focus has been on making all workplaces safer, with considerable success (Figure 17.1).

With the focus now shifting to sustainable use of resources, and protection of the environment; there is potential for a tension to arise between safety and environmental considerations. For example, the current drive to

Challenges for Health and Safety in Higher Education and Research Organisations
Edited by Olga Kuzmina and Stefan Hoyle

Published by the Royal Society of Chemistry, www.rsc.org

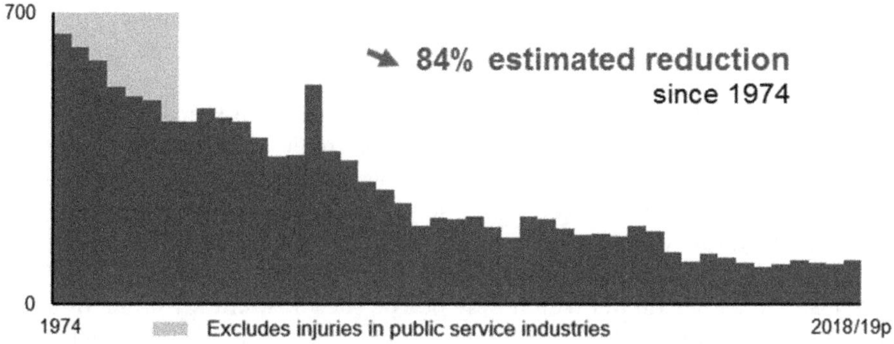

Figure 17.1 Great Britain: Number of fatal injuries to employees. Reproduced from ref. 44 with permission from HSE, Copyright 2019.

replace plastics especially single use plastics with environmentally friendly alternatives has led to an interest in identifying and replacing single use plastic containers with fibreboard and compostable containers. These may not stand up to water damage, have the rigidity and strength of the original so may present new risks of leaking or disintegrating and consequential injury or exposure.

The drive for businesses, including those in the scientific research and higher education sector, to be ethical is a strong motivator for examining the environmental impact of an activity, for example sourcing raw materials, transport, energy, skilled labour, use of discoveries and invention of products and how they are finally disposed of.

With increased external scrutiny and accountability, organisations have had to adapt to embrace the demands of sustainability. For the higher education and research sector, this is necessary if they are to attract students and staff, grants and partnerships and produce products and ideas for commercialisation or service use. Environmental management and sustainability do not just prevent issues to ecosystems. Further the environmental responsibility extends to future generations and can be international in its impact on longer term human and biodiversity health.

17.2 Costs and Demand Areas

Laboratories are considerably more expensive to run than an office due to temperature and humidity controls as well as air changes. Substantially more electrical equipment will be running throughout the year, with the ensuing higher energy and maintenance requirements. The economic equation to be sustainable is more than money alone, it includes a moral dimension of social responsibility as well as the cost of implementation of a change. This may take many years to recoup in any form. Failure to meet these environmental goals, expressed or expected, may lead to poor publicity, reduction in students applying for courses, and failure of grant awards. Some changes bring positive benefits that are straightforward but often the data is not visible, or easily attained, such as the cost to run a piece of equipment against a similar model. Budget pressures and market positioning have been a key incentive to explore where cost savings can be made and making the most of opportunities taken such as grants having criteria linked to sustainable objectives. A common example is the purchase by each researcher of a new −80 °C freezer, for storage of perishable samples, as a matter of course with each new project grant. The majority of research groups will be looking to purchase a cheaper model which may not be the most energy efficient and therefore may contribute to increased energy consumption by the organisation. The lack of institute policies for bulky equipment purchases remains a health and safety and energy use problem, driven by both the attitudes of the research group and centric and cumbersome financial processes.

17.3 Trends

Many organisations have been advancing the sustainable agenda for some time and provide advocacy, examples, data and proformas to assist others wishing to explore the same agenda.[2,3] Many of the projects that have been promoted have been around energy saving, cost-effective systems of recovery and reduction. In turn, there are now an increasing number of organisations that promote, develop or offer cost effective and sustainable services. Increasingly there are initiatives to use alternative substances and materials especially around plastics and solvent recovery. Suppliers and manufacturers are increasingly providing recycle/take back schemes but only for uncontaminated materials such as pipette tips, polystyrene cool boxes, cleaned reagent bottles. This has led to more stringent controls on waste segregation and/or washing facilities. These processes may present new risks as it results in space required for multiple containers for collection, handling issues including manual handling and specialist washing with acids. Further, the overall sustainability equation of collecting, shipping and actual recovery may not be as sustainable as intended but provides a market position and green profile.

17.4 Technology

Advances in electronics and computing have created opportunities to do things that could not have been done before, such as remote control of processes and internet sharing. Modern devices tend to be manufactured as sealed modular boxes, including those used commonly in laboratories, and can be superseded frequently. This leads to whole item disposal of equipment not just repair or replacement of the faulty component. The high cost of disposal tends to encourage keeping faulty equipment and equipment graveyards which present risks associated with storage. Laboratories of today are unrecognisable to those of 50 years ago. The technical understanding of common modern plug and play equipment and instruments is far lower and hence operators are less aware of potential risks. As an example, where a piece of equipment used for automated blood testing develops a fault, which is uneconomic to fix and impossible to repair, then the researcher discards it with limited decontamination of outer surfaces not the contaminated internal parts. Subsequently, there is potential for recycling and waste handlers could become contaminated during invasive stripping of equipment and possibly at risk of infection and anxiety for their health. New different and considerate policies and procedures for decommissioning need to be developed and users trained.

Manufacturers are under a legal obligation to ensure that the products they supply are energy efficient, use proportions of recycled or ethically sourced raw materials, and have recycling information.[4] In research development of prototypes or where existing equipment is used for novel research purposes for which the safety control features are not appropriate.[5] This may increase risk if there is poor understanding of safeguards or need for protective features.

17.5 Standards, Quality and Acceptance

Manufacturing standards are not global, or uniform, with many pieces and parts not being made to known ISO, CEN or BSI standards and difficult to identify, with associated complex chains of supply.

There are many examples of equipment, or components, being marked as CE or other quality standards, but not meeting the requirements. These present health and safety and sustainability and environmental implications as they may not actually meet the stated performance criteria, *i.e.* may be unsafe or constructed of substandard materials. It is important to determine the specification and quality standard of the equipment requirements as part of the procurement process which should include the full lifecycle cost of the items being procured, including waste disposal. Many organisations with centralised procurement have approved product lists including chemicals, personal protective equipment and common equipment. This has the benefits of ensuring the products ordered meet expected requirements. However, this is almost unknown where individuals can order from suppliers' catalogue directly without reference to in-house standards, this is the case in most universities and HE organisations. The choice of suppliers may be limited due to procurement arrangements and thus not permitting the best selection as the range is severely limited. Large organisations where safety critical activities are undertaken (nuclear, electrical, high containment) have prescribed critical safety equipment specifications used in procurement but a more general "choose low energy" requirement is intended to direct selection processes. The informed and responsible selection by researchers of items directly relies on their understanding all aspects of safety and environmental impact, which is not their focus, but it is their responsibility to check conformity to standards.

If sustainable procurement practices are not followed, then it could lead to difficulties in the disposal of the equipment and increased financial impact. On purchase/receipt of the equipment an assessment should be undertaken for conformity and performance. If these are more widely applied, this will ensure safe sustainable operation from the beginning and a clear understanding of the performance, lifespan and consistency of the components and parts, making maintenance and servicing consistent. The historic record of equipment performance and reliability may not persist with replaced components licenced under green credentials. In the laboratory it is essential that standards of cleanliness, calibration and reliability are maintained; this may become more difficult with a wide variety of laboratory tools and chemicals changed to meet green standards and credentials.[3] The regulation of chemicals, including the Registration, Evaluation, Authorisation & Restriction of Chemicals (REACH)[6] and Biocidal Products Regulation (BPR),[7] aims to improve the protection of human health and the environment through the better and earlier identification of the intrinsic properties of chemical substances and to drive progressive substitution of the most dangerous chemicals (referred to as "substances of very high concern") when suitable alternatives have been identified. For example, withdrawal of the disinfectant Hycolin™ led to the need to identify alternatives and a lot of work was subsequently carried out around validation protocols for an effective and approved substance for the disinfection of mycobacteria that is less harmful to the environment. However, the approved products may not be as effective under research conditions and may result in contamination build up and possible infection risk to workers.

17.6 Applications of Green Principles

17.6.1 Selection of Materials and Consumables

Green credentials are important to universities[8] and their students and in applying green policies to estates functions and equipment procurement. Knowledge of energy use and emissions through metering and monitoring is common and for new builds mandatory under building control.

At the level of large equipment there is a lot of information available and schemes to help identify energy efficient products. There is an increasing amount of information about the energy use in laboratory facilities. Also, there is information and selection criteria for bench top and common laboratory equipment, so this can feed into overall energy demands. There are also reference lists of recommended equipment and substances that meet recognised environmental standards.[3] Current practice in HE and research organisations is to place emphasis on sharing equipment, which also requires suitable arrangements to ensure inventory and access do not lead to health and safety incidents. It remains a challenge that procurement and budgets will enable an environmental and sustainable assessment alongside the health and safety assessment. Further, has that change has been risk assessed in a holistic manner that considers both environmental and usability in the laboratory context? For example, purchasing newer refrigeration units which have approved coolants that do not contribute to greenhouse gases such as chlorofluorocarbons (CFCs) which damage the ozone layer but also which operate efficiently. It could be a better choice to replace older refrigeration equipment that is more energy and space efficient.

In 2015, scientists from Exeter University estimated that their bioscience department was generating up to 280 tonnes of plastic waste per year. This extrapolated to over 5.5 million tonnes per year.[9] One manufacturer of high throughput sensitive analytical laboratory equipment has recognised the equipment uses high volumes of plastics so has incorporated a pipette washing and sterilising module to the instrument, so they can be reused with high level assurance of cleanliness. Since 2015, Grenova[10] has helped laboratories in academia and industry reuse 93 million pipette tips, and customers have reported that some tips can be reused 25–40 times.

Plastics that can most commonly be recycled are polystyrene (PS), polyethylene (PET or PETE), polypropylene (PP) and high-density or low-density polyethylene (HDPE/LDPE). Commonly used consumables such as centrifuge tubes are made of PP, whilst culture dishes and flasks are usually made of PS or PETE. Other alternatives to plastic consumables are bio-based plastics[9,11,12] which are made from renewable resources and represent a more promising solution. These products based on "bio-PE" materials can already be found in the life sciences sector, such as laboratory flasks or the base of pipette tip racks.[12]

17.6.2 Repeating Work

Another important aspect of sustainability is around the duplication of experimental studies, underuse of materials generated from a study and various reasons where work must be redone. This is a huge waste of time and resources and demands a less parochial view of science ownership. Research studies should be properly designed to produce as much information as possible with minimum necessary duplication and wastage. Good examples come from the **R**eplacement, **R**eduction and **R**efinement, the 3Rs, principle in the scientific use of animals in studies where the value of the study includes the maximization of information generated for least animal use.[14] This has been taken further in several schemes where an organisation carrying out a study make available tissues and samples or observations to another group. This includes excess animals and control groups being reused or provided to a third party.[15] One specific issue is the transfer of excess bred mice to zoos for food to captive zoo animals. Careful planning to not over breed should proceed this but examples are of where delays in setting up a study have resulted in an animal cohort not being suitable then these are repurposed to another study. The benefits to health and safety are that there is then less exposure to the chemicals or animals arising from repeated work. A properly designed research study will have foresight to the possible outcomes and how this will feed into the bigger picture of the science, so the data is sound and useable in future development.

17.6.3 End Product Development

The design of the research may initially be to answer a question. But as all science may eventually lead to products and services it is a natural part of the design to exclude hazardous chemicals or processes that could not be used in scale-up or manufacture. Major pharmaceutical manufacturing companies have for many years applied this principle and have an "approved" list of chemicals and processes that are to be used in early development work. These lists are revised as knowledge of risks and costs of processing are fed into the modelling. Instilling these procedures into early career students and scientists can improve laboratory safety and protect the environment, with the associated reduction in the costs of wasted resources.

A holistic approach to sustainable and green practices includes all parts of the organisation extending to supply chains, use, waste and operational requirements including scale-up.

There are basic principles in the hierarchy of sustainable use of materials:

- redesign/remove/rethink: do differently
- reduce: less of
- re-source: made of different material or local or renewable supplier
- reuse: use again
- repurpose: use as is but differently
- recycle: recover proportion of the components for another cycle
- recover: selective or partial recovery of energy or substances for reuse
- dispose: range of preferential methods

All of these have health and safety issues around: segregation, storage, handling, transport, disposal. Materials would need to be classified, segregated, stored safely, washed or packaged, contracts established, and assurances given. Staff and students would need specific training and be informed of the various types of waste and their handling requirements. Organisations will need to actively manage these complex and changing situations and understand their liabilities as well as the benefits and costs. It is notable that students and staff are highly motivated to take specific steps if they are straightforward and easy to comply with.

17.6.4 Transportation

Part of the holistic approach to sustainability is to consider the organisations business, including associated processes, such as transport. The transport of goods and services is integral to the supply chain, both on- and off-site, including contractors who work on behalf of the organisation. There are environmentally preferential modes of transport that are comparable to the waste hierarchy[16] and similarly are not without additional health and safety risks. The hierarchy sets out preferences that can be applied to goods delivered to the organisation and policies for purchase and distribution. This applies not only to goods but also to operational practices. For example, depending on institutional policy, students and researchers should travel by public transport, such as trains, which is seen as being environmentally preferable, but other factors should be taken into consideration, such as the of duty of care and provision for vulnerable students and researchers or in neighbourhoods with high crime. Another situation is travel to conferences and meetings, where the operational advantages of face to face conflict with safe and sustainable travel options *versus* remote exchanges where health and safety risk can be met more easily. Selecting the best mode of transport that is sustainable but also considers any conflicting priorities is an important consideration.

17.6.5 Practical Measures

There are many parts of the laboratory environment that need exacting requirements to protect the personnel and the internal environment in which the science and teaching takes place. There are some examples where health and safety as well as environmental protection can also be sustainable.

17.6.5.1 Energy

The biggest sustainability impact on research or scientific education organisations can be found in energy saving and most organisations have gone a long way in implementing schemes of low energy lighting, solar and heat source systems and heat recovery. These types of research or educational facilities are high users of energy, so the benefits are apparent and quantifiable. Economic demands in government and HE organisations have led to the evaluation of the need for facilities, and a number of these facilities, in particular chemistry laboratories, were closed in the 1980s and 1990s. By 2005 there was public concern regarding the lack of academic chemistry departments and consequence for STEM developments.[17] However, since then there has been a rise in students taking up chemistry at Russell Group universities allaying those concerns.[18] The investment in these department is high due to income generation and student income. This in turn has made these universities leaders in energy management.

- Fume cupboards: Fume cupboards have the highest use of water and energy in the laboratory as they use very high volumes of conditioned air and tend to run continuously costing as much as £2000 per year.[20] Current green advice is to educate users to efficient but safe use including only running when required.[19] Modern fume cupboards have many energy and efficiency features, such as the use of horizontal sash openings to limit the volume of air being drawn into the fume cupboard, automatic sash closing mechanisms and low flow settings, that encourage safe use as well as other gains that can be made in the way the fume cupboard is used and maintained.[20] Optimal use includes operating the fume cupboard with the sash closed and opening only to pre-determined set points. This will greatly reduce energy demands whilst maintaining safety. Correct and responsible use of a fume cupboard is also a benefit to safety by reducing exposure, and accidents from explosions or other energetic reactions within the fume cupboard.

- Ultra-low Temperature freezers (ULT): Cold storage of samples is also a high energy demand aspect of laboratories. In my experience researchers have a propensity to retain samples indefinitely. They tend to keep everything. It is therefore important to have a good policy on sample retention, identification and to manage ownership. It is well known that retiring researchers can leave unclaimed materials behind for which the nature and value of the material is unknown. The policy should include detailed sample information to enable future decisions on retention. Procedures to ensure responsibility of the samples is transferred to another manager must also be included. Without these retention policies there are very high costs in keeping samples.[21] It has been reported that insurers insist that claims against losses of refrigerated samples must be known and that they are maintained and part of an emergency plan. This pressure may lead to more sharing of and better managed cold storage.

There is a tendency for cold storage freezers (−20 °C or −80 °C) and other heat-producing equipment to be in the laboratory. There is then a high heat load to the space which in turn must be compensated for by cooling the laboratory space. The additional heat makes for an uncomfortable working environment and occupants then remove laboratory coats and introduce fans and wear light clothing and sandals which are not suitable in a laboratory environment. Good design would ensure there is localised cooling of the cold storage area rather than the room. There are things to consider when installing a ULT, such as, optimal placement of such refrigeration units, the right type, optimal packing, maintenance and use that reduces opening and formation of frost.[20,21,22,23] Generally, laboratory researchers are unaware of the running costs of such a common piece of standard equipment and how to best use these spaces, for example by use of efficient packing systems, sharing freezer space, ensuring the freezer is regularly defrosted and opening only for the briefest time.

Over time the availability of ULTs has encouraged the storage of every biological sample at −80 °C. This is rarely questioned, yet there have been studies that have classified biological samples into different suitable temperatures ranging from ambient to −196 °C depending on what needs to be preserved.[24]

The health and safety benefits of good storage arrangements are the reduction in the number of units, knowing what is held especially for permits and compatibility and reducing duplication and problems arising from unclaimed samples and defrosting issues.

17.6.5.2 *Lighting*

There has been a rapid change in the types and controls of lighting in every workplace. Facilities with a low cost and immediate return on reduced energy costs include:

- Proximity sensing (PIR) *i.e.* only on when someone is present. Problems can arise that the detectors do not recognise someone is working and the lights go off, this can make them less suitable or need careful selection for those at the bench or safety cabinet.
- The various tones, brightness and frequency that LED lights are available in has a wider range of specification than incandescent lights. It is important that for close working these are matched to the work environment including colours of reflective surfaces and the type of activities.

I have known cases where changes to colourways have led to ill health or exacerbated health conditions or raised awareness of poor eyesight when these changes are done without consultation with users or impact of the change on processes.

17.6.5.3 *Chemicals*

Chemicals used in laboratories can be harmful in use but also when not correctly stored or disposed of. As well as assessing the hazards and risk of chemicals used in the laboratory a process of considering the hazards of the chemical to the environment should be made. Many chemicals may be discharged directly to the sewer, through exhaust systems (with or without filtration or capture), in filters or bound to other substrates that are for disposal. Positive impacts for health and safety and environmental protection can be achieved by the selection of the least harmful options and/or inactivation or inert bonding prior to disposal, together with minimal use and purchasing minimum quantities with a clear plan for disposal route and cost.

Inventories and databases for monitoring and use of chemicals can encourage good stock control, sharing and efficient ordering.

There are databases for assessment of environmental impact.[25,26,29,30] Students and researchers should be trained in reducing contamination and using protective devices correctly together with hazard assessment labelling that may include environmental protection steps. Chemicals should be disposed of responsibly and this obligation includes not putting any unapproved chemicals down the laboratory drain not only for environmental reasons but as a duty to others such as maintenance staff and co-workers due to reactions and possible damage to pipe work.

17.6.5.4 Facilities

A good work environment will reduce health and safety risks. Sustainable practices such as the use of natural light and "borrowed light" greatly improve attention and reduces eye strain and lighting costs.

A good work environment includes comfort and ambience using colour and natural ventilation. This has noted benefits for well-being too. The use of natural ventilation has been obscured by the ability of controlled temperatures and humidity *via* air conditioning in buildings. However, cost and energy demands can be very high. In new builds and renovated facilities these can be designed in, they include orientation of the building to take account of solar gain (avoid "hot spots" and glare) and prevailing wind and weather. The design and construction for safety and sustainable buildings is a strong driver for scientific research and for gaining green credentials; many universities and research organisations lead the way and are very active in promoting green research buildings in novel and innovative ways that exceed and extend BREEAM criteria but bear the investment upfront costs.

The design and use of space contribute to sustainability and good design takes account of needs for storage, ventilation, and changes that will arise over facility lifecycle.[20]

Designing for adaptability where changes can be made without huge disruption or need to demolish walls and large areas by the forward planning for knockout walls, large plant space, and ability to modify drainage easily will reduce disruption time but also prevent compromising as a building ages and new science technology and use comes in.

17.6.6 Waste Management

Waste is a large component of managing health and safety in research and HE facilities. Whilst waste cannot be eliminated, by applying various strategies and using the five Rs of waste management hierarchy (reduce, reuse, recycle, recover then residual management), its impact can be reduced.

17.6.6.1 Reuse

Reuse is an attractive way to be green and sustainable, however it involves many partners as there are few opportunities identified for the reuse of items for the same purpose.

Increasingly, research and education facilities are adopting strategies that include a range of reuse methods to minimise wastes. An example is where animal bedding is composted on site and reused by grounds maintenance.

These facilities have extended this to on site composting of office and other materials. Other options include replacing plastics and single use items.[13,39]

There has been an increase in the use of animal carcasses and biological by-product materials such as fats and oils that are uncontaminated through contract disposal. Type-specific collection calls for high attention and care from staff compared to disposal as was customary in the past, as even a small amount of contaminant renders whole batches unrecoverable.

There are several suppliers of small-scale solvent recovery systems that can address laboratory-scale waste solvents on site. These return cleaned solvents back to the laboratory, saving on disposal cost, transport and problems arising from accumulation of solvent wastes awaiting collection.[32] There are on the market laboratory-scale solvent recycling equipment used for xylene, ethyl acetate and toluene, ethanol and cyclohexane.[31] Although replacement with non-toxic processes that do not require high hazard substances should be considered first.

Shared collection between organisations and departments is the only viable route for reclaiming some small volumes of specialist wastes, as there are few processors and scale for recovery systems. One of these is silver and rare metals which are cost effective for recovery agents even for small volumes.

17.6.6.2 Containers

In sustainable laboratories the consideration is not only the chemicals but also the containers and the overpacking of these. One of the most common sustainability complaints from laboratory staff and students is the packaging of small items, *e.g.* dry ice arrived in copious amounts of unnecessary packaging with non-recyclable polystyrene, or a small bottle surrounded by excess of polystyrene packing in a huge cardboard box. Several organisations have looked at their supplier packaging types and materials to rethink and remove them from their waste streams. One university took their concerns about excessive packing of antibodies to the supplier who replaced plastic multilayer packing with much simpler envelopes.[33]

Another example is pipette tip outer box reuse which seems to be working well and is widely adopted. A recent commercial supplier (2BScientific) initiative to take back ice blocks was overwhelmed and did not have enough going out to maintain the scheme, so it was withdrawn. Social media messages were not supportive, and this could be a means to make change happen if there is enough groundswell. Understanding and collaboration is needs to take account of delivery to site, whether there is distribution or onward transport, and requirements for secure packaging.

Specialist conditions for holding prior to laboratory and facility use at the lab must also be considered. Examples are the replacement of polystyrene chips with corn starch chips which have broadly similar thermal properties and are compostable.[28,32] If materials are to be locally stored, then the replacement needs to be suitable to ensure samples are kept stable as well as not putting staff or students at increased risk. For illustration, consideration needs to be given around is the replacement more at risk from rodents, rain and disintegration prior to use, especially if fibreboard or other sustainable products[27] are used?

There are take back schemes, but these are in their infancy and require high levels of management and motivation. One example that has been in place for a prolonged period is the "take back" scheme for Winchester bottles for reuse by the manufacturer. These schemes are not cost effective as there is no tangible cost saving to the user and minimal to the supplier chain due to transport costs and replenishment which needs to be in the overall equation. However, these schemes persist due to the overall drive towards sustainable business.

17.6.6.3 Recycling Schemes

There are increasing initiatives on recycling schemes exist for:

- gloves: nitrile, butyl and latex[34–37,43]
- pipette tips[41] which are increasingly made of less plastic by 40% *i.e.* thickness reduction with no loss of quality
- cool boxes[38,40]
- sample pots[39,41]
- tissue culture flasks[36,42]
- reusable sharps containers[37]

Most of these require uncontaminated materials and arrangements for pick up and are still in their infancy and thus require motivated staff and strict segregation arrangements as well as education and clear instruction. The emerging social responsibility schemes by suppliers are limited and have many barriers such as the scheme size and the need for the location of specific collection containers.

The Rightcycle Program by Kimberley Clark in conjunction with Terracycle[37] in the United States claims to have prevented 300 tons of nitrile and other plastic PPE from entering landfill with just 150 participants. Recovered plastic is used to make garden furniture and other products.

The types of plastic accepted by local councils are variable across the UK and are limited to certain types for which they have contracts for recycling. The types of tube and racks used in laboratories are often composed of different plastics of which only some can be recycled. Plastics that can most commonly be recycled are polystyrene (PS), polypropylene (PP) and high-density or low-density polyethylene (HDPE/LDPE). Commonly used consumables such as centrifuge tubes are made of PP, whilst culture dishes and flasks are usually made of PS. HDPE and LDPE are most commonly found in lids. This makes segregation complex, especially as there may be other components such as seals which have to be separated but are essential to the leak-tightness required in the laboratory activities. With any laboratory item in use and no simple replacement on a like-for-like basis there is a risk in pursuing green products and credentials as other risks may arise. Elimination may not be a safe option, for example not wearing gloves or changing less often, as the risk to the wearer that is more of immediate concern.

17.7 Conclusion

It is possible to apply sustainable and green practices to the laboratory environment in research and higher education. However, it is not as straightforward as home or office environments. Careful consideration is needed to ensure that new risks are not generated by the introduction of sustainable policies and practices. The co-ordination of effort across large organisations and scaling is also a significant barrier. Ranking of HE and research institutes for green, low carbon and sustainability is one method, but there are different schemes and it is optional. There are barriers to adopting and evaluating green agendas – mostly this is cost but also commitment resources and motivation. Full economic and sustainable evaluation and impact of changes is complex. There are regulatory and penalties for ignoring health and safety requirements which are very public and affect reputations and financial stability. The health and safety challenges are always to anticipate the risk and monitor a changing world. The activity needs to be a collaborative activity and information shared to encourage best practice and lessons to be learned.

Abbreviations

3Rs	Replacement Refinement & Reduction of Animals in Research
BSI	BSI Group, also known as the British Standards Institution, is the national standards body of the United Kingdom. BSI produces technical standards on a wide range of products and services and supplies certification and standards-related services to businesses. 389 Chiswick High Road, London W4 4AL United Kingdom
BREEAM	Building Research Establishment Environmental Assessment Method is an assessment undertaken by independent licenced assessors using scientifically-based sustainability metrics and indices which cover a range of environmental issues. Its categories evaluate energy and water use, health and well-being, pollution, transport, materials, waste, ecology and management processes
CEN	The European Committee for Standardization (CEN: *Comité Européen de Normalisation*) is a public standards organization whose mission is to foster the economy of the European Union (EU) in global trading, the welfare of European citizens and the

environment by providing an efficient infrastructure to interested parties for the development, maintenance and distribution of coherent sets of standards and specifications. CEN European Committee for Standardization CEN-CENELEC Management Centre, Rue de la Science 23, B - 1040 Brussels, Belgium. Tel: + 32 2 550 08 11

CE CE marking is a certification mark that indicates conformity with health, safety, and environmental protection standards for products sold within the European Economic Area (EEA). The CE marking is also found on products sold outside the EEA that are manufactured in, or designed to be sold in, the EEA. CE Marking Regulation (EC) 764/2008 Available at: https://ec.europa.eu/growth/single-market/goods/

ISO The International Organization for Standardization is an international standard-setting body composed of representatives from various national standards organizations. The organization promotes worldwide proprietary, industrial, and commercial standards. ISO Central Secretariat. Chemin de Blandonnet 8 CP 401–1214 Vernier, Geneva, Switzerland

Acknowledgements

Thanks for the reviewing assistance of Steve Owens, Lizzy Staincliffe and Brigitte Guile, PHE Environmental and Sustainability team.

References

1. HSE, Historical picture statistics in Great Britain, RIDDOR and earlier reporting legislation Figure 7: Number of fatal injuries to employees in Great Britain 1900-2018/19p Pg. 7, 2019. Available from https://www.hse.gov.uk/statistics/history/historical-picture.pdf.
2. Anonymous, *Safe Successful Sustainable Lab Improving Laboratory Design*, Maintenance and Operation S_Lab University of Bradford, 2015. Available from https://www.effectivelab.org.uk/about-s-lab.html.
3. My Green Lab ACT Environmental Impact Factor Label San Diego, CA 92109 info@mygreenlab.org. Available from https://www.mygreenlab.org/act.html, last accessed Feb 2020.
4. UK Statutory Instruments, No. 871, *The Producer Responsibility Obligations (Packaging Waste) Regulations 2007 (as amended) and 2016 amendments cover recycling and recovery*, 2007.
5. UK Statutory Instruments, No. 2306, *Provision and Use of Work Equipment Regulations (PUWER) 1998*, 1998.
6. EC, *Registration, Evaluation, Authorisation & restriction of Chemicals (REACH) (EC 1907/2006)*, 2006.
7. EU, *Biocidal Products Regulation (BPR, Regulation (EU) 528/2012)*, 2012.

8. People & Planet's University League People & Planet Student Activities Limited – A charitable company (Charity Commission 1140864, OSCR SC043394, company registered in England No. 06417846) Oxford, OX4 1JE. Available from https://peopleandplanet.org/university-league.
9. D. Kuntin, How to... reduce your lab's plastic waste, *The Biologist*, 2018, **65**(6), 28.
10. Grenova Solutions Richmond VA 23224. Available from https://www.grenovasolutions.com/tipnovus/.
11. M. Urbina, A. Watts and E. Reardon, Environment: Labs should cut plastic waste too, *Nature*, 2015, **528**(7583), 479.
12. PlasticsEurope Trade Association for the Plastics Industries, Rue Belliard 40, Box 16, 1040 Brussels – Belgium +32 (0)2 792 30 991.
13. L. Howes, Can laboratories move away from single-use plastic?, *Chem. Eng. News*, 2019, **97**(43). Available from https://cen.acs.org/environment/sustainability/laboratories-move-away-single-use/97/i43.
14. National Centre for the Replacement Refinement & Reduction of Animals in Research NC3Rs Gibbs Building, 215 Euston Road, London, NW1 2BE | Tel: 020 7611 2233 https://www.nc3rs.org.uk/the-3rs.
15. B. Morrissey, K. Blyth and P. Carter, *et al.*, The Sharing Experimental Animal Resources, Coordinating Holdings (SEARCH) Framework: Encouraging Reduction, Replacement, and Refinement in Animal Research, *PLoS Biol.*, 2017, **15**(1), e2000719.
16. Institution of Mechanical Engineers, Transport Hierarchy Institution of Mechanical Engineers 1 Birdcage Walk, London SW1H 9JJ, 2013. Available from https://www.imeche.org/docs/default-source/1-oscar/reports-policy-statements-and-documents/transport-hierarchy.pdf?sfvrsn=0.
17. House of Commons Select Committee for Science and Technology, 2nd Report. Proposed Changes to Chemistry Provision at Sussex University May 2006 Available from: https://publications.parliament.uk/pa/cm200506/cmselect/cmsctech/1011/101104.htm.
18. O. Jones, C. Hendon, A. McNeil, L. Kamerlin, R. Joyner, J. D. Woollins and C. Carmalt, Strong Bonds: The State of Chemistry. Seven Academic Chemists Give their Views on the Field's Health 7 February 2019. Available from https://www.timeshighereducation.com/features/strong-bonds-state-chemistry.
19. A. Arnott, Ventilation Best Practice, Case Study: Ventilation Best Practice Variable Air Volume (VAV) Fume Cupboards University of Edinburgh 2015. Available from https://www.ed.ac.uk/files/atoms/files/lab_case_study_-ventilation_best_practice.pdf.
20. S-Lab, Guide for Laboratories – A Reference Document for the S-Lab Laboratory Environmental Assessment Framework V1.0, 2011. Available from: https://www.ed.ac.uk/files/imports/fileManager/S-Lab_Good_Practice_Guide.pdf.
21. L. Gumapas and G. Simons, *Factors Affecting the Performance, Energy Consumption, and Carbon Footprint for Ultra Low Temperature Freezers: Case Study at the National Institutes of Health*, World Review of Science

Technology and Sustainable Development, 2012, , DOI: 10.10.1504/WRSTSD.2013.050786. Available from https://www.researchgate.net/publication/235342032.
22. M. Howes, Green Labs – Cold Storage 2018, Workshops Cambridge Green Challenge MH – 20 July 2018. Available from https://www.environment.admin.cam.ac.uk/files/20180720_green_labs_cold_storage_workshops_handout_v.1.0.pdf.
23. Green Labs – Lab Energy 2019 – Environment and Energy https://www.environment.admin.cam.ac.uk/files/20190222_green_labs_lab_energy_handout_mh.pdf.
24. University of Colorado Boulder CU Green Labs −70 °C Efforts University of Colorado Environmental Center, 207 UCB, UMC 355, Boulder, CO 80309-0207 303-492-308. Available from https://www.colorado.edu/ecenter/greenlabs/lab-energy-efforts/freezers/70-0c-efforts.
25. My Green Lab, *Guide to Green Chemistry Experiments for Undergraduate Organic Chemistry Labs!*, 2013. Available from https://www.mygreenlab.org/gccg-form.html.
26. *Merck DOZN Quantiative Green Chemistry Evaluator Greener Alternatives Evaluation Matrix*. Available from https://www.sigmaaldrich.com/chemistry/greener-alternatives/matrix-scoring.html.
27. I. Montgomery, *Could These Materials be the Future of Packaging?*. 26 April 2019. Available from https://www.lumi.com/blog/sustainable-alternative-packaging-materials.
28. S. Allen, *10 Eco-Friendly Alternatives to Styrofoam*. Available from https://www.doyou.com/10-eco-friendly-alternatives-to-styrofoam/.
29. Michigan Green Chemistry Clearinghouse Green Chemical Assessment Tools, Available from: https://www.migreenchemistry.org/toolbox/chemical-assessment-tools/?doing_wp_cron=1578493131.9985759258270263671875, last accessed Jan 2020.
30. Accountability, Consistency, and Transparency ACT, The Environmental Impact Factor Label My Green Lab, San Diego, CA 92109424.354.1494 https://act.mygreenlab.org/.
31. T. Zweckmair, H. Slavica, K. M. Klinger and T. Rosenau, *et al.*, Recycling of analytical grade solvents on a lab scale with a purpose-built temperature-controlled distillation unit, *Org. Process Res. Dev.*, 2017, **21**(4), 578.
32. S. Greenway, *13 Plastic Packaging Alternatives, Innovation Excellence*, 2018. Available from https://www.innovationexcellence.com/blog/2018/07/02/13-plastic-packaging-alternatives/, last accessed Jan 2020.
33. A. Paradise, *A Greater, Greener Commitment Lab Manager, Business management*, 2016. Available from https://www.labmanager.com/business-management/2016/03/a-greater-greener-commitment#.XixKrWfgrVi, last accessed Jan 2020.

34. New England Biolabs, *Where the Rubber Glove Meets the Road (to a More Sustainable Lab), Lab Manager, Business Management*, 2017. Available from https://www.labmanager.com/business-management/2017/01/where-the-rubber-glove-meets-the-road-to-a-more-sustainable-lab-#.W8UKBP-mYTRY, last accessed Jan 2020.
35. D. Huynh, Lab glove recycling delivers-results, *In Our Nature*, 2016. Available from https://green.uw.edu/blog/2016-07/lab-glove-recycling-delivers-results, last accessed 20 September 2018.
36. The Spontex® Disposable Gloves Recycling Scheme/The KIMTECH™ Nitrile Glove Recycling Programme. Available from https://www.terracycle.com/en-GB/brigades/spontex.
37. StericycleTM Leeds Yorkshire Bio Systems. Available from https:\\www.stericycle.co.uk/solutions/sharps-waste?campaignid=9486689273&mh_keyword=%2Bsharps&gclid=EAIaIQobChMIu6i_2sS-F6AIViK3tCh1R4QA9EAAYASAAEgL2_vD_BwE, supportuk@stericycle.com.
38. STARLAB (UK) Ltd, 4 Tanners Drive, Blakelands, Milton Keynes MK14 5NA Recycling service. Available from https://www.starlabgroup.com/DE-de/sldocument/download/344900_1, last accessed Jan 2020.
39. K. Zimmer, Life Scientists Cut Down on Plastic Waste Across the US, laboratories are finding creative ways to minimize the amount of plastic they throw away. *The Scientist*, 2018. Available from https://www.the-scientist.com/careers/life-scientists-cut-down-on-plastic-waste-64547.
40. Styropak UK, Styropack Recycling. Available from https://www.styropack.co.uk/recycling/facts-about-recycling-polstyrene-packaging/.
41. Avanti Recycling Centre, Kitling Road Knowsley Business Park, L34 9HL Liverpool Merseyside United Kingdom Laboratory Plastic Recycling. Available from https://www.tradebe.co.uk/laboratory-plastic-recycling.
42. P. Réu, G. Svedberg, L. Hässler, B. Möller, H. Andersson Svahn and J. Gantelius, A 61% lighter cell culture dish to reduce plastic waste, *PLoS One*, 2019, e0216251.
43. Terracycle KIMTech Nitrile Glove Recycling Program. Available from https://www.terracycle.com/en-GB/brigades/gloves.
44. Historical picture statistics in Great Britain, trends in work-related ill health and workplace injury, Health and Safety Executive, 2019. Available from https://www.hse.gov.uk/statistics/history/index.htm, last accessed Jan 2020.

CHAPTER 18

Organise Not Agonise – Getting the Best from Audits and Inspections

JULIA COTTON*[a]

[a]Imperial College London, London, SW7 2BU, UK
*E-mail: j.n.cotton@imperial.ac.uk

18.1 What is an Audit?

An audit is a "systematic, independent and documented process for obtaining objective evidence and evaluating it objectively to determine the extent to which the audit criteria are fulfilled".[1]

18.2 Introduction

This chapter examines the importance of checking and reviewing one's health and safety management system, and discusses different methodologies. It covers concerns, successes and causes of failures around auditing, starting with a focus on interfaces, where there is often uncertainty and lack of knowledge as to who is responsible or in control. Such areas have been the cause of most health and safety management problems, and as such are generally the most fruitful for an auditor.

Audit methodology has remained consistent over the years, and has been the subject of an International Standard since 1990,[2] last reissued in 2018

Challenges for Health and Safety in Higher Education and Research Organisations
Edited by Olga Kuzmina and Stefan Hoyle

Published by the Royal Society of Chemistry, www.rsc.org

as ISO19011, Guidelines for auditing management systems. Modern health and safety auditors must be competent in the audit process, use a process of document review, asking interviewees and operatives open questions, of listening and watching, and visiting or inspecting the work place, ideally whilst normal business is in progress to check whether an established, formal health and safety standard is being met. Once their checks are complete they can provide assurance, alerts and advice to the limit of their competency.

They may find that the standard itself is insufficient (for example a university's code of practice that does not cover all the regulatory requirements) or has not been communicated properly. This may have implications for the entire university. In other instances, a department may not have communicated to its staff and students, or misinterpreted the university's requirements, or only certain sections of the department have complied. It is unlikely that a department or section will simply ignore health and safety emails if they think them irrelevant, but it cannot be discounted.

The Health and Safety Executive (HSE) advise that large organisations should have systems for auditing and reporting health and safety performance, and that senior management (Chief Executive Officer equivalent) should see audits, internal and external, as a necessary and positive tool.

But by what standards should universities' health and safety achievements be measured? Whilst there is a specific university audit standard (USHA's HASMAP system), there are other relevant safety standards such as ISO45001 and HSG65, and other mechanisms for monitoring and reviewing, so an overview is provided.

Finally, how is a successful audit judged? Can a department continue to improve once it has reached its zenith – and what makes an audit succeed or fail?

18.3 Interfaces

I have been auditing health and safety management systems in universities and other research enterprises for the last twenty-five years and throughout this period my greatest concerns have always been around interfaces. Interfaces are any areas, persons or activities not under the direct control of the function being audited. In an academic department, this might mean building curtilage, corridors, roofs, plantrooms and server rooms, and services such as water, sewerage, drains, lighting and heating, fire alarm systems and gas pipelines.

Wastes and waste removal services, maintenance staff and contractors, security and cleaning staff (and their chemicals), are generally controlled through a support service function, and there may well be other academic departments with different risk profiles, or support functions such as catering or retail outlets who share the same building.

Interfaces also include shared spaces and equipment, such as a laboratories and research facilities which may either be controlled by or used by the function. Such equipment tends to be either large or expensive – or both – such as autoclaves, NMRs or even fume cupboards.

The main concern around interfaces is that they are sometimes invisible to those undertaking risk assessments, possibly because areas are hidden within the infrastructure or the activities are taking place outside the researchers' normal working hours.

It is essential that shared laboratories and research facilities have good communications, a good safety culture and formal safety management systems. Without these, such an enterprise is likely to experience high rates of equipment damage, near misses (and under-reporting), and depending on the risk profile, more serious accidents.

18.3.1 Auditing Shared Items of Equipment and Processes

Auditing those shared items of equipment and facilities which are important to a department generally involves careful reading of instructions, and any specific statutory requirements, particularly those around installing, commissioning, maintenance and testing. Which is appropriate providing the provenance of equipment is known and if instructions are still available, but for older items or items gifted or transferred with researchers this can result in an internet hunt.

Training records may not have been kept or there is no designated trainer, so learning outcomes are confused, and even when relevant, associated hazards, risks, controls and emergency procedures may not have been included at all. Process owners may not use the risk assessment as part of training, or there may be little recourse to the risk assessment – if one exists at all. Checking processes are undertaken safely can sometimes only be done by watching people performing the task, but this can make operators nervous and prone to mistakes. However, it may be essential to do this, especially if the process involves movement within and external to the laboratory environment, so putting people at ease in advance of this aspect of the audit is important.

18.4 Communicating for Success

Universities are complex, multi-national organisations that have generally evolved over generations. They have a vast range of competencies, with postgraduate students who do the bulk of the bench work, but who may be inexperienced, sometimes being regarded as the lab manager, and in some instances, staff who may not have been inside a laboratory for decades, supervising them.

Where there are science-based departments, there will also be novel hazards and new ways of using old hazards. Departments which were traditionally low risk, such as computing, are moving into bioengineering and biomedical fields, and may be working with robotics and artificial intelligence, or designing and trialling prosthetics. Researchers who have only ever worked on computers may find themselves visiting or assisting in a laboratory environment but without the historical knowledge of the risks. In this way, departments that were considered low risk unwittingly acquire a high-risk profile, especially where staff and students are no longer working in the low risk home department, but have moved to high-risk environments.

Organisations whose sole purpose is research and development (R&D) might be expected to have safer work environments, with compliance far easier to achieve than at universities. This may be the case in many institutions but not necessarily in all of them, particularly for those which have an international parent company which may operate globally and have a management system that spans tens of different legal requirements and safety cultures. For example, the problems encountered at one high-risk R&D company, whose labs were in the UK, were largely to do with poor communications between senior management, support staff and operatives. The safety management system was written in English, but was so long, tedious and badly written it was almost impossible to make sense of it. The risk assessments and standard operating procedures were almost all written by the same individual and were meaningless to anyone else reading them. The same person was also, bizarrely, responsible for checking and signing off their own documents. Even worse, they did not go into the field to view complex and dangerous operations, and as they had no experience of this work, and didn't consult the operators either, the risk assessments were almost works of fiction.

In the same organisation, the high-risk laboratories (full of reactor vessels being used in fume cupboards), were meant to be off-limits to security staff at night; however, when interviewed, it transpired the security staff were being given *over fifty* emergency escalation procedures a week, for overnight running of the reactor vessels. These were written in technical language, and required security staff to enter the lab and "make the experiment safe". They had no idea what to look for and no explanation as to what a problem might look like. The security staff entered the labs each night as they felt it was a requirement of their role, and nobody had told them otherwise. Given that the laboratory had already suffered a reactor vessel explosion whilst a contractor was present and which practically vaporised the fume cupboard, a problem was indeed likely.

Again, bizarrely, the company's health and safety risk register mentioned the risk of a reactor exploding, and all the controls that they thought they had in place. But when their sister company suffered a reactor explosion, they didn't review their risk register or their risk assessments, and were quite surprised when an identical incident happened on their site.

There are many lessons to be learnt from this company – the main one was that they had spent far too long on writing policies, and not enough time implementing even the most basic safety procedures, such as ensuring security staff didn't enter labs at night as stated in the company's documentation. The lack of consultation and communication with stakeholders such as cleaners, fire, maintenance and security personnel was astounding, given their degree of infiltration into working laboratory areas. The researchers didn't bother to consider or check whether security personnel understood what was required of them, or that things in the labs could kill them.

Complex organisations with so many different safety cultures and levels of competency require a great deal of monitoring to ensure health and safety management systems are implemented and effective. Efficient communications between stakeholders are an absolute requirement; nothing will work without these. But in the Higher Education sector, communications are patchy, mainly because of staff and student turnover and the complex organisational structures. It's the communications equivalent of plate-spinning.

18.5 Health and Safety Audit – Standards to Use in the Research Business

18.5.1 HSG65

HSG65 is a comprehensive health and safety management system produced by the HSE several decades ago, and upon which many universities based their systems. However, it has been thoroughly revised, simplified and reissued and now focuses on the role of senior management, behavioural aspects and a general cyclic system of improvement – Plan, Do, Check, Act, commonly called PDCA (see Figure 18.1). Embedding it within the organisation's management systems can also be highly beneficial as it establishes a mechanism for learning and improvement throughout the business.

18.5.2 ISO45001

ISO45001, introduced in 2018, is a structured and comprehensive certificated health and safety quality management system which has replaced BSEN18001. Adopting its practices and principles as a standard against which to audit is advisable and beneficial as it's far more comprehensive than HSG65. However, it's expensive and labour-intensive for an organisation to implement. To maintain compliance, and then to pay for external

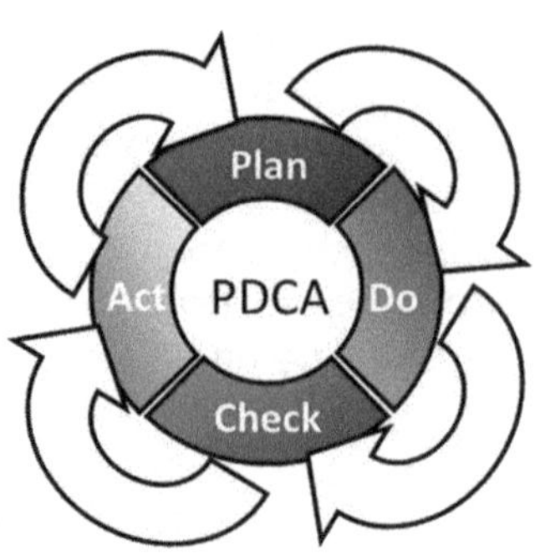

Figure 18.1 The PDCA system.

certification, re-auditing and re-certification, can be a burden where it's not a requirement for the business. However, where research institutions are part of supply chains, they may be required to get certification.

18.5.3 HASMAP

HASMAP is the audit system developed by the Universities Safety and Health Association (USHA). This system enables benchmarking within and between universities, but could easily be applied to other research institutions. It provides departments with incremental standards to be met (basic, substantial and high levels of assurance), over twelve indicators. Auditors and departments have guidance as to the evidence they should seek or provide. HASMAP uses a simple colour scoring system so that improvement or depreciation can be seen easily. Note that HASMAP has been tailored to include all the elements of ISO45001 as well as PDCA.

HASMAP is a flexible concept, and many universities use it for their internal audits. Not so many universities use it to benchmark, maybe because the universities tailor the system to suit their own health and safety management systems, so it's not always possible to compare like with like, although it works well within the organisation. Some universities focus solely on the audited department's safety officer, asking them every question and for every piece of evidence, over one day, with a brief workplace inspection, covering one room. Other universities select two or three indicators, asking for the relevant documentary evidence in advance, and interviewing a selection of personnel questions around this area, followed by an afternoon of inspection and ad hoc discussion in the workplace.

Figure 18.2 represents an example of tailored USHA HASMAP (part of the Leadership element). This has been tailored for use during Year 1 of the HASMAP being introduced to the institution. In the first year, the single leadership element was selected rather than using all twelve elements of the HASMAP. Having achieved a basic standard against each indicator or established a plan to achieve it, progress was checked mid-cycle. If basic had been achieved, the next level of assurance (substantial) was introduced in the following audit cycle and progress against plan checked again. This type of tailoring allows departments who are new to the HASMAP process to focus on achieving one or two key elements to a high standard, or to focus on aspects which are known to be problematic across the institution, such as Planning for Emergencies.

If several departments fail to achieve a specific indicator, this flags a failure or omission at institutional level for this aspect of the health and safety management system.

It is always helpful to manage people's expectations, particularly relating to audit, where potential interviewees can be extremely nervous. In addition, where time is of the essence for busy research departments, understanding the stages of the process and the various requirements helps departments to prepare for the audit and follow up in good time. A flow chart such as presented in Figure 18.3 is often the simplest means of showing a complex process.

LEGEND:

Green	Full compliance
Orange	Partial compliance
Yellow	Non-compliance
	Not achieved

LEADERSHIP: A1 - HEALTH AND SAFETY POLICY

INDICATOR	AUDIT FINDING	HOW WILL DEPARTMENT ADDRESS THIS?	BY WHEN?
BASIC LEVEL OF ASSURANCE – all basic indicators must be green before the next level can be attained			
1.1.1 A function code of practice (CoP) exists and is up to date.	1. Web-based documentation for the department but some is out of date. 2. COP for teaching lab, but see 1.1.5. 3. Not supplied with COP for Room 505		
1.1.2 The CoP is communicated to relevant persons including staff and students.	1. Department has an on-line sign-off for its general policy and there is an induction for the entire student intake which also covers Day One. 2. There is a ground floor teaching lab local rules and associated induction. 3. COP/local rules for Room 505 not supplied; if not in place then it cannot have been communicated. There are no formal arrangements for the use of 505 by Hackspace as described in the Hackspace COP and on its webpage.		
1.1.3 The CoP is signed by the Head of Function.	1. Departmental statement is signed. 2. There are no COP/local rules for Room 505		
1.1.4 The CoP contains a commitment to compliance with College's Health and Safety (H&S) Policy.	Departmental COP (web-based), references the College policy.		
1.1.5 The CoP contains a commitment to the provision of a safe and healthy working environment.	1. There is very little mention of health and safety in the teaching lab documentation, and no mention of risk assessment. An opportunity has been missed to introduce the concept – a key requirement of employers. 2. It is now a requirement for the 3rd year final year project students to provide a risk assessment; but not yet for 4th years.		
1.1.6 The CoP makes a commitment to the prevention of injury and ill health in the workplace.	1. This is lacking, but important in older buildings or labs using hazardous materials where ill-health resulting from exposure might may take years to show (asbestos present within infrastructure, solder and colophony fume).		

Figure 18.2 Example of USHA HASMAP checklist tailored to an organisation.

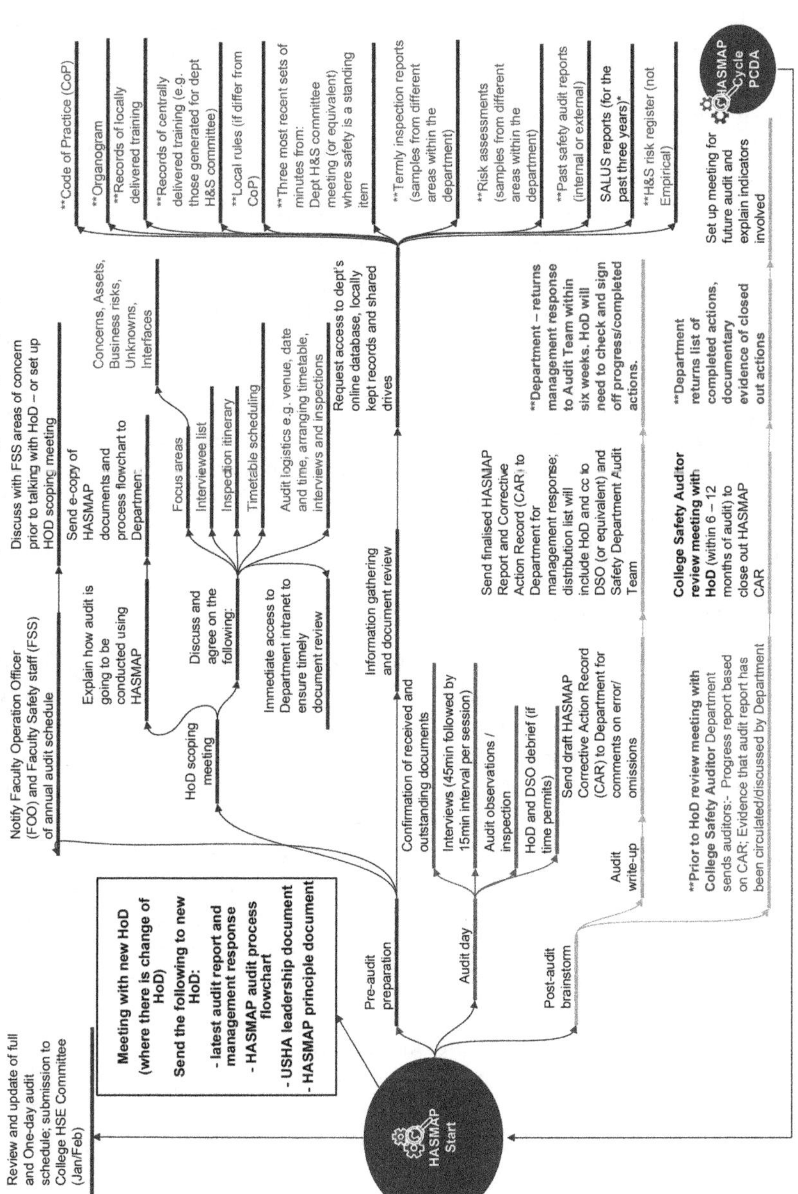

*These documents will be obtained by the Safety Department Audit Team. Notes marked ** require actions/inputs from department.

Figure 18.3 An audit process flow chart.

18.5.4 Other Standards

Research institutions may also have their own organisation-wide Codes of Practice defining their specific local practices which departments are expected to follow, for example on hazardous waste management, fume cupboards, event safety, offsite work, and driving, so these will also require audit to check they are still effective and implemented.

In addition, there will also be fire risk assessments (FRAs) for all buildings controlled by the organisation. FRAs may not always include the specific fire risks arising from the research and teaching activities within the building, and neither do they always capture the shared nature of the buildings, which in any case can be quite dynamic. FRAs are relevant to audit because they can be indicators of poor safety culture – for example if actions have not been closed out in a reasonable timescale. In addition to any remaining open fire-related risks, if new and significant risks have arisen from the research, the FRA may need review, the audit recommendations may need to include this. Fire risk assessments are only as good as the competence of the Fire Risk Assessor, and on whether the organisation acts on any findings in a timely and effective manner – both these aspects are of interest to an auditor.

18.6 Other Checks

18.6.1 Inspections and Spot Checks

Inspections and spot checks may be undertaken at a local laboratory level, at departmental level or faculty level, or by the organisation's internal safety teams. Regardless of where the area for inspection is located, ideally the area owner(s) should be present at the time as well as the safety personnel, so they can record and act on any faults found, emphasise best practices and pass on praise for proactive safety management.

In universities, the success of such spot checks can be quite limited and short-lived as they don't address behavioural or cultural change or shortcomings in the management system. Also, because the constitution of research groups changes every couple of years, staff focus on the work and the next grant rather than on tidying laboratories and changing the culture they inherit from previous incumbents.

I believe inspections and spot checks are more successful in industry because labs tend to be purpose built for an activity, or are designed to be completely re-configurable so a fresh start can be made each time activities change. Obsolete equipment and out of date reagents are not permitted, so there is an expectation of good housekeeping, and that spot checks and inspections are there to help this process.

In universities, some academics may have had tenure for decades in the same laboratory. Even though the activities have changed, the safety culture has failed to move with the times, obsolete equipment takes up valuable storage space, just in case it is ever needed again, and as the labs will often also have poor ergonomics as well, inspections and spot checks are more likely to fail. In other instances, a research group will be so busy with grant deadlines that they fail to notice the chaos around them. A further key factor that can assist in the success of inspections is having a proactive and diligent laboratory manager, something that many universities can no longer afford.

18.6.2 Spot Checks

Spot checks are unannounced, short, informal inspections, which are generally not formally recorded and which may not involve the person responsible for the area. Heads of Department may find this a useful or divisive approach to take depending on their management skills; spot checks are intended to look at a specific area or concern rather than the entire department. As part of the drive to increase visible leadership and drive positive safety culture change, this is also an ideal opportunity for senior management to find out more about the research, and to ask the all-important "What would happen if...?" questions. Such questions are a very useful way for the management at all levels to get to know staff, and to monitor their responses to check if they fully understand emergency procedures for example. It's also an opportunity to praise best practice, and to highlight non-compliance, and to show that health and safety is taken seriously.

18.6.3 Recording Inspections

The method used to record inspections may depend on the type of inspection. Some may be of a few workrooms and led by the local representative who may have no experience in spotting problems or have become blind to them. For these, a checklist approach would suitable. Others inspections may cover many areas of a complex workplace or specific topics, and be led by a team of safety personnel, experienced in identifying flaws in infrastructure, research and support activities such as waste streams, general housekeeping, or repeat findings.

Adopting a practice where an inspection team leads, records and highlights priorities, but where the area owner accompanies the team and is responsible for identifying action owners and for recording close-out, helps to foster ownership. The allocation of roles needs to be clarified with all involved in the inspection before its start – and the roles and responsibilities of area managers should include "monitoring" for health and safety.

18.6.4 Online Recording Systems

Locations, photos and comments can be uploaded and shared easily, their adoption can be made simpler by the purchase of sufficient end-user licences (where needed), and appropriate hardware and software on which to run them. Some universities use phone apps (such as iAuditor) where one can create one's own specialist inspection checklists, and corrective actions and reports will be automatically emailed to area owners, but there are many others. Software may require tailored inspection questions sets to reflect the types of activity within the areas being inspected along with inputting and updating any contact details, managing licences and versions as well as operating systems, all of which require resource and accurate information – particularly for room numbers. For buildings where room configurations may have changed many times over the years, accurate room numbers can present quite a problem. Small software companies or in-house programmes may allow for greater flexibility in design, but using highly tailored, purpose-built software could also result in business continuity issues. There are further disadvantages to using software and checklists, whether online or on paper, as no matter how accurately they reflect workplace hazards and risks, one can focus so much on them that something out of the ordinary is missed.

Over-familiarity with an area and being blind to hazards of long-standing is a real risk so area owners are discouraged from conducting their own formal inspections alone, but to invite a peer from a different area to act as a critical friend. Checklists however, if properly designed, are extremely useful for departmental inspections conducted by less experienced safety personnel and representatives.

18.6.5 Daily Checks

Prior to the start of the working day, if the organisation has the luxury of a laboratory manager, either they or persons designated by the senior staff should check the status of wastes, electrical and other high-risk equipment, and in high-risk workplaces, the status of control and emergency measures such as eye-wash stations.

18.6.6 Shut Down Checks

At the end of the working day, or when a specific item of equipment has been used it may be necessary to ensure areas or equipment have specific shut down procedures followed. For example, the last person leaving a lab may be expected to perform certain checks or ensure an item of equipment is turned off and in a safe state to ensure the next user is not at risk.

18.6.7 It's Not My Area!

Common areas used by the department should be included in inspections, including corridors, storerooms, cold rooms, kitchenettes, cleaners' cupboards, wastes, gas and cryogenic stores and access and egress on the building's curtilage. Ideally, formal university-level or head of department-led inspections should invite/notify the Building Manager, but failing that they need to be sent any adverse reports for action or advice.

18.6.8 Competency

It may take several inspections or several years to become competent in the basic inspection process of an entire department, depending on its risk profile. Some university auditors invite members of departments' safety personnel to join them for mentoring to improve their inspection skills and obtain practice. Local safety officers should be trained in the reasons why inspections are essential as well as in the use of checklists, what to look for and record; what questions need to be asked, and how to ask them; and how to approach people who may be more senior than themselves.

18.6.9 Reporting and Escalation of Inspection and Audit Reports

Where checklists are used, they need to be appropriate and tailored to the departmental activities, and the frequency of checks tailored to a higher-level escalation/reporting procedure. For example, inspections might be carried out immediately prior to department staff or safety meetings, so they can be discussed at the time.

18.6.10 Follow-up and Close-out

If a department or section of a department undertakes any type of inspection or monitoring and follows it up with actions, this is seen to be a positive indicator. But the escalation, follow-up and close-out process is key; without assigning people to actions, and then communicating actions to them with a required timescale, the same problems are likely to be found during the next inspection, which might be from the HSE.

18.7 Inspection *Versus* Audit

The most successful inspection processes are led from the top. The Head of Department forewarns personnel of the inspection but not the exact date, informs them that the inspection team will include themselves, and if it is topic-based, what they are going to be inspecting. Those being inspected should also understand that any actions from previous inspections will be reviewed. This avoids the concept of completely unannounced inspections and spot checks which can have a surprisingly negative effect on personnel if the inspection process has only been recently introduced, and they are not mentally prepared for the concept of senior management descending on them.

Sometimes planned inspections can be less effective than spot checks. If the inspection date is known in advance, the lab staff may be tempted to tidy up the laboratory or workplace beforehand. However, the laboratory will quickly relapse into non-compliance and poor housekeeping as the root causes of any problems may have been hidden from the inspection team, and therefore not addressed.

For an inspection to be successful, the inspection team must be competent, firm but sympathetic, and have a robust reporting, escalation and close-out process. Inspection teams are not always comprised of senior management and not always independent of the area they are inspecting. Therefore, the ideal team should also comprise people unfamiliar with the area "a new pair of eyes" and must be empowered either to act or for senior management to act on their recommendations. If not, at its worst, an inspection will be a complete waste of time resulting in no improvements at all.

Whilst inspections are a very important part of an audit, an inspection is not an audit. An inspection is where people look at things that are present or not, whether good or bad. It does not analyse why things have occurred, whereas an audit looks at documentary evidence, records, verbal reports, and may even watch how people behave or conduct a task.

When a department knows it is to be audited and evidence of a management system is being sought, it generally hastens to establish those management systems before the auditors arrive, which can result in lasting change. However, for stand-alone inspections, if a department knows it is to be inspected, it may only "tidy up", thereby giving the inspector a false sense of what is "normal". Staff may also disappear for the duration of the inspection, making it difficult to check behaviour or compliance with personal protective equipment for example.

The time and effort spent in preparation for audit to design and implement working systems, write risk assessments and SOPs, train and induct, create and file training records, update web pages, replace out of date signs and posters and have a major clear out, is perceived as an investment because the workplace has improved; whereas an inspection may result in a temporary improvement which reverts the following day. In this way as the new system brings immediate and visible benefits, an audit can result in lasting change. Although inspections have their place as stand-alone monitoring, they are an important part of the audit process, to provide a visible check that described systems are in place and activities are undertaken as described.

18.8 Measuring Successes and Failures

Departments' attitudes to being measured publicly against other departments may not always be positive. There are several reasons for this, sometimes because people don't like airing their dirty washing in public but also because in a large department the audit is often only a snapshot of a small part of its function, and if this is the area scrutinised, then a false impression may be taken away. Quite often departments just don't have the resources to stop the research for long enough to establish and maintain a good safety culture; they may have overcrowded, old laboratories and no permanent lab managers, or there may be restructuring in the air, or simply bad management.

In other instances, it would be like comparing apples with bananas *e.g.* small chalk and blackboard departments with large, multi-risk engineering ones.

How can one score, and advise, a department – with all of this to contend with?

18.9 Why Does an Audit Succeed or Fail?

18.9.1 Scoping the Audit – Help Ensure Buy-in

To help ensure interest and buy-in, the audit needs to be scoped with the Head of Department. Introduce them to the concept that even the best run and safest department may be reliant on an interface such as a support service to keep it operating and, as they don't control interfaces, there may be false assurance in such areas. If the Department depends for income and reputation on a key research facility or item of equipment or even a control measure, then the concept of an audit to provide assurance and alert and advice suddenly becomes realistic. This is a business risk-orientated safety audit, and protecting the expensive area, facility or piece of equipment from harm, means the Head of Department will readily accept the need to evidence training, monitor and supervise, maintain and test, report incidents, protect from floods and fire and other health and safety issues.

Other aspects of interest are activities and associated hazards with which they are personally unfamiliar; or areas which are already known to them to be of concern, but about which they can do little. It may be an interface or a colleague. This latter is arguably the most difficult to deal with given the short tenure of most heads of department.

In this way, the audit findings are far more pertinent to the senior management, and recommendations more likely to be transformed into actions.

18.9.2 New Brooms and Old Lags

In the last decade, many universities have adopted leadership training programmes for their senior management teams and heads of academic departments (also advocated by USHA). Some universities have also established such training for their heads of support services; this training has made a marked difference to the level of engagement with health and safety at all levels, particularly with middle management (for example, principal investigators and research group leaders).

Part of the problem with establishing safety culture through an audit is that it may involve the Head of Department or Section Head or equivalent having to convey or implement less palatable audit findings. However, many heads of departments and section heads, whether in universities or other institutions, often only have a short term in office; five years or less. After this, they must return to their fellows in the academic pond. If they have made unpopular changes, been less than sympathetic, or too radical, then their return may not be welcomed, and as research departments function best with collaboration, thoughts of the eventual return may haunt a Head of Department or Section Head, and hinder them in carrying out their safety management roles. These problems are more likely to be of concern to new heads of departments who may lack management experience or experience of university culture, which can be slow to change.

But consultation, communication and collaboration should help with change, providing it is done sympathetically, in a reasonable time scale and for mutually perceived benefits.

For heads of departments who have been in post for years, the problems may be different, in that they just don't see the problems, particularly if a department or function is one that has evolved with them, and planning for health and safety has never taken place. They may also have not bothered to attend "Leadership" training, or if attendance was mandatory, may have attended but not listened!

Problems may also arise if a head of department has moved from a higher risk area into what they perceive as a lower risk area, and where if they are dismissive of the lower risks or of the entire safety management systems, their attitude will permeate the department. At worst, an entire department can be poisoned by poor safety culture at the top; incidents and accidents may go unreported, staff may blow the whistle; there may be a higher than normal turnover or unexpectedly high stress levels.

The opposite is also possible, where a head of a lower risk department has become a head of a higher risk department, and may not understand the risks at all. In most instances this makes them extremely risk aware, and they are happy to seek assurance *via* the audit process, and by active participation in it.

Encountering a disinterested head of department is part of the challenge of health and safety auditing, where one is not just measuring against a standard, but also selling health and safety. The message is simple: personal benefits for the individuals and a business benefit to the department – or a business risk if health and safety standards are not met. Leadership training has made an enormous difference, as it includes an element of health and safety law and legal accountability, so the modern head of department is generally well-aware of what is required of them. Occasionally the auditor might need to remind them.

18.9.3 Audience Participation

One of the key performance indicators in the USHA HASMAP scheme is that of leadership and visible participation in standard setting and monitoring. If possible, heads of departments or the function being audited, should be invited to join the workplace visit or inspection element of the audit, and to participate in asking the "what would you do if" questions to random staff and students, particularly to those in shared laboratories where roles and responsibilities are often not formalised. Heads may need to be briefed to take an empathic approach, and warned that if something unexpected or non-compliant is found it is likely because there has been a communication or resource failure and not a deliberate attempt at sabotage! Otherwise there is a danger that either the Head of Department or the person being audited may become defensive, or even aggressive, which is not constructive or conducive to the audit process, and the auditor will need to ensure this is dealt with appropriately and sympathetically.

18.9.4 Competence of the Auditor and Audit Team

The competence of the auditors is key to an audit's success.

Auditors come in all shapes and sizes; but Myers-Briggs or Belbin or other character-assessment methods will generally uncover a layer of judgemental behaviour in most auditors. Lead auditors need to understand this and manage themselves and their teams accordingly. Wherever possible, the auditor and audit team attend some form of behavioural safety training, to help them to ask the right questions in the right way and to listen effectively and learn how to convey sometimes unpleasant messages.

The danger of using tick-box audit methods is that if in fact a high level of analysis is required, the brilliance of a high audit score may overshadow more subtle findings. It is up to the auditor to look for indicators and anomalies, despite high audit scores. For example, one department had a manager who bullied and micro-managed their staff in all aspects of their roles, including in health and safety. All staff had body language and behaviour which showed high levels of nervousness and stress. These departments had extremely high audit scores for those areas within the immediate control and visibility of the managers. Areas outside of this appeared invisible (interfaces) and were dismissed by the manager as being unimportant, but on scrutiny, major non-compliances were found throughout. In this instance, as has occurred in other audits, the more serious audit findings were just part of a collection of indicators of poor management, and the manager resigned.

The Lead Auditor has a specific role to play in the audit team, principally to recognise the strengths and specialism of the audit team and to utilise them fully, and where specialist subject matter knowledge is required, provide this from elsewhere in the organisation (a temporary audit team member), or depending on the level of audit required, buy-in an external specialist.

Guest auditors, particularly when mentoring those who may be inexperienced in auditing, must be fully briefed by the Lead Auditor before the audit begins. The worst audit experience I ever had was at the start of my audit career, when I hadn't realised just how important this was, and my guest auditor began accusing the interviewee very aggressively of non-compliance, until the interviewee became sufficiently angry to throw us both out of their office. I had to return later to smooth things over! To ensure this does not happen, a briefing should include the process to be used by the auditor and audit team, including where to sit, watching body language, active listening, and waiting to be given the lead from the Lead Auditor. During interviews, it's very easy for the auditor and the auditee to lose the thread, and miss an important detail if the auditors are interrupting one another. An auditor briefing should also include who will lead on which aspects; a reminder to keep to time, keep notes, and remain professional, and keep everything confidential, even if the audit doesn't go as planned.

Auditors also must be managerially independent of the area being audited, with no conflict of interest. Where internal expertise is sought from within academic departments or external universities, this needs to be considered, particularly if there are competing departments involved. As one would normally undertake a document review prior to interviews and the site inspection, one needs to be careful about what documents leave the organisation or department to be sent to a guest auditor, just in case competition might be an issue.

Assure, alert and advise:[3] these broad audit objectives may be referred to as the 3As – to provide assurance to management, alert them to problems and advise on corrective action. For general management system audits, the advice would simply be to implement the relevant aspect. But where audits are focussing on topics, it's important to ensure that the team is comprised of a subject matter expert, so that the advice is appropriate. Auditors may also be able to share best practice found elsewhere as potential solutions. If an auditor doesn't have the answer that is perfectly acceptable.

18.9.5 The Most Important Interview

The first interview is the most important of the audit, because if it goes well, and the interviewee has felt supported and relaxed rather than interrogated, the news will spread rapidly to other interviewees, and they will be primed to be helpful. Therefore, it's always important to put the interviewee at ease by smiling at them, thanking them for coming, and then asking some neutral questions.

Neutral questions might range from asking their name, how long they have worked there, what they do, how they travel to work, and even if they are comfortable, but the important thing is to appear not just professional, but also friendly and interested. Ask them, too, if they know what the audit is about and why they have been invited to be interviewed. Interviewees can find the thought of being audited a terrifying experience, particularly if they feel they are of low status in the organisation, and nobody has ever spoken to them before about their job or role.

After this you can explain to them why you are there, and why you are interviewing them, and possibly what the management intention is behind the audit. For some audits, one may need to assure the interviewee that any data they provide will be anonymised; and that all audit interviews are confidential. After this, the audit can begin.

At the end of the interview, remember to thank the interviewee again.

"Could I just stop you there?" If you mishear or need clarification, or the auditee is going off subject, or time is running out, or if you simply haven't been able to record data quickly enough, then an interjection may be required. It can be helpful to use an appropriate hand gesture – raising one's hand – at the same time, just to break the flow of conversation.

Open and closed questions: one gets far more information out of people by asking open questions – for example, those that begin with what, why, who, which, when and how as they should elicit a far larger response than a closed question, *i.e.* anything to which the answer can only be yes, no or maybe. Closed questions are necessary to confirm a piece of information, but auditors need to be aware of what they are saying and of how they are asking questions.

18.9.6 Debriefing: Give a Man Something to Live for – Delivering Bad News

This was possibly the single most important advice I ever received regarding auditing, and occurred during my first ever audit debrief with a senior management team – how to deliver bad news. The audited department was excellent in almost every aspect of the audit and had clearly worked extremely hard to put management systems in place, but there were a few notable exceptions. Unfortunately, I began the debrief by focussing on the gaps, so any good news that followed on from this was effectively obliterated. The Head of Department listened calmly, whilst his management team were clearly very upset, and at the end of the presentation, he very kindly told me how I could have better presented the information by presenting the good news first, then the bad, and then ending on the overall performance being positive "the sandwich technique". I never forgot that lesson or his phrase "Give a man something to live for!"

The opposite can also be true, whilst one should present positive findings first, one must be careful not to give the department unrealistic expectations during the audit that everything is as it should be, and then hit them with bad news during the debrief. Better not to say anything at all.

During the initial scoping meeting, one should establish a "no surprises" mechanism, whereby if serious problems are found during the audit's course, that senior management will be informed immediately.

18.9.7 External *Versus* Internal Audit

There are pros and cons to external audits. They can be expensive as a capital cost, and as the amount of staff time taken up with accompanying the external team, setting up and attending interviews. If not properly scoped, or if the University management are not sympathetic, or if the external team have not been properly briefed or given the relevant evidence, the findings and recommendations may be dismissed, not implemented, and the report buried.

When commissioning an external audit, as one cannot be sure how well the audit report will be written, it is advisable to ask for a sample of a recent report (redacted). Also, an external auditor should have had experience of auditing within universities and an understanding of how they are financed, sometimes by charity grants. Whilst all UK institutions must comply with legislative requirements, the money and infrastructure is not always readily available, so audit reports will need to highlight any requirements for compliance, reference legislation and discuss timescales.

Internal audit, whilst it has the benefit of knowledge of the organisation, may be under-resourced. Many universities have auditors, but many are also safety advisors as well, and their other roles and responsibilities may make timetabling difficult.

External scrutiny is essential to the health of any health and safety management system, but it should be done with care – if too frequent, institutions can spend too much time reacting to successive audit findings, with insufficient time to close them out properly. This can lead to poorly designed systems which are doomed to failure and to a loss of faith in the value of audit as a process.

18.9.8 Resources – Audit Team, Department and University or Organisation

Planning an annual or two-yearly audit schedule may seem straightforward until the time comes to set the dates, and arrange the interviews. Dates are also needed for scoping, follow-up and close-out meetings. Whilst this only takes a few hours of computing and telephone time, this is spread over three or four weeks or more. The audit programme needs to be flexible as academic departments may be contacted at short notice by other mandatory external audit requirements, and so ask for their internal safety audit to be deferred; or the entire department may be involved in chasing a grant programme, or in attending a conference.

One audit was deferred for a year as the building had suffered a major flood and needed total refurbishment.

In universities, the internal audit team will generally be unable to set up departmental interviews as it has no jurisdiction over staff and students in other departments. This needs to be made clear to the Head of Department during the setting up of the audit and the initial scoping meeting with them, that their department will need to provide logistical support in setting up interviews and inspections and providing a suitable interview room. This may or may not be the case in other organisations.

Sufficient time must be given to write up audit reports, check and agree drafts for errors of fact, receive the management response and action plan, follow-up progress – and check the department is monitoring and recording its own progress, and then to close-out the audit findings.

Problems arise if the audit programme is overly-ambitious, as time is needed between audits to write reports, and if the programme has had one or two or more cancellations, then a twelve-month programme could end up being squeezed into a six-month period, to the detriment of the audit team. Deferred audits can get pushed back until the following year, and all such changes make scheduling and publishing an accurate schedule in advance an almost impossible task.

18.9.9 Unrealistic Time Scales for Implementation

Departments, generally those who have not been audited before, and who are subjected to large audits, may have too many recommendations to be able to resource. Ensuring that departments have a realistic time scale in which to implement their action plan is key to success; one such head of department undertook to complete a five-year action plan in six months, the resulting pressure causing ill-health for several members of staff.

Large departments also tend to have far more interfaces in shared labs and buildings, on other campuses as well as offsite, so the practicalities of communicating and facilitating a large and complex action plan can be impossible. Few departments have the luxury of permanent or part-time laboratory managers and technical staff, so implementing and monitoring an action plan in the work place is difficult because other lab staff who may be on short-term contracts, are actively involved in the research and are expected either to produce results or fail in a highly competitive market. They may not wish to be bothered with health and safety – and understandably may not see implementing a laboratory-wide system as part of their role or responsibility.

Where an audit finding has had implications for the University itself, resources will need to be found. To obtain such resources the report will need to be summarised, succinct and to the point, with needs, costs and outcomes clearly stated, and will need to be presented to the appropriate level of senior management by the most appropriate and effective messenger. A follow-up progress report is advisable including a review of whether financial needs have been met. For example, in one university, an audit found that the results of fume cupboard testing had not been understood or acted on by those in charge or by fume cupboard users, and subsequently the installation, commissioning and testing processes were also found inadequate, there were major cost implications across the entire university as so many of its building contained fume cupboards.

Several meetings were held with support service departments, before funds were approved and an action plan established.

18.9.10 Evolution *Versus* Planning

Like many other systems, health and safety management systems can either be planned or evolve. If they evolve, they are likely to be informal, not documented, with no training and knowledge of only parts of the system held by several individuals or by one person. If they leave, then all knowledge of the overall system and its aims and intended purpose will be lost. This seems to be a feature of universities' laboratories, particularly in large, shared areas, where there is nobody working within the laboratory who is responsible or accountable for health and safety.

Some Research Group Leaders and Principal Investigators (line managers) may not step into a laboratory for months, and some may not set a good example on those occasions when they do (often without the required personal protective equipment), and they may not notice non-compliances in the form of bad housekeeping. Furthermore, in shared labs, line managers may feel unable to say anything to those they don't manage.

Research appears to be getting increasingly expensive to undertake. Consequently, laboratories are becoming bigger, and there is even more shared space and equipment. How can such large interfaces be managed effectively?

Ideally there should be a planned health and safety management system which is reviewed whenever the organisation or section is about to undertake a significant change in its risk profile. This might be due to a takeover, relocation, enlargement, or the introduction of new risks. However, such management systems take time to construct, write, consult upon, and then to implement, but once in place, review should be straightforward, providing stakeholders are consulted with and resources are available.

18.9.11 Auditing Safety Management in Shared Labs and Workspaces

Universities may no longer have the luxury of technical staff and laboratory managers; but research staff don't necessarily have the time, skills or underpinning knowledge to manage a large shared laboratory, and they are not necessarily funded by their grant to take this role. Although many researchers do take on this role, because it is an informal one, as they move on, the knowledge will be lost. In large shared laboratories, therefore, there must either be a researcher(s) who has been delegated to undertake a clearly defined, formally documented role, for which they will need to be trained – or there needs to be a local laboratory/safety manager (or safety advisor depending on their remit), who understands the hazards and risks, controls and emergency procedures, and the underpinning laboratory management needs of the area such as waste streams.

The extents of their authority need to be defined and communicated directly to all the lab users and their line managers; for example, if they find a non-compliance, what steps will they be required to take? The reporting line for any researcher undertaking this health and safety role (who may already be doing it or who has been selected by their peers), should be formally agreed with the line managers and head of department, as having their delegated authority.

Health and safety management systems audits will always focus on who is responsible for health and safety, but an informal volunteer "lab manager" who may be responsible for themselves or their own research group may be in a difficult position legally if they provide "safety management" for others in the same shared space. They don't necessarily have the training to spot non-conformities, give safety advice or inductions, or undertake and review risk assessments for other groups, neither do they have a formally delegated duty to make other lab users comply.

In terms of audit, the best means of safety management in large shared laboratories is to have all researchers trained in the basic laboratory health and safety management systems and in risk assessment; safety advisors and professionals with relevant underpinning knowledge to support them when necessary; and an excellent communication and reporting structure, with the relevant aspects formally documented.

18.9.12 Auditing Safety Management of Other Interfaces

Audits need to cover all aspects impacting or which could be impacted by a laboratory's practices and processes. For example, an audit focus might be how wastes are managed, including packaging, storage and transport, security, labelling, emergency arrangements for spillage and contamination. This might involve the cleaning staff, contractors, and movement through the building, which might need input from the Building Manager, or if it were chemical waste, possibly from the Fire Officer as well. For more specialised processes, the audit standards would not be the USHA HASMAP, but the University's own Codes of Practice, along with the required legislative and local authority requirements, and the audit team would include the Building Manager and Fire Officer. Identifying and involving stakeholders is key to the success of any audit.

18.9.13 Safety Culture

Leadership and safety culture, good or bad, stems directly from heads of department and line managers. Routine visits from senior management asking "what would you do if" questions, and encouraging all laboratory users to also make this their practice, eventually leads to self-policing undertaken hopefully in the pleasantest manner possible, not as a rebuke but as a gentle reminder. Persistent non-compliance in the face of this, may require further investigation and a reappraisal of the workspace, technique, equipment *etc.*

It is equally important never to allow non-conformance or bad practice to pass apparently unremarked; otherwise it may be a sign of approval. In the same way, making a point of praising best practice or where effort has been made to improve, will also be a sign of approval followed one hopes by continuing effort.

18.9.14 Business Continuity – Loss of Knowledge

Often an audit is requested after a serious incident – taking a reactive rather than proactive approach can itself be indicative of poor safety culture. But there is no doubt that areas where earlier incidents have been reported, investigated and have recurred are also indicative of a culture that fails to properly analyse and learn from its earlier mistakes.

One of the most depressing audits I conducted was where a series of incidents and breaches had taken place within the department. The report was written, but during the audit follow-up period, the head of department resigned. When a new head was finally appointed and I came to discuss the audit with them, I was as surprised as him to find that his line managers had forgotten to brief him!

The entire area of safety and business continuity was poorly understood by the Higher Education sector until a few years ago. As more safety and business risk professionals have been employed (often because of a serious incident or near miss), it is now common to have safety as a risk on an organisation's risk register, and to have a specific health and safety risk register at departmental level.

Even after an audit has highlighted the need for formalising roles and responsibilities, and recording daily duties and local processes, there is sometimes also insufficient resource to undertake the task, and it gets left for another day. Inevitably, loss of knowledge will occur when personnel leave, and the department may remain unaware of what it is they've lost until something breaks.

18.10 Summary – the Shape of Success

For audits to be successful, organisations must have bought into the concept of audit at the most senior level, and have provided systems and resources to support implementation. In addition, they must be prepared to act on audit findings. Equally, they should be prepared to formally repute findings giving their evidence-based justification, or formally report that they are leaving the findings as an open risk, giving reasons why there is a delay or why it is not possible to close them out.

Auditors and audit teams must be competent – not just in the subject matter being audited, but in the audit process itself, particularly in their listening and non-verbal communications skills.

The audit process should be robust and documented so departments know what to expect in advance.

Careful scoping and having heads of departments to choose which aspects of their business they wish to scrutinise promotes ownership, particularly if these are essential to the operation of the department – and especially if they are also interfaces like a shared laboratory or workspace. The scoping meeting is also possibly the only opportunity to discuss logistics including the length of time to be spent on site, setting up interviews, possibly with people outside of the department, interview rooms, areas to visit, mechanisms for sharing documents, confidentiality *etc.*

Audits need to be scheduled and planned and documents reviewed in good time, so that an appropriate audit team can be mustered. This includes identifying and communicating with stakeholders, possibly involving them in the audit inspection (such as building managers, fire safety advisors, cleaning supervisors).

Once the audit is complete, ideally on the day, there should be a comprehensive debriefing and subsequent ownership by senior management, follow-up and close-out; all essential parts of the audit process. The audits which are most successful are those where senior management request a written progress report at quarterly staff or safety committee meetings, and then take further action as required.

The auditor's written report must be produced within a reasonable timescale. It also needs to be comprehensive yet succinct, thorough and easy to understand, and not to bury the department in recommendations. Departments should have an opportunity to check errors of fact, (names, locations, dates), but not to comment on them – that should be reserved for their management response.

Where resources are limited, areas could be audited on a risk-based approach, with seemingly higher risk departments being audited more frequently or more comprehensively.

As well as internal audits, external audits and reviews of the health and safety management system, are essential for keeping an organisation on track, providing they have the appetite and resources to make the necessary changes.

Success depends on effective safety culture – and this is dependent on having informed, responsible and ethical leadership who can get the best out of people in a supportive and respectful way. If senior management is seen to consider health and safety important and to act accordingly, then others will follow.

Further Reading

1. Leading Health and Safety at Work, 2008 (UCEA and USHA).
2. A guide to measuring health and safety performance, HSE December 2001.
3. Health and safety management in higher and further education: guidance on inspection, monitoring and auditing, HSC 1991.
4. Health and safety benchmarking – improving together – HSE 1999.
5. Managing for health and safety – HSG 65, HSE 2013.

6. Leading health and safety at work, INDG417, HSE 2013.
7. Involving your workforce in health and safety, HSG263, 2015.
8. Health and Safety, Environment and Quality Audits: A Risk-based Approach, Stephen Asbury, Elsevier Butterworth-Heinemann 2018.
9. Responsible Research – Managing Health and Safety in Research – guidance for the not-for-profit sector, UCEA 2012.
10. ISO 19011:2018 – Guidelines for auditing management systems.

References

1. Guidelines for Auditing Management Systems: ISO 19011:2018 – BSI Standards.
2. ISO10011-1:1990 Guidelines for Auditing Quality Systems.
3. S. Asbury, *Health and Safety, Environment and Quality Audits: A Risk-based Approach*, Elsevier, Butterworth-Heinemann, 2018.

Subject Index

Page numbers in *italic* indicate a figure; page numbers with a T indicate a table.